SPACE TELESCOPE SCIENCE INSTITUTE

SYMPOSIUM SERIES: 2
Series Editor S. Michael Fall, Space Telescope Science Institute

QSO ABSORPTION LINES: PROBING THE UNIVERSE

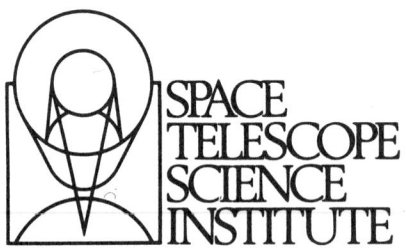

QSO Absorption Lines: Probing the Universe

Proceedings of the QSO Absorption Line Meeting
Baltimore, 1987 May 19-21

Edited by

J. CHRIS BLADES
Space Telescope Science Institute

DAVID A. TURNSHEK
Space Telescope Science Institute

COLIN A. NORMAN
*Department of Physics and Astronomy
Johns Hopkins University
and
Space Telescope Science Institute*

Published for the
Space Telescope Science Institute

CAMBRIDGE UNIVERSITY PRESS

Cambridge
New York   New Rochelle
Melbourne   Sydney

Published by the Press Syndicate of the University of Cambridge
The Pitt Building, Trumpington Street, Cambridge CB2 1RP
32 East 57th Street, New York, NY 10022, USA
10 Stamford Road, Oakleigh, Melbourne 3166, Australia

© Cambridge University Press 1988
Individual papers in this edition are © Space Telescope Science Institute 1988

First published 1988

Printed in Great Britain at the University Press, Cambridge

British Library cataloguing in publication data
QSO Absorption Line Meeting (1987, Baltimore, Md.)
QSO Absorption lines
1. Quasars. Spectra. Lines
I. Title   II. Blades, J. Chris
III. Turnshek, David A.   IV. Norman, Colin A.
V. Space Telescope Science Institute
523

Library of Congress cataloguing in publication data
QSO Absorption Line Meeting (1987 : Baltimore, Md.)
QSO absorption lines : probing the universe : proceedings of the
QSO Absorption Line Meeting, Baltimore, 1987 May 19-21 / edited by
J. Chris Blades, David A. Turnshek, Colin A Norman.
p.   cm. - (Space Telescope Science Institute symposium series 2)
"Published for the Space Telescope Science Institute."
Includes bibliographies.
ISBN 0-521-34561-8
1. Quasars-Spectra-Congresses.   2. Absorption spectra-
-Congresses.   3. Red shift-Congresses.   4. Galaxies-Spectra-
-Congresses.   I. Blades, J. Chris.   II. Turnshek, David Alvin.
III. Norman, Colin A.   IV. Space Telescope Science Institute (U.S.)
V. Title.   VI. Series.
QB860.Q76 1987
523-dc 19   88-816   CIP

ISBN 0 521 34561 8 Hard covers

CONTENTS

Preface ... ix

Participants ... x

QSO Absorption Lines: Introduction ... 1
W. L. W. SARGENT
Introduction ... 1
Properties of the Heavy-Element Redshifts ... 2
Physical Properties of the Ly-α Clouds ... 3
Recent Developments on the Ly-α Forest ... 4
Origin of the Heavy-Element Redshifts ... 5
The Origin of the Ly-α Forest Lines ... 6
New Results on Clustering ... 6
QSO Absorption Lines and Large-Scale Structure ... 8
The QSO Pair Q2343+1229, Q2344+1225 ... 8
Conclusions ... 9
References ... 9
Discussion — Colin Norman, Chair ... 10

BAL QSOs: Observations, Models and Implications for Narrow Absorption Line Systems ... 17
DAVID A. TURNSHEK
Introduction ... 17
Observations ... 18
Model Properties ... 26
Relation to the Narrow Absorption Lines ... 36
Future Work ... 43
References ... 44
Discussion — E. Margaret Burbidge, Chair ... 46

QSO Absorption Systems with $z_{abs} \simeq z_{em}$ ... 53
C. B. FOLTZ, F. H. CHAFFEE, R. J. WEYMANN, S. F. ANDERSON
3C 191 – An Historical Digression ... 53
Is There a Statistical Excess of Absorption Systems Near QSOs? ... 54
Do Strong Associated Systems Occur Preferentially in Radio Loud QSOs? ... 56
Is There a Dichotomy Between the Occurrence of BAL and Associated Absorption? ... 60
3C 191 Revisited ... 61
Speculations ... 62
References ... 65
Discussion — E. Margaret Burbidge, Chair ... 66

The Ly-α Forest: Observational Evidence for Line Evolution — 71
RICHARD W. HUNSTEAD
Introduction — 71
Historical Background — 72
Recent Determinations of γ — 75
The Inverse Effect — 78
Distribution of Equivalent Widths — 80
Evolution of Individual Clouds — 82
Issues Requiring Further Observations — 84
Conclusions — 86
References — 87
Discussion — John Bahcall, Chair — 89

Properties of the Ly-α Clouds — 91
R. F. CARSWELL
Introduction — 91
Deriving Distributions — 92
Cloud Properties — 97
Conclusions — 98
References — 99
Discussion — John Bahcall, Chair — 100

Absorption Lines and Galaxy Formation — 107
MARTIN J. REES
The Systems of High Column Density — 107
Equilibrium and Confinement of Clouds — 111
Photoionization of Clouds; z-Dependence of the UV Background — 115
The Distribution of Column Densities f(N) — 116
References — 117
Discussion — John Bahcall, Chair — 118

Properties of the Heavy-Element Absorption Systems — 127
JACQUELINE BERGERON
Introduction — 127
The MgII and CIV Samples — 128
Average Size of the Absorbers and Cosmological Evolution — 129
Ionization Level — 131
Photoionization Models — 135
Heavy Element Abundances — 138
Properties of Galaxies Giving Rise to MgII Absorption Systems — 139
References — 141
Discussion — Alec Boksenberg, Chair — 143

High Resolution Observations of Heavy-Element Absorption 147
J. CHRIS BLADES

Introduction	147
Low Redshift Absorption: QSO-Galaxy Systems	149
Intermediate Redshift Systems: MgII Absorption	158
Absorption Systems at High Redshift	162
References	170
Discussion — Alec Boksenberg, Chair	173

In Search of a Unified Description of the Narrow Absorption Line Systems 179
DAVID TYTLER

Introduction	179
Abundances	180
Column Densities	181
Ionization	182
Sheets or Spheres?	183
Evolution	184
Clustering	184
Interpretations	185
References	187
Discussion — Alec Boksenberg, Chair	188

The Properties of the Gaseous Galactic Corona 195
BLAIR D. SAVAGE

Introduction	195
Distribution of the Gas	196
Ionization Characteristics of Milky Way Halo Gas	209
Elemental Abundances in Milky Way Halo Gas	211
Kinematics of Milky Way Halo Gas	212
The Origin of Galactic Halo Gas	214
Absorption by Milky Way Halo Gas and the Absorption Line Systems Seen in the Spectra of QSOs	217
References	220
Discussion — Donald Morton, Chair	223

Analogies at Low Redshift for QSO Absorption Line Regions 227
D. G. YORK

Introduction	227
Description of Absorbers	227
Constraints on Halos	229
Star Forming Regions	230
Observational Tests	231
Galaxy Formation	233
References	234
Discussion — Donald Morton, Chair	235

The Extent of Galaxies in 21 cm 241
RENZO SANCISI
Interacting Systems 241
Isolated Systems 244
References 250
Discussion — Donald Morton, Chair 251

Evening Panel Discussion 257
Discussion — Peter Shaver, Chair 257

21 Centimeter Absorption Lines 275
F. H. BRIGGS
Introduction 275
The Nature of the 21 cm Line 276
The Redshifted 21 cm Line Absorbers 278
Unique Observations in the 21 cm Line 284
Conclusions 289
References 289
Discussion — David Jauncey, Chair 291

Damped Ly-α Absorption Systems 297
ARTHUR M. WOLFE
Introduction 297
The Lick Survey 298
Properties Directly Deduced from Accurate Spectroscopy 299
Other Physical Properties 302
Implications for Galaxy Formation 304
References 305
Discussion — David Jauncey, Chair 306

QSO Absorption Lines and the Universe at Redshifts Between 1 and 4 319
J. P. OSTRIKER
Introduction 319
The Ly-α Clouds 320
The Proximity Effect 323
Shells, Voids and Correlations in the Ly-α Cloud Distribution 324
Metal Line Systems and Dust Obscuration 324
References 325
Discussion — David Jauncey, Chair 326

Final Panel Discussion 331
Discussion — Arthur Wolfe, Chair 331

PREFACE

The Universe at redshifts between 1 – 4 is now an area of intense observational study and theoretical modelling. It is in or near this redshift range that many fundamental processes and events occur that determine the structure and evolution of galaxies and their environments, the formation and rapid evolution of active galaxies and QSOs, and the thermodynamic state and metallicity of the intergalactic medium. Much of what we know at these redshifts is based on the study of the absorption lines in QSO spectra. There is a vast and complex body of work here, but a general picture is starting to emerge about how the various phenomena are related.

The study of the physics of QSO absorption lines is based largely on two observational areas: (1) ground-based optical and radio studies of redshifted absorption, and (2) space-based ultraviolet studies of the spectra arising in the disk and halo of our Galaxy and in the environments of nearby galaxies. As an adjunct to the spectroscopic work, high-resolution imaging is required to detect and study the objects which produce redshifted absorption. The whole area will benefit significantly when the Hubble Space Telescope is operating.

We decided to hold the ST ScI 1987 May Workshop on QSO Absorption Lines because the community had not previously had a meeting devoted to this subject, the Hubble Space Telescope will be very important for future work in the area, and, most importantly, fascinating astrophysics could be discussed with a real observational basis.

Many scientists contributed to the meeting either by giving reviews, presenting posters, or participating in the discussions, and most major Observatories and groups in the field were represented. A large number of posters of new results were presented, which were published in July by the ST ScI. The discussion and debate following each review talk was of great value and is indeed the hallmark of our Workshops. We have included the discussions in the book; they are topical and show the flavor of the meeting. Future researchers can wonder how perceptive—or otherwise—we were! The chapters of the book reflect the order that was used for the review talks and discussion sessions during the Workshop.

Once again, Barbara Eller arranged the smooth running of the meeting and associated events with her usual flair and standard of excellence. We thank Karen Vrana and Loretta Willers for help with the meeting, Jonathan Wheatley, Marylin Bell and Carl Grillmair who recorded the talks and discussions, and Lausanne Lee and Melissa Whittington-Riebau for transcribing the discussion tapes. We thank Stu Simpson and Carl Schuetz for assembling the figures in the book, and Duncan Forbes who helped in the editing. Finally, we thank Dorothy Whitman who performed the remarkable feat of turning a pile of inhomogeneous scientific manuscripts and discussion transcripts into a beautiful looking book.

Chris Blades

Dave Turnshek

Colin Norman

PARTICIPANTS

Elise Albert
Ski Antonucci
John Bahcall
Neta Bahcall
Stanislaw Bajtlik
Steven Balbus
Xavier Barcons
Peter Barthel
Marylin Bell
Jacqueline Bergeron
John Black
Chris Blades
Ralph Bohlin
Alec Boksenberg
David Bowen
Frank Briggs
Robert L. Brown
Geoffrey Burbidge
Margaret Burbidge
Richard Buss
Robert Carswell
Adeline Caulet
Roger Chevalier
Ross Cohen
Paul Coleman
Philipe Crane
Stefano Cristiani
Arlin Crotts
Oved Dahari
Laura Danly
Arthur Davidsen
Michael Davis
Ewine van Dishoeck
Michael Disney
Megan Donahue
Robert Duncan
Ian Evans
Michael Fall
Jim Felten
Craig Foltz

Duncan Forbes
Riccardo Giaconni
Prab Gondhalekar
Richard Green
Carl Grillmair
George Hartig
Cyril Hazard
Tim Heckman
Julia Heisler
Mark Henriksen
Richard Hunstead
Garth Illingworth
David Jauncey
Edward Jenkins
Helmut Jenkner
Mike Jewison
Vesa Junkkarinen
Anne Kinney
David Koo
Anuradha Koratkar
Vinod Krishan
Phil Kronberg
Ken Lanzetta
Barry Lasker
Felix Lockman
Knox Long
C. J. Lonsdale
Limin Lu
Olivia Lupie
Duccio Macchetto
George Miley
Palle Moeller
Simon Morris
Donald Morton
Colin Norman
Bill Oegerle
Jeremiah Ostriker
Leonid Ozernoy
Yichuan Pei

Larry Petro
Peter Quinn
Martin Rees
Otto Richter
Gordon Robertson
Hans de Ruiter
Allan Sandage
Renzo Sancisi
Wallace Sargent
Blair Savage
Allen Schiano
Irwin Shapiro
Paul Shapiro
Mike Shara
Prajval Shastri
Peter Shaver
Mike Shull
Eric Smith
Graeme Smith
Harding Smith
Bill Sparks
Lyman Spitzer
Pete Stockman
Christopher Thompson
David Turnshek
Neil Tyson
David Tytler
Marc van der Valk
Nolan Walborn
Joe Wampler
John Webb
Jonathan Wheatley
Richard White
Gerry Williger
Arthur Wolfe
Bruce Woodgate
Adam Wysota
Donald York
Steve Zepf

QSO ABSORPTION LINES: INTRODUCTION

W. L. W. Sargent
Palomar Observatory
California Institute of Technology

Abstract. A brief account is given of the physical properties of the heavy-element and Ly-α forest absorption redshift systems observed in QSO spectra. Theories for their origin are summarized. It is emphasized that the two types of absorption system have qualitatively different properties and that they are almost certainly not components of a single population as Tytler has recently claimed. Extensive new data on the heavy-element redshifts reveal clustering on velocity scales unlikely to have been produced by the motions of clouds within galaxies. Extensive absorption systems observed in the relatively wide QSO pair Q1037–2704 and Q1038–2712 and in the spectra of other QSOs in the surrounding field are described. Observations of this kind appear to represent a new way of studying the large-scale distribution of galaxies at earlier epochs in the expansion of the Universe.

1. INTRODUCTION

Since I want to describe some new results, I shall not dwell on the history of the subject as is customary on these occasions. Instead, I shall confine myself to a brief summary of the present state of affairs, without saying how we got there. Thus, I shall largely resist the temptation to dispense remarks of a vaguely philosophical character of the kind which M. J. Rees once described to me as "unspecific wisdom". Nevertheless, it is clear from the impressive number and variety of the talks and posters at this Workshop that the field of QSO absorption lines is just beginning a new phase, one in which the lines are actually being used to obtain information on the distant Universe. Accordingly, we shall hear reports on the properties and distribution of distant galaxies, on the intergalactic medium and the meta-galactic radiation field, all obtainable in no other way.

The QSO absorption lines are currently divided into three distinct categories:
1. the heavy-element redshifts;
2. the Ly-α forest lines; and
3. the Broad Absorption Line (BAL) redshifts.

The heavy-element redshifts include two important sub-categories: the damped Ly-α systems, which Wolfe (1988) will discuss, and the Lyman limit systems.

It is almost universally agreed that the Broad Absorption Line redshifts, whose properties will be described by Turnshek (1988), are caused by expanding gas which has been ejected from the QSO in question. I shall not say anything further about these systems. On the other hand, the origin of the first two categories has been more controversial. There are several reasons for believing that the majority of the sharp absorption lines are formed in cosmologically distributed, intervening objects. The strongest pieces of evidence are as follows.

1. The distribution of number of redshifts observed per object is close to Poissonian, as would be expected. (Later in this paper it will be shown that the distribution is not exactly Poissonian and that this is to be expected when the complex nature of the large-scale distribution of galaxies is taken into account.)
2. The presence of common absorption systems in the spectra of QSOs widely separated on the plane of the sky (Shaver and Robertson 1983a,b).
3. The energy problems which arise if the absorption systems are supposed to be produced in gas ejected from the QSO (Goldreich and Sargent 1976).

An exception to this generalisation is provided by certain heavy-element absorption complexes which are found close to the emission redshift in certain categories of radio–QSOs (Foltz et al. 1986; Barthel 1987; Foltz et al. 1988). These complexes, which may be related to the broad absorption troughs, almost certainly arise in gas which has been ejected by the galaxy containing the radio source.

2. PROPERTIES OF THE HEAVY-ELEMENT REDSHIFTS

The typical neutral hydrogen column density is $N(HI) \approx 3.0 \times 10^{18}$ cm^{-2} and the Doppler parameter is in the range $5 < b < 25$ km s^{-1}. Excited fine structure lines of CII* and SiII* are usually not seen, implying an electron density in the absorbing gas $n_e \leq 10$ cm^{-3}. There is a wide range of ionization since $N(CII) \approx 10^{15}$ cm^{-2} and $N(CIV) \approx 10^{15}$ cm^{-2}. From this we can deduce that the gas is probably photoionized by a flat-spectrum source (for example, the meta-galactic QSO flux), that the ionization parameter is $\Gamma = n(H)/n(\gamma) = 6.5 \times 10^3$ and that a significant fraction of the gas is neutral. The heavy-elements are somewhat underabundant; $Z \approx 0.1 Z_\odot$. Only crude limits can be placed on the sizes of the absorbing clouds. The line depths show that the clouds cover the QSO emission region which is estimated to have a size $D \approx 10^{19}$ cm. The Jeans length in the gas is estimated to be 2.4×10^{21} cm. The two lines of sight to the gravitationally lensed QSO, Q0957+551 ($z_{em} = 1.40$), show the same C IV absorpion redshifts at $z_{em} = 1.12$, but the detailed absorption profiles are different (Boksenberg and Sargent 1983), implying individual cloud sizes of $D < 2.7 \times 10^{22}$ cm. On the other hand, this observation together with the observation (Shaver and Robertson 1983a; Boksenberg and Sargent 1983) of common absorption redshifts in the QSO pair UM680/681 whose separation is 1 arcmin or 300 kpc, implies that the absorption occurs in relatively small clouds which are embedded in much larger structures. The heavy-element redshift systems often exhibit velocity structure scales extending from 20 km s^{-1} (the lower limit that has been resolved) up to 600 km s^{-1}. The value 150 km s^{-1} is typical. The number density of typical strong CIV redshift systems is $dN/dz \approx 1.5$ at $z = 2$. As we shall see, there is no strong cosmological evolution in the density of absorbers.

3. PHYSICAL PROPERTIES OF THE Ly-α CLOUDS

There are considerable differences between the heavy-element and single Ly-α redshift systems. The typical column density is N(HI) $\approx 10^{15}$ cm^{-2}. There is a power-law distribution of column densities of the form $n(\text{HI}) \approx N^{-1.5}$. (This is the area-weighted observed distribution which translates into a true distribution $n(\text{HI}) \approx N^{-3.5}$.) The line widths are in the range $10 < b < 40$ km s^{-1}; $b \approx 25$ km s^{-1} is typical. There are several limits on the sizes of the absorbing clouds. The observation of common Ly-α absorption lines in the spectra of the two brighter components of the gravitationally lensed QSO, Q1115+080, implies that $D > 0.4$ kpc. An upper limit of $D < 400$ kpc is inferred from the absence of common Ly-α forest lines in the spectra of the QSOs UM 680 and UM 681 referred to above. In a critical measurement, Foltz et al. (1984) showed that most, but not all, Ly-α forest lines were observed in both components of the pair Q2345+007 which is thought to be a gravitational lens, although the lensing galaxy has not been observed. Making a plausible assumption about the redshift of the lensing galaxy, Foltz et al. deduced that the minimum size of the Ly-α clouds was 8 kpc; this is likely to be close to the actual value if it is true that some strong lines are not seen in both lines of sight. If the pair is not a lens then the value of D would be about 15 kpc. Statistical methods have been used to set limits on the abundances of any heavy-elements associated with the Ly-α clouds. Norris et al. (1983) found marginal evidence for absorption in the OVI λλ 1034, 1038 doublet to be statistically associated with Ly-α absorption lines in the spectra of several high-redshift QSOs. They derived an abundance ratio O/H $\approx 10^{-2}$(O/H)$_\odot$ on the assumption (see below) that the Ly-α clouds are photo–ionized by the meta–galactic flux of radiation from QSOs. Later, Chaffee et al. (1985) conducted an analysis of a single high column density system in the QSO Q0014+81 and deduced that they had detected SiIII λ1206 line. The problem is that this line lies in the Ly-α forest. More detailed calculations of the expected ionization conditions in the clouds indicated that SiIII λ1206 would be the strongest observable heavy-element line in their spectral range. The inferred abundance was (Si/H) = 0.001 (Si/H)$_\odot$. A Lyman limit system with no detectable lines of heavier elements was discovered in the QSO Q2126–158 ($z_{em}= 3.30$) by Boksenberg and Sargent (1983). This system has $z_{abs} = 2.99$ and the HI column density is well determined to be N(HI) $= 3 \times 10^{17}$ cm^{-2}. The limit on the possible heavy-element abundance, as reinterpreted by Chaffee et al. in the light of their improved calculations, was $Z = 0.001 Z_\odot$. In summary, it is still not established in my opinion that any lines of heavier elements have been detected associated with the Ly-α forest lines. This is true for both the weaker lines when added together and for single strong systems. In this respect the Ly-α forest redshifts are clearly different from the heavy-element redshifts.

Only weak clustering has been detected in the 2–point correlation function of the Ly-α clouds (Webb 1987, personal communication). An earlier study by Sargent, Young and Schneider (1982) revealed no clustering either in the correlation function of a single QSO, Q1623+268 ($z_{em} = 2.55$), or in the cross–correlation between the Ly-α forest lines in this QSO and a second object, Q1623+269, separated from Q1623+268 by 3 arcmin (or ≈ 1 Mpc) on the plane of the sky. If the local, galaxian correlation function is written as $dP = N_0[1 + \xi(r)]dV$ where $\xi(r) = (r/r_c)^{-1.77}$ and where $r_c = 5$ Mpc, the simplest hierarchical clustering model predicts that r_c changes with redshift according to the relation $r_c(z) = r_c(0)(1+z)^{-5/3}$. In fact, the limit obtained by Sargent et al. was $r_c = 0.2$ Mpc at z= 2.5 as compared with $r_c = 1$ Mpc predicted by the simple theory. The marginal clustering detected by Webb is consistent with the upper limit obtained earlier. Rees and Carswell (1987) also pointed out that there is no evidence for voids in the Ly-α cloud redshift distribution.

The Ly-α clouds exhibit strong evolution in redshift (Peterson 1978; Young, Sargent and Boksenberg 1982; Murdoch et al. 1986). According to the latest work, the number density of the absorption lines evolves as $dN/dz \approx (1+z)^{2.3\pm0.1}$ and that for lines with rest equivalent width above 0.36 Å, $dN/dz \approx 60$ at $z = 2.5$.

Thus, the Ly-α lines differ qualitatively from the heavy-element redshifts in at least three important respects: composition, clustering and evolution in number density.

4. RECENT DEVELOPMENTS ON THE Ly-α FOREST

Tytler (1987a, 1988) has recently derived the distribution of the number density of Ly-α absorption lines as a function of column density and has found that the observed distribution may be represented as a single power law over a wide range in column density from $N(H\ I) = 10^{12}$ to 10^{21} cm^{-2}.

The low end of this range is dominated by lines of the Ly-α forest, while the high end is dominated by Ly-α lines associated with redshifts containing heavy-element lines. Tytler points out that the single power law makes it hard to believe that the lines are produced by two distinct kinds of objects. Moreover, the absence of a break in the distribution at a column density of $N(HI) \approx 10^{17}$ cm^{-2} makes it hard to believe that the objects which produce the lines are highly ionized as is commonly believed. Instead, Tytler proposes that the clouds are neutral at all column densities; for this to be true they must be small, dense objects in order not to be ionized by the meta-galactic QSO flux. In fact, Tytler advocates a typical size of about 3 pc. This is hard to reconcile with the observations of common lines in the spectra of gravitationally lensed QSOs. Moreover, the claim of one population of objects is in stark disagreement with other observed regularities—in particular, the differences in composition, cosmological evolution, and clustering which were summarized above.

In addition, some authors have questioned Tytler's claim that a simple power-law distribution actually applies (Bechtold 1987; Wolfe 1988); the distribution of raw equivalent widths derived by Sargent et al. (1980) certainly showed different slopes for the Ly-α forest and heavy-element Ly-α lines. Therefore, it seems unwise at this stage to pursue the consequences of Tytler's radical suggestion until the facts are more clearly established.

A second development, which observationally is on firmer ground, concerns the behavior of the Ly-α forest in the vicinity of the Ly-α emission line. Carswell et al. (1984) pointed out that, contrary to a previous analysis by Sargent et al. (1980), there is statistical evidence for a deficiency of absorption lines in the blue wing of the Ly-α emission line. They interpreted this as being due to the effect of the QSO in effecting the ionization of the Ly-α clouds in its immediate vicinity. In a related development, Murdoch et al. (1986) pointed out that whereas in QSO spectra overall there is a marked tendency for the density of Ly-α forest lines to increase with redshift, in a given QSO spectrum there is a flat or even declining tendency with increasing redshift (see also Hunstead 1988). This result, named the inverse effect by Murdoch et al. was shown to be due to a statistical deficiency of Ly-α absorption lines close to the emission line, as Carswell et al. had found. Tytler (1987b) extended this result, which now seems to be firmly established. Bajtlik, Duncan and Ostriker (1987, 1988) have quantified Carswell et al.'s suggestion and have shown that Tytler's data may be fitted very well if it is assumed that close to the QSO the local radiation field dominates over the metagalactic QSO flux.

The discovery of the inverse effect, or the proximity effect as it is called by Bajtlik et al. (1988), is very important because it allows a direct determination to be made of the total meta-galactic ionizing flux at high redshifts, where the contribution of lower luminosity sources such as star-forming galaxies and Seyfert nuclei cannot be observed. Interestingly, Bajtlik et al. found a good fit to the observations with a meta-galactic flux at $z = 2.5$ of $I_\nu = 10^{-21}$ ergs s^{-1} cm^{-2} ster^{-1}, a value which has been estimated as the total contribution of QSOs alone. Moreover, the success of Bajtlik et al.'s calculation probably enhances one's confidence that the Ly-α clouds are indeed highly ionized by the meta-galactic UV flux as Sargent et al. (1980) proposed.

5. ORIGIN OF THE HEAVY-ELEMENT REDSHIFTS

The heavy-element redshifts are thought to be due to intervening galaxies. The lines of evidence in favor of this hypothesis are as follows.
1. The discovery of CaII and NaI absorption lines in the spectra of QSOs close to nearby galaxies on the plane of the sky (Boksenberg and Sargent 1978; see also the review by Blades 1988).
2. The discovery of galaxies at the same redshift as absorption systems observed in the spectra of QSOs with redshifts up to $z \approx 0.8$. Several impressive cases of such galaxies are detailed in the survey article by Bergeron (1988).
3. The presence of lines of heavy-elements which make it unlikely that intergalactic objects are responsible.
4. The energy arguments referred to in §1 which rule out ejection from the QSO.
5. The existence of common absorption systems in the spectra of widely separated QSOs which can be interpreted as being due to galaxies in clusters or superclusters.

The frequency of absorption systems of a given type may be used to infer the required mean cross-section for absorbers of different types, assuming that the Schechter (1977) function describes the distribution over galaxy luminosities and that the radius of a galaxy is related to its luminosity in the same manner as the Holmberg radius, namely $R \approx L^{5/12}$. With these assumptions the effective radius, R^*, of an L^* galaxy comes out to be 75 kpc for the Lyman limit systems, 50 kpc for the MgII systems, 90 kpc for the CIV systems and 480 kpc for the Ly-α clouds, supposing that they originate in galaxies. These numbers, which were given to me by Tytler (1986, private communication), are evaluated for a mean redshift of $z = 2.5$, for $H_o = 100$ km s^{-1} Mpc^{-1} and include only spiral galaxies. As is well known, these effective cross-sections are much larger than the optical sizes of galaxies at the present epoch. Also, according to Tytler (1987a) there are 6 times as many Lyman limit absorbers with $N(HI) \geq 3.0 \times 10^{18}$ cm^{-2} at $z = 2$ than are seen locally in 21 cm measurements. In a similar vein, Wolfe (1988) has deduced that there are five times as many damped Ly-α absorbers (presumed to be galactic disks) with $N(HI) \geq 2.0 \times 10^{20}$ cm^{-2} at $z = 2.6$ than there are locally per unit co-moving volume. These observations may constitute evidence for substantial evolution of galactic disks in relatively recent epochs.

A critical question concerning the heavy-element redshifts is the origin of the velocity structure. Three types of possible origin have been proposed:
1. motions of clouds within galaxies which would be limited to about 200 km s^{-1} in

order to be gravitationally confined;
2. explosive phenomena in galaxies such as supernova remnants; and
3. the tip of the galaxian correlation function, which should extend out to velocities beyond 1000 km s^{-1}, well beyond the velocities which would be expected for gravitationally confined clouds. We shall return to this question in §VII.

6. THE ORIGIN OF THE Ly-α FOREST LINES

The exact nature of the Ly-α clouds which are responsible for the Ly-α forest discovered by Lynds (1971) is still controversial. It is clear that they are cosmologically distributed, intervening objects which are poor in heavy-elements. Given the present uncertainties in their properties, the clouds could be dwarf irregular galaxies (Fransson and Epstein 1982), primordial intergalactic clouds (Arons 1972; Peterson 1978; Sargent et al. 1980; Black 1981), intergalactic shocks (Chernomordic and Ozernoy 1983) or even the extended primordial halos of normal galaxies (Bahcall and Spitzer 1969). More recently, Rees (1986) has proposed that the clouds are confined by 'mini–halos' of the cold dark matter whose existence has been invoked to reconcile the lack of observed anisotropy in the microwave background radiation with the existence of galaxies, clusters and superclusters today (see also Rees 1988). Hogan (1987) has taken Tytler's arguments in favor of small, dense clouds seriously and has suggested that the Ly-α forest results from shocks generated by collisions between thermally unstable proto–galactic clouds. On the other hand, Bond, Szalay and Silk (1987) see the clouds as pressure–driven relics of primordial density fluctuations. On this scenario, while more massive clouds are confined by the cold dark matter, the Ly-α clouds are unstable entities whose expansion has been triggered by the onset of QSO and galaxy formation at a presumed redshift of around $z \approx 4$.

If the Ly-α forest lines are indeed intergalactic in origin, there are three possibilities regarding their stability:
1. they are pressure-confined by a general intergalactic gas;
2. they are freely expanding; and
3. they are self–gravitating.

As Ostriker (1988) elaborates, freely expanding clouds are ruled out because their expansion velocities are too big for the observed line widths while for self-gravitating clouds the collapse times are too short.

This leaves pressure–confined clouds as proposed by Sargent et al. (1980) and Ostriker and Ikeuchi (1983). These latter authors devised a model in which the clouds originate from a shock–heated intergalactic medium containing fragmentary shells. On this hypothesis, the most massive condensations form galaxies while the lowest mass ones evaporate. This leaves a narrow range of 10^7 to 10^8 $M_\odot$ in equilibrium at $z = 2.5$. The clouds evaporate and dissolve as the Universe expands and the pressure of the intergalactic medium goes down.

7. NEW RESULTS ON CLUSTERING

Together with A. Boksenberg (Royal Greenwich Observatory) I recently completed observations of the absorption spectra of 60 QSOs with redshifts in the range $1.8 < z_{em} < 3.56$, primarily in order to study the clustering and evolution of heavy-element redshifts as represented by the easily identified CIV $\lambda\lambda$1548,1550 doublet.

The spectra were obtained with Boksenberg's IPCS detector mounted on the blue camera of the Double Spectrograph at the Cassegrain focus of the Hale telescope. The on-line display of the accumulated count level was used to obtain a uniform S/N ratio of 20:1. The resolution was 1.4 Å. Most of the spectra covered the wavelength range from the CIV $\lambda 1549$ emission line down to the Ly-α emission line. In several cases the coverage extended into the Ly-α forest and in some cases the resolution was 0.7 Å. The spectra have all been reduced, measured and redshifts have been determined. The results are now being assembled for publication (Sargent, Boksenberg and Steidel 1988). The results extend those obtained in similar, but less ambitious, surveys carried out by Young, Sargent and Boksenberg (1982) and by Foltz et al. (1986).

The main results of the CIV survey are as follows.

A total of 202 CIV absorption redshifts were identified in the 60 QSOs. A uniform sample was isolated of 130 redshifts with rest equivalent widths for both CIV lines greater than 0.15 Å. Some of these systems fall in clumps which are clearly not independent entities. Therefore, an additional sample (called the 'Poisson sample') was formed in which clumps containing more than one system on a scale of less than 1000 km s^{-1} were treated as one system. The Poisson sample contains 107 absorption redshifts.

In this sample, which is composed primarily of radio-quiet QSOs, there is no significant tendency for absorption redshifts to cluster around the emission redshift of the QSO. This behavior is in marked contrast to that exhibited by radio QSOs as Foltz et al. (1986) have clearly shown.

The distribution of number of absorption redshifts per QSO does not conform to a Poisson distribution as is expected for randomly distributed intervening absorbers (Bahcall and Peebles 1969). This is true at the 90 percent confidence level even for the Poisson sample. A likely explanation of this result is that over the small ranges in redshift ($\Delta z \sim 0.6$) available in any given QSO, the details of the large scale distribution of galaxies along the line of sight play an important role in the statistics.

The density of absorption systems per unit redshift range dN/dz decreases with increasing redshift in the range $1.4 < z_{abs} < 3.4$ according to an approximate law $dN/dz \propto (1+z)^{-1.2 \pm 0.4}$. All three samples show the same qualitative behavior. On the other hand, a constant co-moving density of absorbers with constant cross-section should have $dN/dz \propto (1+z)$ if $q_o = 0$ and $dN/dz \propto (1+z)^{1/2}$ if $q_o = 1/2$. The observed behavior of the C IV doublet density is quite different to that shown by the Ly-α clouds which increase in density as $dN/dz \propto (1+z)^{2.3}$ over the same redshift range (Murdoch et al. 1986). Moreover, the 'Lyman limit' absorption systems and the MgII redshift systems are observed to increase with redshift at something like the rate expected for $q_o = 0$ (Tytler 1986, private communication).

The 2-point correlation function for the CIV doublets has been generated. It is generally flat as is expected for a randomly distributed sample of absorbers. However, there is significant clustering on scales less than 500 km s^{-1} in the rest frame of the absorbers. Analysis shows that the observed clustering is very unlikely to be due to motions of clouds within galaxies but is more likely due to the galaxian correlation function. These new results, which are still being analyzed, emphasize the differences between the heavy-element redshifts and the Ly-α forest lines, which display little or no clustering tendency.

8. QSO ABSORPTION LINES AND LARGE–SCALE STRUCTURE

In March 1986, Steidel and I obtained spectra ($\sim$ 2 Å resolution, S/N ratio $\sim$ 40) at the Las Campanas Observatory of two QSOs, Tol1037–2704 ($z_{em} = 2.193$) and Tol1038–2712 ($z_{em} = 2.331$), to which attention had been drawn by Jakobsen et al. (1986). The QSOs are separated by 17.9 arcmin on the sky. Confirming and extending the work of Jakobsen et al., it was found that there are at least five absorption systems in the spectrum of each QSO, occupying the same narrow range in redshift from 1.89 to 2.14. There is a statistically significant excess of absorption systems in the spectra of both QSOs; however, the tendency which was pointed out by Jacobsen et al. for the systems in one QSO to match a corresponding system in the other to within 1800 km s^{-1} was judged not to be significant (Sargent and Steidel 1987). For a Universe with $q_o = 0.5$, the separation of the two lines of sight is $\sim$ 4.3 h_{100} Mpc, by far the largest distance over which correlated absorption has ever been seen. In addition, we obtained low-resolution ($\sim$ 4 Å) spectra of a fainter QSO Tol1038-2707 (only 5 arcmin from Tol1038-2712) which has a strong absorption system ($z_{abs} \approx 1.887$) which corresponds to one of the common absorption systems in the other two QSO spectra, as well as a spectrum of Tol1037-2742 containing a possible strong absorption system at $z \sim 1.86$. In April 1987, another nearby QSO, Tol1035-2737, was found to have a redshift of $z_{em} = 2.168$ and to have absorption redshifts at $z_{abs} = 2.125$, 2.040 and 1.905, respectively. Collectively, these observations suggest that this general direction in the sky contains a large supercluster extending from a redshift of around $z = 2.14$ to 1.86. Tol1038-2712 lies behind the supercluster, Tol1037-2704 and Tol1035-2737 lie at its most distant edge, while Tol1038-2707 is close to the nearer edge. The supercluster has an extent along the line of sight of $\sim$ 50 Mpc at the epoch corresponding to $z \sim 2$. Its boundaries on the plane of the sky are not known as yet; however, the extent in this direction is at least 10 Mpc. Such extended structures, if elongated, would produce a non-Poissonian distribution of number of absorption redshifts per QSO of the kind referred to in the discussion of the results of the CIV survey.

9. THE QSO PAIR Q2343+1229, Q2344+1225

This pair was discovered on a U.K. Schmidt objective prism plate by C. Hazard and confirmed by Boksenberg and myself. Q2344+1229 has a magnitude of 17 and a redshift of $z = 2.53$, while Q2344+1225 has magnitude of 17.5 and a redshift of $z = 2.77$. There is a common absorption system in both objects at $z_{abs} = 2.42$; in Q2344+1229 this system has a damped Ly-α absorption line, while in Q2344+1225 the system is weaker and more complex. To add to the coincidences, there is a damped Ly-α absorption in the higher redshift QSO, Q2344+1225, at $z_{abs} = 2.53$, similar to that of the lower redshift object. The separation of these two QSOs is 5.7 arcmin, corresponding to 2 – 4 Mpc at $z = 2.5$, depending on the cosmological model. This is yet another case of a pair of common or associated absorption in relatively widely spaced QSOs. It will be interesting to explore the limits on the physical properties of the object responsible for the damped Ly-α system in Q2344+1225 which are imposed by its proximity to Q2344+1229. Interestingly, there is a third, fainter QSO about 2 arcmin from Q2344+1225 with a similar emission redshift, $z = 2.75$, but with no common absorption systems.

10. CONCLUSIONS

It will be seen that the QSO absorption lines have many potential uses in cosmology. We have real prospects that they can be used to study the gross properties of the interstellar gas in galaxies back to the time of their formation. The counts of absorption lines as a function of redshift will eventually provide an important cosmological test. Observations of correlated absorption, both in redshift and in nearby QSOs on the plane of the sky, are leading to important information on the evolution of the large-scale distribution of galaxies in space. Finally, the enigmatic Ly-α clouds offer a unique opportunity to directly study the evolution of the intergalactic medium and the meta-galactic ionizing radiation flux and to provide information on primordial abundances of the elements.

Acknowledgements. I thank my co-workers, A. Boksenberg and C. Steidel, for allowing me to quote the results of our joint work in advance of publication. I also thank D. Tytler for providing me with details of his unpublished work. The work described in this paper was supported in part by the National Science Foundation under Grant AST84-16704.

REFERENCES

Arons, J. 1972, *Ap. J.*, **172**, 553.
Bahcall, J. N., and Peebles, P. J. E. 1969, *Ap. J.*, **156**, L7.
Bahcall, J. N., and Spitzer, L. 1969, *Ap. J.*, **156**, L63.
Bajtlik, S., Duncan, R. C., and Ostriker, J. P. 1987, *QSO Absorption Lines: Probing the Universe, A Collection of Poster Papers*. eds. J. C. Blades, C. A. Norman, D. A. Turnshek, (ST ScI Publication), p. 49.
_____. 1988, *Ap. J.*, submitted.
Barthel, P. 1987, *QSO Absorption Lines: Probing the Universe, A Collection of Poster Papers*. eds., J. C. Blades, C. A. Norman, D. A. Turnshek, (ST ScI Publication), p. 19.
Black, J. H. 1981, *M.N.R.A.S.*, **197**, 553.
Blades, J. C. 1988, *This volume*.
Bechtold, J. 1987, *Ap. J.*, submitted.
Bergeron, J. 1988, *This volume*.
Boksenberg, A., and Sargent, W. L. W., 1978, *Ap. J.*, **220**, 42.
_____. 1983, in *Proc. 24th Liège Symposium, Quasars and Gravitational Lenses* (Liège: Institut d'Astrophysique), p. 500.
Bond, J. R., Szalay, A., and Silk, J. 1987, *Ap. J. (Letters)*, submitted.
Carswell, R. F., Morton, D. C., Smith, M. G., Stockton, A. N., Turnshek, D. A., and Weymann, R. J., 1984, *Ap. J.*, **278**, 486.
Chaffee, Jr., F. H., Foltz, C. B., Roser, H. J., Weymann, R. J., and Latham, D. W., 1985, *Ap. J.*, **292**, 362.
Chernomordic, V. V., and Ozernoy, L. M. 1983, *Nature*, **303**, 153.
Foltz, C. B., Chaffee, Jr., F. H., Weymann, R. J., and Anderson, S. F., 1988, *This volume*.
Foltz, C. B., Weymann, R. J., Roser, H. J., and Chaffee, Jr., F. H. 1984, *Ap. J. (Letters)*, **281**, L1.

Foltz, C. B., Weymann, R. J., Peterson, B. M., Sun, L., Malkan, M. H. and Chaffee, Jr., F. H. 1986, *Ap. J.*, **307**, 504.

Fransson, C. and Epstein, R. 1982, *M.N.R.A.S.*, **198**, 1127.

Goldreich, P. and Sargent, W. L. W. 1976, *Comments on Ap. and Space Sc.* **6**, 133.

Hogan, C. B., 1987, *Ap. J.*, **316**, L59.

Hunstead, R. W. 1988, *This volume*.

Ikeuchi, S. and Ostriker, J. P. 1986, *Ap. J.*, **301**, 522.

Jakobsen, P., Perryman, M. A. C., Ulrich, M. H., Macchetto, F., and Di Serego Alighieri, S. 1986, *Ap. J. (Letters)*, **303**, L27.

Lynds, C. R. 1971, *Ap. J. (Letters)*, **164**, L73.

Murdoch, H. S., Hunstead, R. W., Pettini, M., and Blades, J. C. 1986, *Ap. J.*, **309**, 19.

Norris, J., Hartwick, F. D. A., and Peterson, B. A., 1983, *Ap. J.*, **273**, 450.

Ostriker, J. P., 1988, *This volume*.

Ostriker, J. P., and Ikeuchi, S. 1983, *Ap. J. (Letters)*, **268**, L63.

Peterson B. A. 1978, in *Proceedings of IAU Symposium 79: Largescale structure of the Universe* ed. M. S. Longair and J. Einasto (Dordrecht: Reidel) p. 390.

Rees, M. J. 1986, *M.N.R.A.S.*, **218**, 25P.

————. M. J., 1988, *This volume*.

Rees, M. J. and Carswell, R. F. 1987, *M.N.R.A.S.*, **224**, 13P.

Sargent, W. L. W., and Boksenberg, A., 1983, in *Proc. 24th Liège Symposium, Quasars and Gravitational Lenses* (Liège: Institut d'Astrophysique), p. 518.

Sargent, W. L. W., Boksenberg, A., and Steidel, C. C. 1988, *Ap. J. Suppl.*, in preparation.

Sargent, W. L. W., and Steidel, C. C. 1987, *Ap. J.*, **322**, 142.

Sargent, W. L. W., Young, P. J., Boksenberg, A. and Tytler, D. 1980, *Ap. J. Suppl.*, **42**, 41.

Sargent, W. L. W., Young, P J. and Schneider, D. P., 1982, *Ap. J.*, **256**, 374.

Schechter, P. 1976, *Ap. J.*, **203**, 297.

Shaver, P. A., and Robertson, J. G., 1983a, in *Proc. 24th Liège Symposium, Quasars and Gravitational Lenses* (Liège: Institut d'Astrophysique), p. 598.

————. 1983b, *Ap. J. (Letters)*, **268**, L57.

Turnshek, D. A. 1988, *This volume*.

Tytler, D. 1986, *private communication*.

————. 1987a, *Ap. J.*, **321**, 49.

————. 1987b, *Ap. J.*, **321**, 69.

————. 1988, *This volume*.

Wolfe, A. M. 1988, *This volume*.

Young, P. J., Sargent, W. L. W., and Boksenberg, A. 1982, *Ap. J. Suppl.*, **48**, 455.

DISCUSSION

G. Burbidge: Well, having just come from the Arp Symposium in Venice, I would like to remind the audience that there still is a considerable body of evidence suggesting that many objects' redshifts are not associated with distance, consequently one would be forced by this hypothesis to argue that all of the absorption in these objects is intrinsic. Now, you started off by giving the normal statement that the Broad Absorption Line systems are indeed associated with intrinsic properties, in essence I think because no one knows how to make them intervene, and that the narrow line systems are intervening. All I want to point out is that, it seems to me, there is still a good deal of evidence of various kinds suggesting that significant numbers of sharp lined systems may be intrinsic, and you mentioned that at the end of your talk in one sense. There is also this next question of the line-locking hypothesis which Margaret first proposed in the early 1970's and which various people attempted to dispose of in the middle 1970's, but indeed is still around. If you look for example in Janet Drew's sample of 1978, or in even more recent work, you can see that certain ratios, particularly the ratio of 1.11, are very clearly present. Now, since some of the systems involved in that ratio are the so-called damped Ly-α systems, which are claimed to be intervening galaxies, one would have, if one believed in the reality of a line-locking ratio, to argue that this is intrinsic, and therefore that the absorption was indeed associated with the objects and not with intervening galaxies. The final point I would like to make in this connection is, even if you go to the mixed picture, in which you allow many of these objects to have redshifts associated with the expansion of the Universe, if you do indeed argue that a significant fraction of the sharp absorptions as well as the broad absorptions are intrinsic, then that does probably relieve many of the problems associated with the demands on the halos and galaxies and all the other things which, as far as I can see, lead to a rather contrived picture, and have ever since evidence for large halos were first found.

N. Bahcall: In those cases where you see the two QSOs in a large structure with absorption lines, is there a way that you can put a limit on the possible velocity within that large structure by comparing detailed wavelengths of the absorption lines?

Sargent: You can in principle. Of course, along the line-of-sight you can choose to call a given redshift either distance or velocity. That's a choice you have to make. On the other hand, perpendicular to the line-of-sight you can make some statements of that kind. If you look at the pairs of absorbers in the two objects that I've shown you, that you think might really be the same pair, the sort of velocity differences you get are below 1500 km s^{-1}. There are a lot of statistics that go into trying to figure out whether those pairs along the spectrum are really pairs or not, or whether they are just randomly distributed within a particular interval. As I said, two different calculations of that effect give very different probabilities, and I think it would require another case to be sure, and maybe several other cases. I wanted to make the point that we are in the situation to be able to study superclusters eventually using techniques of this kind. I should say, in partial reply to what Geoff said, that I don't see how the ejected hypothesis does you much good in looking at common absorbers in several QSOs in the same part of the sky unless they're very close.

G. Burbidge: That may be true, but that's a very complicated scenario, and I haven't fully digested it.

Webb: My question concerns the clustering result of the CIV systems and also the inverse effect that you discussed. Supposing for your 60 QSOs you had used, rather than the CIV lines, the Ly-α absorption lines associated in the same systems, but without CIV information. What do you think the correlation function might look like?

Sargent: Based on other cases it would have looked the same as for CIV. In these particular objects we generally did not observe the Ly-α region because the sort of resolution that you need to get some sense out of the Ly-α region is higher than you need just to split the CIV.

Webb: However, the Ly-α lines are stronger than the CIV lines, and so I'm trying to reconcile the differences found (in the correlation function) between the Ly-α only clouds and the CIV clouds. Don't you think the amplitude would maybe be smaller for the metal line systems if you had not used the CIV doublet, but instead the corresponding Ly-α line?

Sargent: I don't know. We can do this work eventually without using just CIV doublets, but they were the easiest things to pick out.

Wolfe: Wal, in the damped Ly-α system that you showed can you say anything about the velocity difference between the two objects when you look at the narrow metal lines?

Sargent: This is where I show a damped Ly-α in a QSO which is close on the sky to another one. The difference crudely measured is less than .01 in Δz, and I think that can be refined by better measurements of the emission redshift. You may have noticed that both QSOs in that pair have rather shallow lines, and they will really require a lot more work to get the emission redshifts correct, but it's less than .01.

Wolfe: So it's not clear yet whether it's a common object or two objects?

Sargent: I think they're very close together, whatever it is.

Ozernoy: Have you made an attempt to see structure in the peak of the correlation function at 600 km s^{-1}? For instance, you can compare column densities of adjoining lines and see if they have the same column densities.

Sargent: No. That hasn't been done. I did the simplest thing, which was to put all the lines together without regard to strength. Eventually we shall have to study them as a function of equivalent width to see how the correlation looks.

Chen: I always worry about selection effects in any statistical result. Because observation of a QSO at rather high resolution takes a lot of time, there may be a tendency to center the wavelength range of the spectrum on absorption seen at low resolution. So is there any worry that based on the presence of absorption in a low resolution spectrum, the decision is then made to take a higher resolution spectrum?

Sargent: Are you saying that's what I do, or what other people do? (Laughter)

Norman: What do you do, Wal? (Laughter)

Sargent: If you chose the objects because you knew they had absorption in the first place you would be completely wasting your time, so you obviously don't do that. In fact you may have noticed that although the spectra have enormously higher signal-to-ratios (around 20 to 1) than the discovery spectra typically have, 6 out of 60 turned out to have no heavy-element absorption at all. Therefore, if I was trying to cheat, I did it extremely badly. Of course, it is also the case that in order to study clustering you have to have at least two absorbers in any one spectrum.

Foltz: Of your 60 QSOs, how many were BAL QSOs?

Sargent: None.

Foltz: Did you throw them out?

Sargent: Yes.

E. M. Burbidge: Are you intrigued by the inverse effect?

Sargent: Yes.

E. M. Burbidge: I wondered also in your very rich Ly-α forest, how many of the lines can be accounted for by Ly-β, Ly-γ, etc. Can you sort those out?

Sargent: Yes. To start with, I found this inverse effect about five years ago, independently, and then tried to explain it before publishing it. Of course, the first possibility is that there is a problem with Ly-α lines and Ly-β lines. However, if you look at Tytler's plot you see that he only goes from Ly-α down to Ly-β, so there's no possibility that it's Ly-α lines and Ly-β lines.

E. M. Burbidge: How may QSOs were in that sample?

Sargent: I don't know. Tytler may know.

Tytler: Nineteen QSOs or so, and 400 lines.

Sargent: There is a way, in principle, of testing the idea that this effect is caused by the ionization of the QSO blasting all the Ly-α clouds around it, which would happen to all close by things in the same line-of-sight. I showed you one case, the case that Artie was just talking about, where the line-of-sight from one QSO went close to another QSO. Now the Ly-α lines, which were in the line-of-sight to the more distant one, which were in the vicinity of the even closer QSO, also ought to have been blasted. Right? So you should see a defect in close pairs where the line-of-sight has passed another QSO. I've looked for that effect, and in the first example I looked at, which is a paper I published with Young and Schneider (1982, *Ap. J.*, **256**, 374), there seemed to be such an effect. However, in other cases, there's nothing. I think in principle that that would enable you to tell whether this blasting idea is correct. If it is, then it will complicate the statistics of the Ly-α forest enormously because they will depend on how close to another QSO the line-of-sight went which, of course, it will be possible to tell, but nevertheless it will be complicated to take into account.

York: Geoff mentioned the pressure the statistics put on this idea of halos, and you quoted four different radii for different measurements—Lyman limits and so forth. Can you say to what extent those have been cleaned for the local effect, the radio QSO type of effect? Is that a sample that doesn't have any radio sources, and they're all consistent with each other?

Sargent: It doesn't need to be cleaned for that, because it was mostly stuff that we have done that didn't have very many radio QSOs.

York: So the Lyman limit numbers, are those yours?

Sargent: Those are the ones that Tytler has obtained. I think that usually radio sources are not common in that sample.

P. Shapiro: Was there any evidence in the one example you have of a close damped Ly-α absorber of standard metal line absorbers at the same redshift that could be interpreted as halo of the damped Ly-α system?

Sargent: Yes. There were associated heavy-element lines which were sharp and low ionization which is commonly the case in these high column density systems.

Rees: I have a question which perhaps preempts Bob Carswell's about the size of the clouds. Does the gravitational lens double image really tell you about the size of an individual cloud or just some correlation scale; and what statement are you prepared to make dogmatically about cloud sizes in light, for instance, of the fact that they must cover the continuum or maybe the line emitting region of the background QSO?

Sargent: Well, I'm not a theoretician, so I never make dogma. (Laughter) Ray Weymann himself has pointed out that getting two identical absorbers in two images does not mean you're seeing the same thing in both. I tend to think that you are seeing the same thing in both because there seems to me to be a correlation between the strength of the lines in both, and the velocity differences are small. I think that with really good spectra you can measure the velocity differences accurately and determine a mass by multiplying the velocity difference squared by the separation, and then compare that with the mass that you infer from the column density itself and the ionization fraction. We're already on the limits of being able to do that, and it would only require somewhat higher resolution on these very difficult objects to make such calculations, which I think would then allow one to tie in the properties of the Ly-α absorbers better.

Rees: Because I think you can almost say directly then that they are gravitationally bound, right?

Sargent: Almost, yes, but not quite. The gravitationally bound hypothesis is not the hypothesis that I have been pushing.

N. Bahcall: On the correlation function of the Ly-α clouds, I thought I understood you to say this correlation is weaker than that of galaxies, but I didn't hear you say by how much. Is it consistent, for example, with the trend that we are finding of going down with luminosity or estimated mass?

Sargent: I don't know. It's about a factor of five weaker than you would guess for the case where you extrapolate back the local correlation function to $z = 2.5$ using the simplest extrapolation. For example, if the correlation length here is 5 Mpc, at a redshift of 2.5 it ought to be 1 Mpc and the limit is less than 0.2.

BAL QSOs: OBSERVATIONS, MODELS AND IMPLICATIONS FOR NARROW ABSORPTION LINE SYSTEMS

David A. Turnshek
Space Telescope Science Institute
Baltimore

Abstract. The current status of observations pertaining to BAL QSOs is reviewed. Evidence which motivates a simple disk-like geometric and kinematic model for the BAL region is presented. This includes constraints on distance, covering factor and the acceleration/deceleration of the BAL region flow. Evidence which suggests that BAL region abundances are enhanced relative to solar values is also reviewed. The relation between this model and models for QSO broad and narrow emission line regions is discussed. Finally, evidence which suggests that some of the narrow absorption line systems seen in QSO spectra are intrinsic or caused by the BAL phenomenon is considered.

1. INTRODUCTION

The subject of BAL QSOs could be addressed either at a workshop on QSOs and AGNs, where discussion about the emission line region is emphasized, or at a workshop such as this one, where the origin of the narrow absorption lines seen in QSO spectra is an issue. In my opinion there is strong evidence that the BAL region gives rise to a component of the broad emission line region and, at the same time, that the phenomenon producing the BALs may give rise to narrow absorption line systems. There have been four other review papers dedicated to the subject of BAL QSOs (Weymann and Foltz 1983; Turnshek 1984b; Weymann, Turnshek and Christiansen 1985; Turnshek 1986). This current review is not designed to replace the previous ones, but is designed to complement them. I am aware of no serious errors in the previous reviews. BAL QSOs characteristically exhibit ultraviolet resonance line absorption due to highly ionized species *typically* extending from outflow velocities near 0–5000 km s^{-1} up to 10,000–30,000 km s^{-1}. There are a great variety of BAL profile types, but they occur in otherwise relatively normal QSO spectra.

The first two BAL QSOs were discovered by C. R. Lynds (1967) and E. M. Burbidge (1970), but the identification of these types of QSOs did not become routine until objective prism surveys became a common way of discovering QSOs. Given the fact that BAL systems are widely accepted to be intrinsic to the QSO, there are basically two extreme interpretations for them which should be reviewed at the outset. One possibility is that the BALs may occur in only a peculiar subset of QSOs comprising about 3–10 % of the QSO population. In this interpretation the BAL region covering factor must be large, implying that an observer would see BALs in a BAL QSO's spectrum from all lines-of-sight. The early model of Scargle, Caroff and Noerdlinger (1970) adopted this interpretation. The other extreme possibility is that the value of the BAL region covering factor is not near unity, but is approximately equal to the fraction of QSOs that have BALs. In this case an observer would see BALs in a QSO's spectrum only if he were looking in a preferred line-of-sight which intersected BAL clouds. This interpretation would be in agreement with the early model put forth by Lucy (1971). Of course, the consequence of this latter interpretation is that all QSOs may have BAL regions.

As I noted earlier, it is reasonable that a talk on BAL QSOs be given both at a conference discussing QSO/AGN emission line regions and at a conference on (narrow) QSO absorption lines. Therefore, for this conference I discuss both of these topics. First, I review the observations. Then I review the properties that a model for the BAL region must possess. Finally, I consider evidence which may suggest that the narrow absorption lines seen in QSO spectra are related to the BAL phenomenon. I suspect that this latter topic may be of greater interest to the audience of this workshop.

2. OBSERVATIONS

In the past 5–10 years it has been recognized that a significant fraction of QSOs exhibit BALs and a considerable body of data concerning their properties now exists. Below I briefly review these observations.

2.1. Fraction of QSOs with BALs

Optical discovery programs for QSOs show that at moderate to high redshifts BAL QSOs comprise about 3–10 % of the QSO population (Hazard et al. 1984). Since CIV absorption must usually be present to discover a BAL QSO, they are much more difficult to discover at emission redshifts less than about 1.25. Nevertheless, based on observations with IUE or the rare occurrence of MgII BALs, low redshift examples of BAL QSOs have been identified. This work suggests that the fraction of QSOs with BALs at low redshift is not significantly different from the fraction observed at moderate to high redshifts (Turnshek and Grillmair 1986). To stress this point, I show in Figure 1 two low redshift BAL QSOs recently discovered on the basis of MgII BALs during QSO survey work being conducted at the MMT (Foltz et al. 1987b). In this study we report on the discovery of 192 QSOs on four UK Schmidt objective prism plates. This work is part of a much larger survey in which UK Schmidt objective prism plates are digitized with the Automated Plate Measuring machine in Cambridge. Objects brighter than $m_B = 18.5$ are then examined using software algorithms to automatically select QSO candidates. These candidates are observed spectroscopically to confirm their nature. Simulations will be used to consider in detail the question of selection effects in our survey. Such work will eventually lead to a clear determination of the fraction of QSOs with BALs. However, deferring the issue of selection effects to the future, I simply note that of the 192 QSOs initially discovered, two out of 80 QSOs with $z_{em} < 1.25$ have MgII BALs and probably 8 (but possibly 11) out of 112 QSOs with $z_{em} > 1.25$ have BALs. Finding two BAL

QSOs in a sample of 80 on the basis of MgII BALs was unexpected given the current estimate that MgII BALs are only visible ~ 5 % of the time in BAL QSO spectra.

At the very highest redshifts ($z_{em} > 3$) Hazard et al. (1986) suggest that the fraction of QSOs with BALs is notably larger than 3–10 %. It has been pointed out that in the well-studied SGP region, 4 of the 24 QSOs known to have $z_{em} > 3$ are BAL QSOs (Hewett 1987, private communication). In fact, as of this writing, the QSO with the highest known redshift $z_{em} = 4.43$ is a BAL QSO (Warren et al. 1987).

Finally, in terms of the luminosity dependence of the BAL QSO phenomenon, an examination of 54 low redshift ($z_{em} < 1.25$) PG QSOs/AGNs in the IUE archives show that 5 of 34 objects with QSO-like luminosities exhibit BALs, while none of 20 objects with Seyfert-like luminosities exhibit BALs (Turnshek and Grillmair 1986).

2.2. Characteristics of BALs

BAL QSOs exhibit a great deal of variety in absorption characteristics in terms of outflow velocity, level of ionization, velocity structure in the absorption troughs and strength of the absorption. The BALs often appear detached from the emission peak, but usually set in by the time an outflow velocity of 5000–10,000 km s^{-1} is reached. Maximum outflow velocities normally lie in the range 10,000–30,000 km s^{-1}. Cases where maximum outflow velocities are lower include RS23 (Turnshek et al. 1988; see also Figure 2) and Q0302+170 (Foltz et al. 1984; see also Figure 6). Cases where maximum outflow velocities may be higher include UM141 (Turnshek et al. 1980) and Q1414+087 (Foltz et al. 1983; see also Figure 5). In the latter case a maximum outflow velocity of ~ 60,000 km s^{-1} is suggested. Resonance line transitions due to CIV, SiIV, NV and OVI are commonly observed in absorption. Some Ly-α absorption is usually present, but is often weak. Sometimes CIII and AlIII BALs are observed. In rare instances, about 5 % of the time at moderate redshifts, MgII and CII are observed. As discussed in posters presented at this workshop, there are also good cases of BALs due to resonance line absorption of FeIII or PV, SIV and SVI (Turnshek et al. 1987; see also Figure 13) and AlII and FeII (Hazard et al. 1987a,b). Hazard et al. also report absorption from excited states of FeII and possibly FeIII. The velocity structure of the BAL troughs ranges from being very smooth for either shallow or deep absorption to very broken-up for either strong or weak absorption. Sometimes absorption-absorption line-locking is present and at least for the SiIV absorption lines in Q1303+308, has been shown to be statistically significant, with only about a 1 % probability of being due to chance (Foltz et al. 1987a; see also Figure 10).

2.3. Correlation Between BAL and Broad Emission Line Profiles

There is some indication that the properties of BAL profiles are correlated with the properties of the corresponding broad emission lines. This possibility is most easily seen if one parameterizes the CIV broad emission line according to peak emission line height to continuum height ratio (line-to-continuum ratio) and FWHM velocity, and then examines the BAL profile types that can occur as a function of broad emission line profile type. In order to illustrate the variety of BAL profile properties and demonstrate the possible existence of this correlation, I have prepared four figures on similar scales, each showing the Ly-α to CIV region in the QSO rest frame for three BAL QSOs. BAL QSOs having relatively narrow broad emission lines with

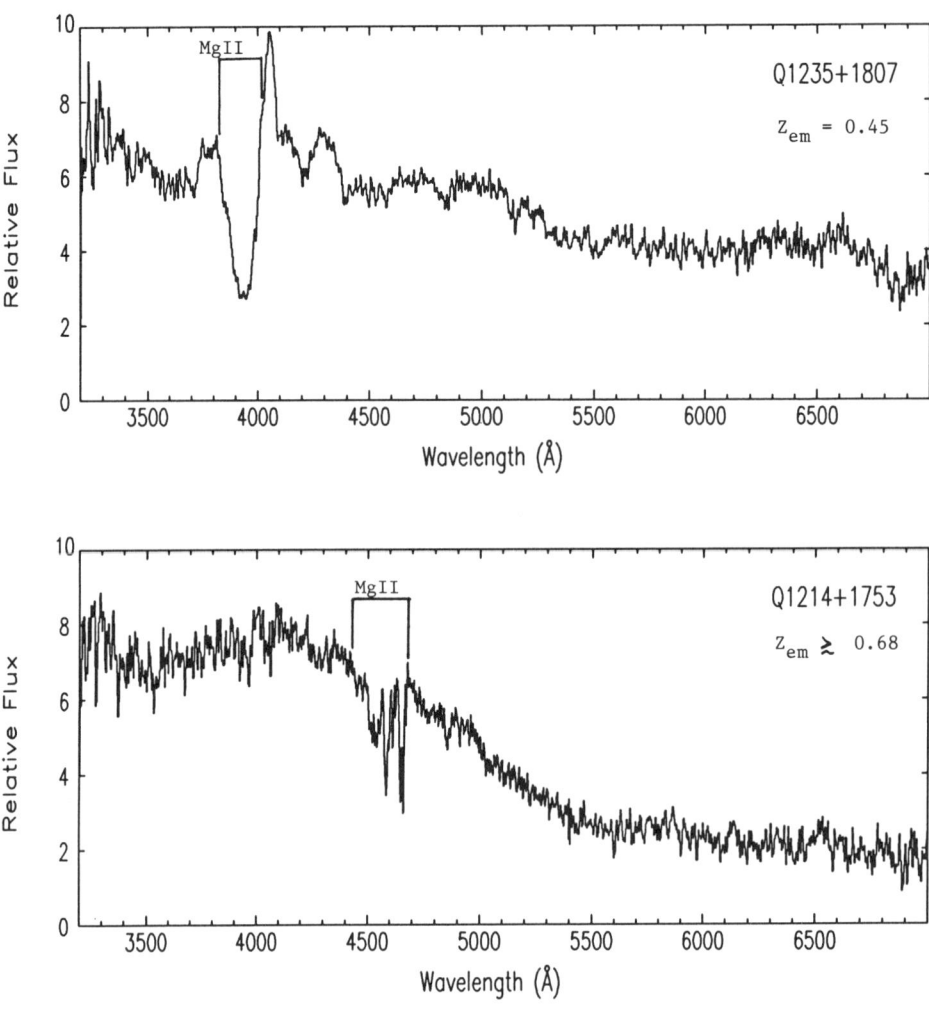

Figure 1. *Spectra of 2 new low redshift BAL QSOs discovered in the APM QSO survey.*

large line-to-continuum ratios are shown in Figure 2. They have smooth single-trough absorption which sets in at very near zero outflow velocity. These are the so-called PHL5200-type BAL QSOs (see Turnshek et al. 1988). BAL QSOs having relatively narrow broad emission lines, but smaller line-to-continuum ratios are shown in Figure 3. They have multiple trough absorption which also sets in at very near zero outflow velocity. BAL QSOs with broader broad emission lines, but still moderate line-to-continuum ratios are shown in Figure 4. They have BALs which exhibit some smoothness, but which are detached from the broad emission line peak, not setting in until outflow velocities of 3000–5000 km s^{-1} are reached. Finally, BAL QSOs with the weakest broad emission lines (FWHM hard to measure) are shown in Figure 5. They tend to be the ones having BAL profiles with the most structure. The process of trying to quantify such a correlation is difficult because so many parameters are involved and because good data are sparse. Finally, additional types of BAL/emission line profiles do exist, but it is not clear how they may fit into this suggested correlation. Hazard et al. (1984) and Hartig and Baldwin (1986) have suggested other possible correlations.

2.4. Emission Line Properties of BAL QSOs

The question of whether the distribution of emission line properties of BAL QSOs are the same as those of non-BAL QSOs is an important one because this tells us something about the BAL/emission line region geometry and the value of the BAL region covering factor for the case where all QSOs are taken to have BAL regions. Selection bias in discovery programs presents a real obstacle to getting a clear picture of the distribution of emission line properties (e.g., see §2.1). If the models for the emission line regions and BAL regions of QSOs are spherically symmetric, then given a small BAL region covering factor which is roughly constant, the distribution of emission line properties in BAL QSOs and non-BAL QSOs should be similar. However, there is evidence that, although the range of emission line properties for BAL QSOs and non-BAL QSOs may be similar, the distribution of emission line properties are not. In my opinion, in BAL QSOs as compared with non-BAL QSOs, four effects are fairly clearly established:
1. CIV emission is weaker;
2. NV emission is stronger;
3. AlIII emission is stronger; and
4. FeII emission is stronger.

See the work of Turnshek (1984b), Hartig and Baldwin (1986), Surdej and Hutsemekers (1987) and Junkkarinen, Burbidge and Smith (1987) for details and concerns. As discussed in §3.4, if BAL region covering factors are small and all QSOs have BAL regions, this must imply that the BAL region geometry is non-spherically symmetric.

2.5. Radio Properties of BAL QSOs

Another important issue concerns the radio properties of the BAL QSOs compared to other optically selected QSOs. Stocke et al. (1984) have shown that bonafide BAL QSOs are radio quiet. Multi-frequency observations at radio wavelengths of new BAL QSOs discovered since the Stocke et al. observations and previously known BAL QSOs have confirmed this (P. Coleman 1986, private communication). On the other hand, there are radio-loud QSOs which have an excess of absorption near the emission

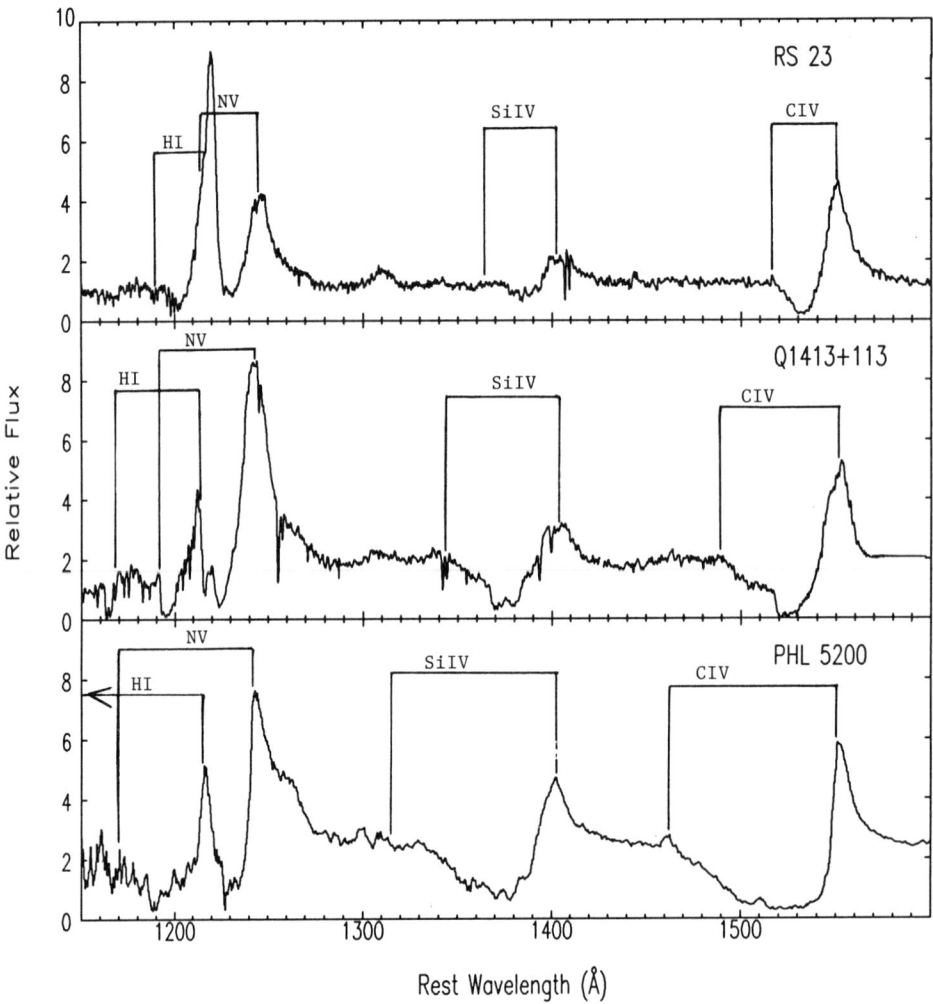

Figure 2. *Spectra of 3 QSOs exhibiting relatively narrow broad emission lines with large CIV emission line-to-continuum ratios. Objects which have emission lines exhibiting these characteristics tend to have smooth single trough absorption which sets in very near zero outflow velocity.*

redshift (Briggs, Turnshek and Wolfe 1984; Morris et al. 1986; Anderson et al. 1987), and the absorption seen in these objects may have some relation to the BALs. I discuss the similarity between this type of absorption and BALs later in §4.1. Foltz et al. (1988) discuss the properties of this class of absorption line systems in the paper following this one.

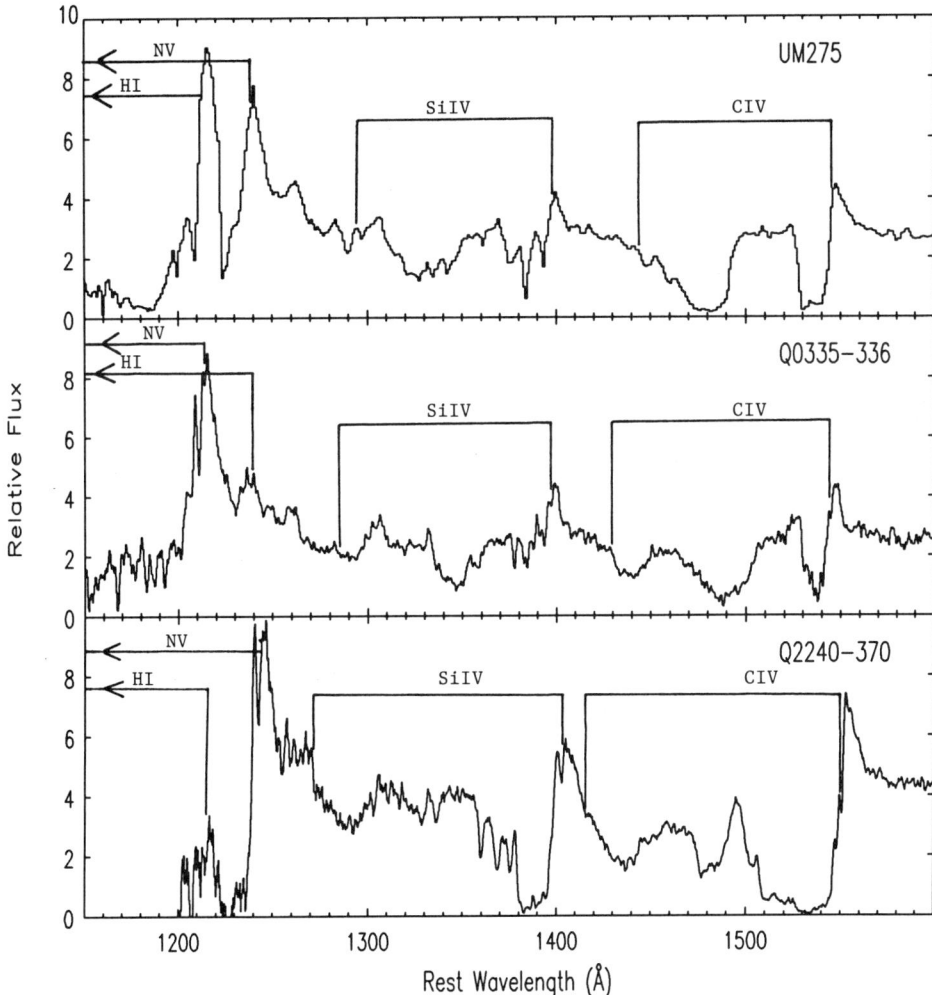

Figure 3. *Spectra of 3 QSOs exhibiting relatively narrow broad emission lines with moderate strength CIV emission line-to-continuum ratios. Objects which have emission lines exhibiting these characteristics tend to have multiple trough absorption which also sets in very near zero outflow velocity.*

2.6. Optical Polarization and Variability of BAL QSOs

The optical polarization properties and possibly the optical variability properties of BAL QSOs appear to differ markedly from other radio quiet QSOs. If BAL region covering factors are small and all QSOs have BAL regions, this must also imply that BAL region geometry is non-spherically symmetric. I briefly summarize the observations. A broad-band polarization survey of optically selected QSOs has indicated that BAL QSOs have a fairly high probability of exhibiting significant polarization, unlike the situation for non-BAL QSOs (Moore and Stockman 1984; G. Schmidt 1987, pri-

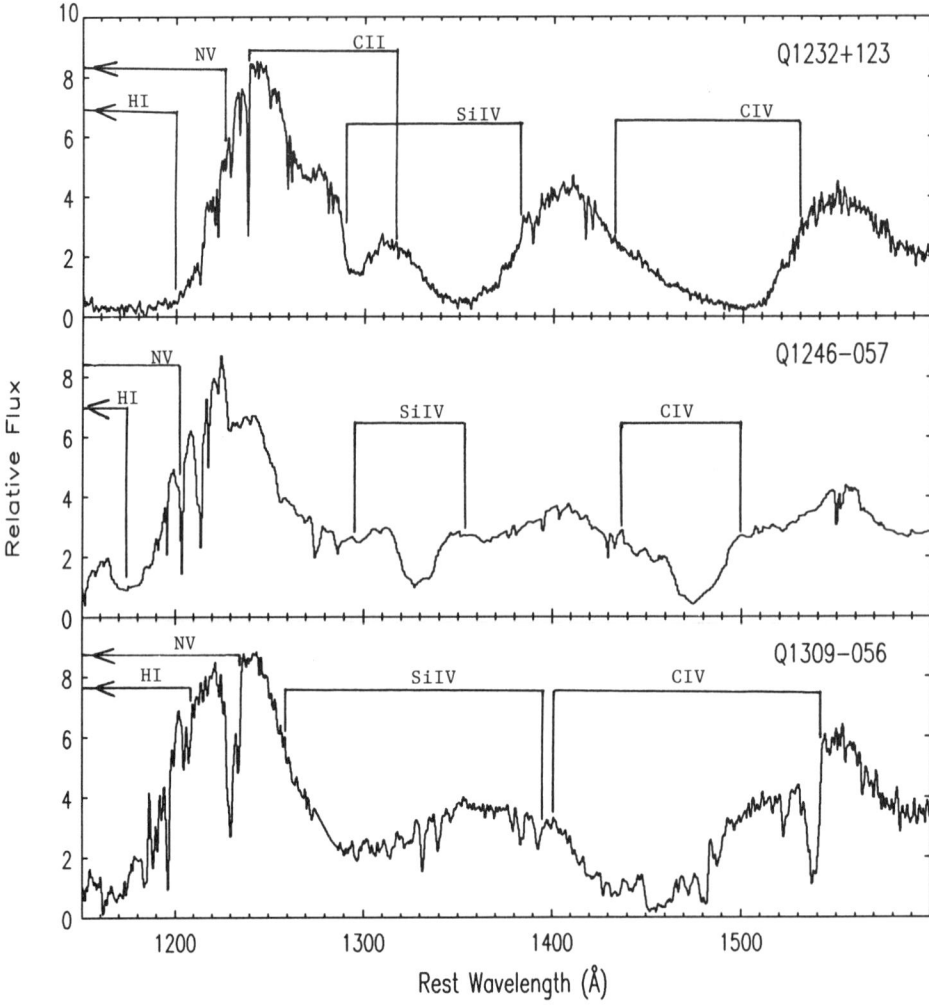

Figure 4. *Spectra of 3 QSOs exhibiting more extended broad emission lines with moderate strength CIV emission line-to-continuum ratios. Objects which have emission lines exhibiting these characteristics tend to have BALs which show some smoothness, but which are detached from the emission peak, not setting in until outflow velocities of 3000–5000 km s^{-1} are reached.*

vate communication). In their survey of 19 BAL QSOs, 5 BAL QSOs exhibit a fairly high optical polarization of 1.5 % or more. The 5 objects exhibiting significant polarization are Q0135-400, Q0932+501, Q1246-0578, Q1413+113 and PHL5200. The PHL5200-type BAL QSOs may be the ones that typically have the highest optical polarizations. For one of the objects (PHL5200) spectropolarimetry has been published (Stockman, Angel and Hier 1981).

In view of the fact that radio-loud optically violent variable QSOs exhibit signifi-

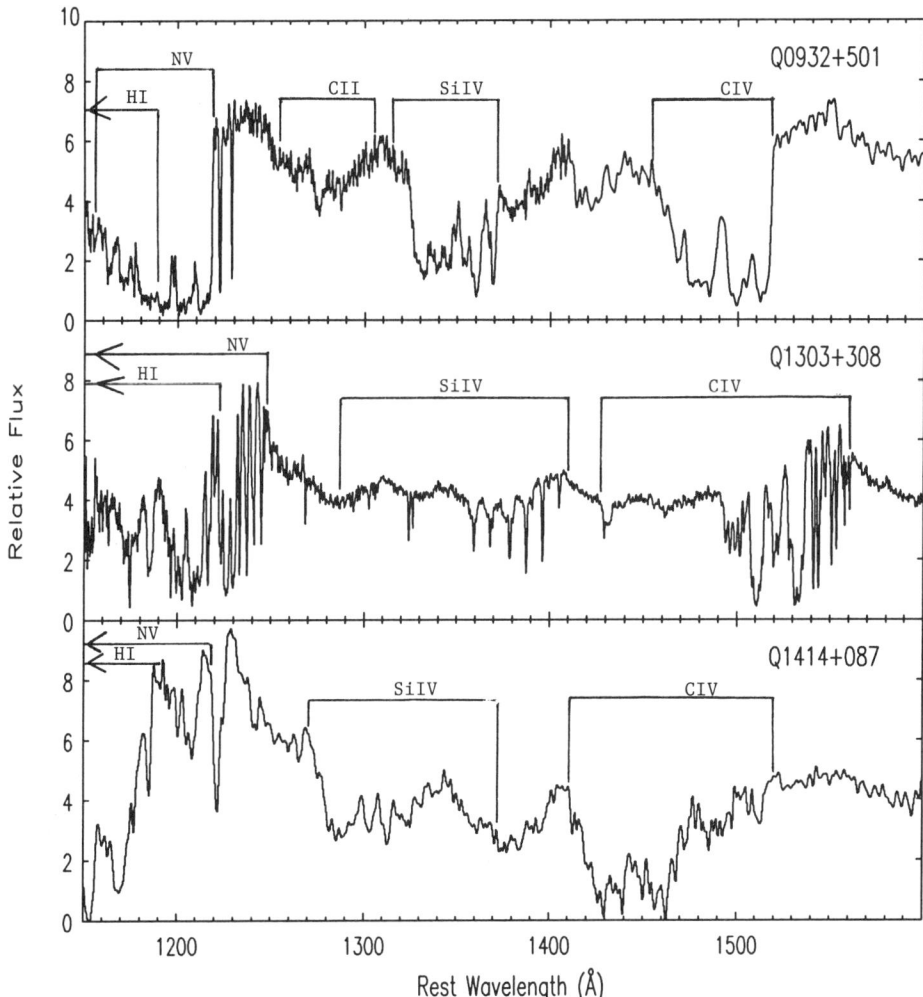

Figure 5. *Spectra of 3 QSOs exhibiting broad emission lines with very small CIV emission line-to-continuum ratios. The emission line widths are hard to measure accurately in these objects. Objects which have emission lines exhibiting these characteristics tend to have BALs which show the most structure.*

cant optical polarization, a possibly related topic is the question of optical variability of BAL QSOs. A photographic optical variability study of 64 optically selected moderate to high redshift QSOs by Netzer and Sheffer (1983) accidentally included 5 BAL QSOs. On two plates taken 31 years apart, the non-BAL QSOs as a class did not appear to exhibit variability in excess of ~ 0.2 mag rms scatter which was about the precision of the photographic photometry; however, 3 of the BAL QSOs showed evidence for variation at levels of ~ 1.1 mag, ~ 0.8 mag and ~ 0.5 mag. In another study, the BAL QSO 1E 0104.2+3153 discussed by Giola *et al.* (1986) has varied approximately ~ 0.5 mag between the beginning of 1982 and the beginning of 1984. To study this phenomenon in more detail, my collaborators and I have undertaken a program of photometric monitoring and we have just recently begun to obtain multi-epoch data.

2.7. BAL Trough Variability

Just recently, three reports have been presented indicating that the strength of the BAL troughs may have changed with time. The study by Foltz et al. (1987a) indicates that the strength of the highest velocity CIV absorption in Q1303+308 may have decreased over a two year period, while Turnshek et al. (1988) suggest that the SiIV absorption trough in Q1413+113 may have increased in strength over a 4 year period. Observations of Q1246-057 over a 2 year period are also reported to show variation in absorption line intensity (Penston and Smith 1988), but these same authors found no convincing evidence for variations in Q1309–056. Confirming observations for all of these cases, preferably coupled with photometric monitoring, need to be obtained. Junkkarinen, Burbidge and Smith (1983) report that the strength of the absorption in PHL5200 has not changed over a 6 year period. Also, there is no evidence for outflow velocity variations in BAL QSOs (e.g., Junkkarinen, Burbidge and Smith 1983; Foltz et al. 1987a). In general, studies of the possibility of BAL trough intensity and outflow velocity variability are rare.

2.8. Imaging of BAL QSOs

In concluding my brief review of the observations, I summarize the important results on imaging of low redshift BAL QSOs. Three low redshift BAL QSOs have so far been imaged and have been shown to be extended. Hickson and Hutchings (1987) barely resolved the image of the $z_{em} = 0.29$ BAL QSO Q1700+518. Recently, I have obtained a limited amount of data in collaboration with Burrows, Hutching and Weymann on the $z_{em} = 0.13$ BAL QSO Q1416-129 and this object is also resolved. In both cases the images are rather circular, inconsistent with the presumed host galaxy being edge-on or nearly edge-on. The data for the $z_{em} = 0.089$ BAL QSO Q1411+442 are much better. Accurate images taken by Malkan, Green and Hutchings (1987) show this BAL QSO's host galaxy to be a large spiral with a very long arm or tail which has an inclination angle of 57°. The IUE spectrum shows Q1411+442 to be a BAL QSO of the PHL5200 type, being very similar spectroscopically to RS23 and Q0302+170. These observations have implications for aspect angle effects and the nature of the BAL region geometry.

3. MODEL PROPERTIES

In this review I do not discuss mechanisms for producing the observed level of ionization in the BAL region nor do I discuss the details of mechanisms proposed for accelerating the BAL region clouds to the high outflow velocities that are observed. These topics are covered in some detail in the review by Weymann, Turnshek and Christiansen (1985). They review the substantial problems involved with reconciling the observed BAL region column densities with the results of photoionization models which adopt solar abundances. These problems are still not solved, probably because

BAL region abundances are very enhanced relative to solar abundances (see §3.6). They also review acceleration mechanisms and show that a thermal wind model provides a very effective way of accelerating the BAL region clouds given the constraints that observations impose on models. Despite evidence for absorption-absorption line-locking, the force due to radiation pressure is likely to play only a minor perturbing role in the cloud acceleration process.

My main goal in this section is to present a geometric and kinematic picture of the BAL region which follows from the observations. This is a necessary first step for developing an accurate complete model of the BAL region which eventually can explain the origin of the BAL and emission line regions, the ionization, the abundances, the acceleration and the details of the observed velocity structure. To accomplish this, first I discuss the starting point for developing such a model for the BAL region. Next, I discuss constraints which place both upper and lower limits on the size-scale of the BAL region and discuss evidence which indicates that the global BAL region covering factor is small. Observational considerations which place constraints on the large and small scale geometry of the BAL region and which provide evidence that the BAL region flow is decelerating in PHL5200-type BAL QSOs are then presented. While the geometric and kinematic model that is developed for the BAL region is consistent with the observations, a considerable amount of observational and theoretical work needs to be done in the areas of evaluating selection effects in discovery programs and model calculations in order to explore the details. I also review evidence suggesting that BAL region abundances are enhanced relative to solar values. After presenting this geometric and kinematic model, I then discuss how this view may fit into the framework of models for the QSO's broad and narrow emission line regions.

3.1. The Starting Point of a Model

There are a number of issues which need to be weighed when trying to arrive at a starting point for a model of the BAL region. One issue involves assessing the success of previous models which may provide the basis for building a model of the BAL region. Given that existing models which postulate the existence of many small clouds are fairly successful within the range of parameter space that they have explored, a basic question is: can the constraints provided by observations of BAL QSOs be incorporated into existing models without altering the existing models too much? In the case of the BAL region the answer is a qualified 'yes'. One finds that the existing models can be built upon, but only at the expense of reinforcing the multi-component picture of the QSO emission line region which has been reviewed and motivated in some detail by Davidson and Netzer (1979) and more recently by Osterbrock and Mathews (1986). On the other hand, an equally valid approach is to take a fresh look at the problem and decide whether new information provided by the observations of BAL QSOs suggests a completely different starting point for a model. For example: do the data warrant treating BAL QSOs as a separate population of QSOs which are largely unrelated to more normal non-BAL QSOs? Do the data provide some constraint which would completely invalidate a basic assumption of existing models or which is at odds with constraints imposed by non-BAL QSO observations? Since the answer to these last two questions is 'no', the approach that has been taken here is to try to fit the BAL region into the framework of existing models. This is done, while at some level realizing that this is not an ideal situation, because it is always apparent that constraints on model parameter space are not very extensive.

Notably, some other workers have taken the opposite approach (e.g., Surdej and Swings 1981; Drew and Gidding 1982; Drew and Boksenberg 1984; Surdej and Hutsemekers 1987). Generally, these authors present models for the BAL regions of QSOs that are not consistent with the framework of existing models and they adopt the view that BAL QSOs and non-BAL QSOs represent two essentially different populations of QSOs. It is fairly clear that, within the context of the model which will be developed here, this cannot be true. As discussed in §3.3, within the framework of existing models, observations constrain the maximum value of the global BAL region covering factor to be fairly small, implying that certain lines-of-sight to QSOs with BAL regions must show fairly normal non-BAL QSO spectra. It remains to be seen whether additional constraints which are derived from observations of BAL and non-BAL QSOs become incompatible, thereby demanding two separate populations of objects. In the interest of not trying to invent new populations of objects without cause, it would seem that the present course of trying to fit the BAL region into the framework of existing models is the most profitable one.

3.2. The Size Scale of the BAL Region

Based on imaging studies of low redshift BAL QSOs (see §2.8), it is fairly clear that the BAL region does not arise many kiloparsecs from the central source QSO as a result of the acceleration of the host galaxy's interstellar medium due to the presence of an active nucleus. This possibility was originally proposed by Scott, Christiansen and Weymann (1984) and presumably would require the QSO's host galaxy to be viewed edge-on for BALs to be seen. This is definitely not the case for the BAL QSO Q1411+442. On the contrary, the more recent possibility that BAL profile properties are correlated with corresponding broad emission line profile properties suggests that the BAL region is intimately related to the broad emission line region which is thought to be on the order of ~ 1 pc from the central source. In addition, the observation of BALs from excited states of FeII and possibly FeIII with lower level excitations in the 0.12–3.7 eV range reported by Hazard et al. (1987a,b) would also suggest that the BAL region is nearby the central source.

In the context of existing models, observations which place both a lower limit and an upper limit on the BAL region location with respect to the central source have been discussed (Turnshek et al. 1985; Turnshek and Grillmair 1986; Turnshek et al. 1988). Good agreement with the observed data in the region of the Ly-α/NV profiles can be obtained if one assumes that the BAL region is separate and distinct from and completely occults the region producing the Ly-α broad emission line. Therefore, as a direct lower limit on BAL region location, one knows that it lies farther from the central source than the region producing the Ly-α broad emission line. This is consistent with the conclusion that MgII broad emission lines cannot be produced in the BAL region (Turnshek et al. 1984). This lower limit is therefore dependent on the model for the broad emission line region (see §3.7). At the same time, constraints imposed by the photoionization process can be used to infer a lower limit on $n_e r^2$ (the BAL region electron density times the distance to the BAL region squared). This is because the photoionization parameter is $\xi = L_\nu/n_e r^2$ and the luminosity, L_ν, is known. Thus, there is a mininum value for $n_e r^2$ because photoionization will make the BAL region level of ionization too high if $n_e r^2$ is too low. The minimum amount of emission arising from collisional excitation in the BAL region can thus be estimated given suitable values for the global BAL region covering factor, the temperature of the BAL region, the column density of the ion of interest, and so on (Turnshek 1984a). The results generally indicate that observable high ionization broad emission lines may arise due to collisional excitation in the BAL region, even for very small assumed values of the global BAL region covering factor. This is especially true for [OIII]λ5007 broad emission. In fact, the lack of observed broad [OIII] emission in low redshift QSOs has been used to constrain the global BAL region covering factor to be less than 0.01 for the case where $n_e < 10^6$ cm^{-3} in the BAL region. Since the fraction of QSOs observed to have BALs is greater than 0.01, this latter result generally leads to the conclusion that $n_e > 10^6$ cm^{-3} in the BAL region, which suggests for upper limits that the BAL region lies within $\sim$ 500 pc for the most luminous QSOs and within $\sim$ 30 pc for the least luminous QSOs. However, due to the uncertainties regarding abundances and the level of ionization of the BAL region (see §3.1 and §3.6), it is very important at this stage to remember that *these deductions are not based on self-consistent calculations*; the emissivity of the BAL clouds due to collisional excitation is not very tightly constrained.

3.3. The Global BAL Region Covering Factor

The global BAL region covering factor for a QSO is simply the fraction of the QSO covered by BAL region clouds integrated over all possible lines-of-sight to the QSO. As discussed by Turnshek (1984a), an upper limit on the size of the global BAL region covering factor can be set by assuming that no more than the entire observed NV broad emission line may be created by NV resonance line scattering of inner continuum plus Ly-α photons in the BAL region. Using a constraint based on NV instead of CIV has advantages. For the case of CIV, an upper limit on the global BAL region covering factor is usually set either by checking the residual intensity at the bottom of the CIV absorption trough or by checking for excess extended emission on the red side of the CIV broad emission line. However, these upper limits on the global covering factor must apply only for the case where the BAL region flow is accelerating. As I note in §3.5, this may not be the case for some BAL QSOs. Furthermore, in cases where CIV is utilized, one usually assumes that most of the observed CIV broad emission line *does not* come from resonance line scattering, but rather arises from collisional excitation. This is consistent with models for the broad emission line region and the fact that collisional excitation must contribute to the CIV broad emission line because a CIII] broad emission line is observed. Of course, the CIII] broad emission line cannot come from resonance line scattering since CIII] BALs are not observed. However, since models for the broad emission line region often predict little NV emission due to collisional excitation, it is not *ad hoc* to assume that most of the observed NV emission comes from resonance line scattering.

In order to set a limit on the global covering factor, one must first assume that the NV absorption troughs are extensive enough in wavelength coverage to scatter most inner Ly-α broad emission line region photons. This is seen to be true in most BAL QSOs by an examination of the spectra. Examples of where this does or does not occur are shown in Figure 6. As a result of the fairly low BAL region outflow velocities in the BAL QSO Q0302+170, Figure 6 shows that the NV broad emission line is not very enhanced by resonance line scattering; however, in the more normal case where BAL region outflow velocities are large, the NV broad emission line appears enhanced as in the spectrum of Q1413+113 also shown in Figure 6.

In order to place limits on the maximum value of the global BAL region covering factor, it is convenient to initially assume that the equivalent width of the scattered NV emission produced will be independent of observer aspect angle. As I discuss in §3.4 and note below, the assumption of spherical symmetry is probably erroneous, but the assumption gives a firm upper limit on the value of the global BAL region covering factor. Finally, one must also assume that none of the NV scattered radiation is destroyed. From these considerations, observations of BAL QSOs allow us to infer that the global BAL region covering factor is less than 0.2–0.4 for most BAL QSOs, regardless of whether the flow is accelerating or decelerating. Of course, the covering factor is probably lower than this because some of the NV broad emission line may arise from collisional excitation and because the global BAL region geometry and BAL cloud shapes may be such that NV photons are *preferentially* scattered in the direction of the observer in BAL QSOs (see §3.4). If the BAL region flow is accelerating, the depth of troughs and the strength of any red emission extension for CIV would generally require global BAL region covering factors less than 0.1–0.2 (Turnshek et al. 1980; Junkkarinen 1983). These estimates of the maximum value of the global BAL region covering factor are typical for most BAL QSOs. The exceptions are cases where the NV and/or CIV broad emission lines are much weaker. In these cases the upper limits on the value of the global BAL region covering factor can be considerably lower.

Surdej and Hutsemekers (1987) have recently argued that the results of this type of analysis may be in error and that, as noted in §3.1, the simplest interpretation of the BAL QSO phenomenon is that BAL QSOs and non-BAL QSOs form at least two distinct populations of QSOs. This would imply that the global BAL region covering factor was near unity, suggesting that one's line-of-sight would nearly always intercept a BAL region when observing a QSO that had BAL regions, regardless of the observer's aspect angle. Surdej and Hutsemekers suggest that three mechanisms may reduce the net amount of emission created by resonance line scattering, thus invalidating the results discussed above. First, they note that turbulence in an expanding atmosphere can reduce the emission to absorption ratio in a resonance line scattering model. Second, they note that occultation by the central core is particularly important for decelerating flows and can reduce the net emission from resonance line scattering. Third, they claim that collisional de-excitation in the BAL region can destroy the resonance line scattered photons. They further argue that separate populations of BAL and non-BAL QSOs are consistent with the data because the (underlying) broad emission line properties of the two classes of objects differ. While it is true that the broad emission line properties of BAL QSOs and non-BAL QSOs probably differ, aspect angle effects in non-spherically symmetric geometries and small global BAL region covering factors may also explain these differences since they are not large. This is not surprising because such a model has extra free parameters. The main point which should be made here is that a large global BAL region covering factor, which Surdej and Hutsemekers (1987) favor, is apparently inconsistent within the frame work of existing models because the mechanisms they propose for reducing the emission from resonance line scattering probably will not operate (Turnshek et al. 1988). However, the validity of any criticisms should be considered in the context of choosing an appropriate starting point for a model (§3.1).

Finally, one remaining argument, implicit from the discussion in the previous section, suggests that the BAL region has a small global covering factor. Namely, if the global BAL region covering factor were not small, one would expect substantial emission due to collisional excitation (e.g., from CIV) arising in the BAL region.

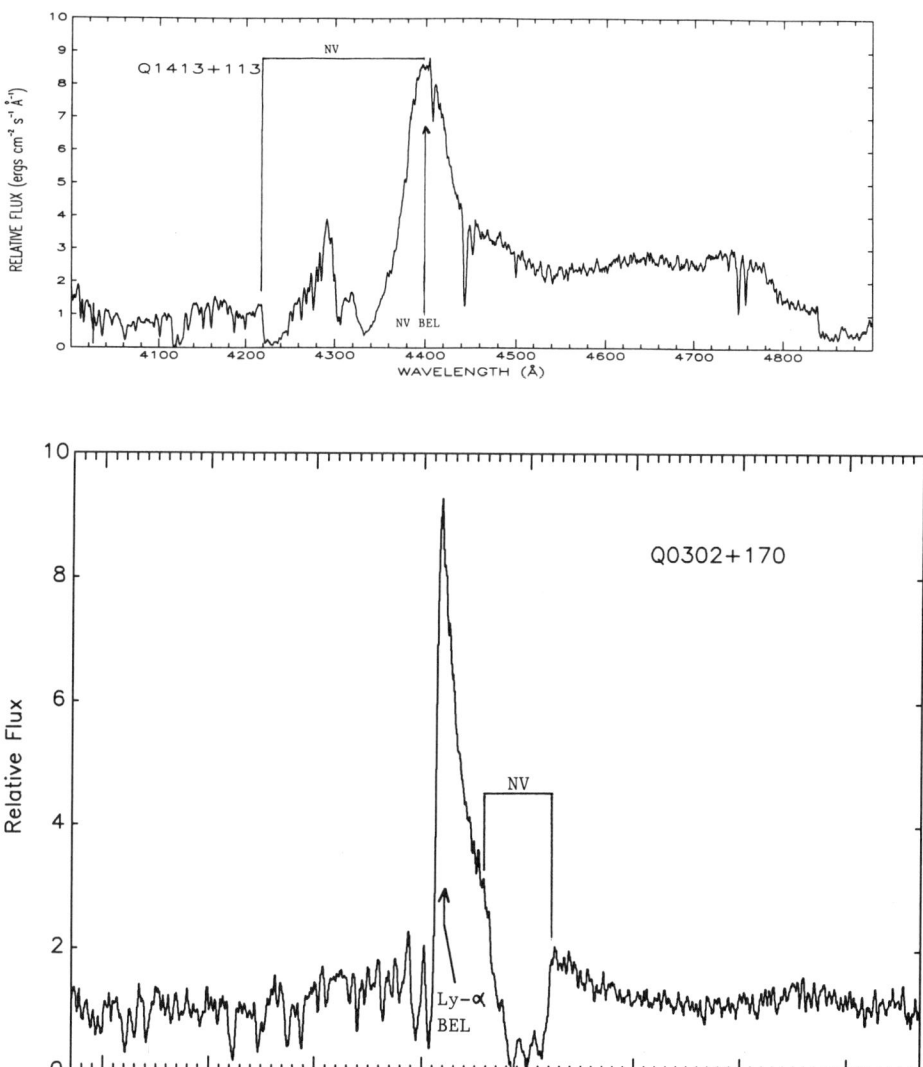

Figure 6. Spectra of the Ly-α and NV region in 2 BAL QSOs. In the top figure, Q1413+113 is seen to have very strong NV emission to the blue of where the NV absorption trough sets in. Presumably this emission is strong because the velocity width of the absorption is extensive, allowing NV emission to be enhanced due to NV resonance line scattering of inner Ly-α emission and continuum photons. In the bottom figure, Q0302+170 is seen to have weak NV emission to the blue of where the NV absorption trough sets in. Presumably the emission is weaker here because the velocity width of the absorption is small; therefore, NV emission will not be very enhanced by NV resonance line scattering of inner Ly-α emission photons.

Turnshek (1984a) has suggested that even for BAL region covering factors as low as 0.03 the amount of emission produced by collisional excitation in the BAL region may be substantial.

3.4. The Large and Small Scale Geometry of the BAL Region

The discussion so far has concentrated on understanding the location and lateral extent of the BAL region clouds relative to the region emitting the broad Ly-α line, how this relates to broad line emission that may arise in the broad absorption line region and how the observations constrain the value of the global BAL region covering factor. It seems clear that, due to the location of the BAL region and the normally extensive outflow velocities that are observed, resonance line scattering in the BAL region leads to efficient production of NV broad emission lines in QSO spectra. Taking this to be true, one can consider the question of what properties the large and small scale BAL region geometry would have to be in order to explain the observations that:
1. the NV broad emission lines seen in QSO spectra are symmetric; and
2. the NV broad emission lines seen in BAL QSO spectra are generally stronger than those seen in non-BAL QSO spectra.

In order to explain only the first observation, one must assume that the BAL region has some symmetry about the central source. Therefore, on the large scale the BAL region is unlikely to have a single jet-like geometry, and is more likely to have either a spherically symmetric geometry, a disk-like geometry or a geometry in which oppositely opposed BAL region jets exist. In order to explain the second observation, a mechanism must exist to have the NV resonance line scattered photons preferentially scatter in the direction of the BAL region clouds. The model must be such that an observer would see stronger NV broad emission lines when his line-of-sight passed through the BAL region (i.e., in a BAL QSO) as opposed to when his line-of-sight did not intersect a BAL region (i.e., in a non-BAL QSO). A geometry in which NV photons are scattered in such a preferential direction cannot be a spherically symmetric one. However, NV photons can be made to scatter in a preferential direction (i.e., in the direction of BAL region clouds) both in a disk-like geometry and in a geometry with oppositely opposed jets, providing some small scale geometric constraints are imposed on the shapes of individual BAL region clouds. A geometry in which the BAL region is confined to a disk-like plane around the central source QSO is considered here. In order to have photons preferentially scatter along the plane of the disk, the small scale BAL region geometry would have to be such that individual BAL clouds were pancake-shaped with the thinnest part of the cloud being in the direction of the BAL region outflow. With this small scale geometry, photons would preferentially escape perpendicular to the pancake surface. Figure 7 schematically illustrates the large and small scale BAL region geometry postulated. An additional reason why the large scale BAL region geometry is likely to be non-spherically symmetric rather than spherically symmetric has been given by Turnshek (1986), where it is argued that the multiplicity of troughs seen in many BAL QSOs indicates that, although the global BAL region covering factor is small, the local BAL region covering factor is fairly large. This suggests that a non-spherically symmetric geometry, possibly a disk-like one, holds. Based on observations of low redshift BAL QSOs (e.g., Malkan, Green and Hutchings 1987), the orientation of such a BAL region disk is unlikely to be aligned with the disk of any host galaxy.

Large scale geometric picture of BALR clouds at outflow velocity v_0

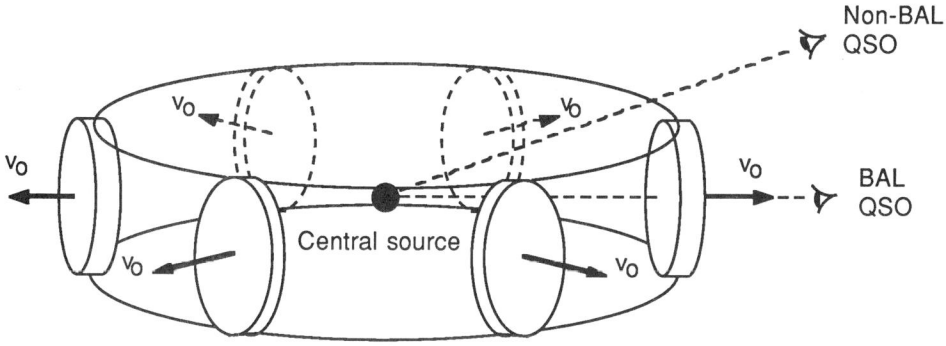

Small scale geometric picture of individual BALR cloud

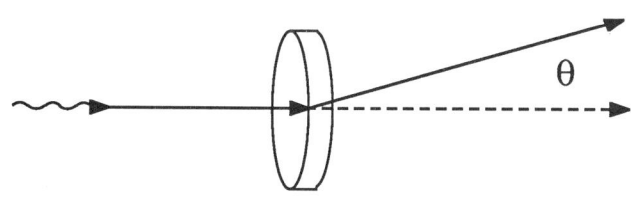

Figure 7. *Postulated large and small scale geometry for the BAL region. Photons find it hard to escape perpendicular to the plane of the large scale BAL region disk (upper figure) because individual clouds are pancake shaped (lower figure).*

If NV broad emission lines are typically enhanced by a factor of ∼ 2 when the line-of-sight passes through the BAL region disk, then the global BAL region covering factor, as deduced from NV, more likely must lie in the 0.1–0.2 range, rather than the 0.2–0.4 range noted earlier. Such an enhancement is consistent with the observations given our knowledge of the selection effects in the samples that have been used to infer that NV broad emission lines are enhanced in BAL QSOs.

3.5. Is the BAL Region Flow Accelerating or Decelerating?

Recently, Turnshek et al. (1988) made a detailed study of PHL5200-type BAL QSOs. These QSOs show rather narrow (e.g., CIV FWHM $\sim$ 2500 km s^{-1}) broad emission lines with line-to-continuum ratios of $\sim$ 2 and smooth absorption starting at outflow velocities close to zero (e.g., see Figure 2). Their analysis of one BAL QSO in particular, Q1413+113, shows that the regions giving rise to all of the broad emission lines in that spectrum are likely to be completely covered and occulted by the BAL region. This leads to a problem for models of the BAL region which have the region accelerating on the large scale. The problem is that accelerating models for the BAL region would produce a NV broad emission line that appears not to be occulted by the BAL region. One way around this apparent contradiction is to prefer a model for Q1413+113 in which the BAL region outflow velocity decreases with increasing distance from the central source, i.e., a decelerating flow. Several investigators have shown that in a decelerating region, apparent occultation of emission will occur even for the case where emission arises in the decelerating flow (Grachev and Grinn 1975; Marti and Noerdlinger 1977; Surdej and Swings 1981). Although at the present time deceleration can only be inferred from evidence for occultation, it is important that future work actually try to fit the data using a resonance line scattering model with a decelerating flow. As will be discussed in the next section, this evidence for deceleration is not at odds with some of the results that have been obtained from narrow and broad emission line region studies.

3.6. Abundances and Ionization in the BAL Region

In a poster for this Workshop, Grillmair and Turnshek (1987) present an algorithm for determining BAL region column densities as a function of outflow velocity. This technique has been used to determine column densities for a number of BAL QSOs, and in particular for two BAL QSOs which lend themselves to interesting abundance determinations in a second poster by Turnshek et al. (1987). I summarize the results on abundance determinations presented in these posters. In the BAL QSO Q0932+501 the C/H abundance ratio is inferred to be enhanced 10–100 times solar values. In the BAL QSO Q1413+113 the Fe/C or P/C and S/C abundance ratios are inferred to be enhanced at least a factor of 100 over solar values. Data on BAL QSOs as a class suggest that the abundance of metals is generally enhanced relative to solar values.

3.7. Relation Between the BAL Region and the Emission Line Regions

First I should point out that, irrespective of the correctness of existing models for the broad and narrow emission line regions, small global BAL region covering factors generally demand that a single model must explain the broad and narrow emission lines and the BALs. Based on the framework of existing cloud component models and the constraints on BAL region geometry and kinematics that I have discussed, it seems most natural to interpret the BAL region as one of the cloud components in existing models for the emission lines. It is appropriate for theorists, therefore, to use results from BAL QSO studies to place additional constraints on existing models. Osterbrock and Mathews (1986) have recently given an excellent review of emission line regions of AGNs and QSOs. I do not discuss in detail results reported in their review. Instead I discuss how results from BAL QSO studies must be interpreted in the context of models which Osterbrock and Mathews (1986) discuss.

The main property of the BAL region which must be incorporated into the framework of existing models is that it lie beyond and be larger in total lateral extent than an inner region producing the broad Ly-α emission line. Therefore, the BAL region and the broad Ly-α emitting region may be quite separate and distinct and they may actually have a different origin. Additional evidence which supports this view comes from the facts that the BAL region can give rise to extensive observable absorption, whereas the broad Ly-α emitting region cannot, and that abundances inferred from BAL region studies (see §3.6) are higher than what is normally inferred from emission line region studies. At the same time, it is clear that the BAL region may give rise to a component of the broad emission line region. For example, the BAL region almost certainly gives rise to the broad NV emission line.

As Osterbrock and Mathews (1986) review, if the emission line region is assumed to be due to an outward flow of clouds, then there is considerable evidence which would suggest that the clouds closer to the central source have higher velocity than clouds farther out. This can be inferred, for example, from observations showing that the FWHM velocity of emission lines is larger for lines formed in higher density regions closer to the central source. In fact, the data on emission lines can be used to argue that the simple interpretation that the broad emission line region and the narrow emission line region are completely distinct from one another is an inadequate description of the observations. Osterbrock and Mathews (1986) suggest that in the narrow emission line region either the flow is decelerated outward, or else individual clouds are accelerated outward but those which form at larger distances reach smaller terminal velocities. They suggest that the broad emission line region merges with the narrow emission line region, even though the flow pattern in the broad emission line region is not as clear. As I reported in §3.5, in one type of BAL QSO there is evidence that the BAL region flow decreases outward, so results from BAL QSOs may be in qualitative agreement with this view. However, since BAL region densities are thought to be considerably higher than narrow emission line region densities (see §3.2), the BAL region apparently must be embedded somewhere within this normal broad/narrow emission line region flow.

One speculation is that a disk-like distribution of massive stars undergoing supernovae-type explosions within the inner tens of parsecs may give rise to a local region which can cover and occult some inner region where practically all of the Ly-α broad emission line is produced. Such a scenario might meet the requirement of a small *global* BAL region covering factor and a large *local* BAL region covering factor, producing the kind of aspect angle effects that are inferred from comparisons of BAL and non-BAL QSO properties. In the presence of a rapidly expanding thermal wind, $z_{abs} > z_{em}$ from the BAL region may not be seen, because the wind would quickly sweep up clouds injected into the flow. Some other source for the clouds giving rise to the broad Ly-α and some other emission lines would probably be necessary. This would be an advantage, however, if one wanted to construct a model in which only the BAL region abundances were significantly enhanced.

The actual distance between the central source and the region producing the broad

Ly-α emission line is important to know, especially in the context of constraining the size scale of the BAL region. In the original single component models for the broad emission line region, the Ly-α emitting region was constrained to lie $\sim$ 1 pc from the central source, because CIII]λ1909 also had to be produced in the same region and this suggested that the density was $< 10^{10}$ cm^{-3}. However, it is now believed that very inhomogenous conditions exist in the emitting region and no such density constraint should be rigorously imposed. Therefore, it is quite possible that the Ly-α emitting region lies closer to the central source.

4. RELATION TO THE NARROW ABSORPTION LINES

There are two general issues concerning the possible relation between the BALs and the narrow absorption lines that in my opinion should be addressed. The first is the question of any observed similarity between the properties of the BALs and the properties of the so-called narrow absorption line systems. Is there any link between these two categories of systems? It would seem only natural given the variety of the BAL systems, their velocity extent and their frequency of occurrence, to have some lines-of-sight passing through the edge of a QSO's BAL region giving rise to relatively narrow absorption line systems in objects which are classified as non-BAL QSOs. To illustrate this, I discuss some spectra which are not of the standard BAL type, but which have properties which are reminiscent of the BAL type. I also discuss the question of whether the BAL flow itself may contaminate the intergalactic medium and give rise to intervening narrow absorption line systems. Both Briggs, Turnshek and Wolfe (1984) and Hazard *et al.* (1984) have suggested that a scenario in which BAL remnant ejecta contaminate the intergalactic medium may be important.

4.1. Similarity Between the BALs and the Narrow Absorption Lines

In studying the different narrow absorption line systems that are observed in QSO spectra, four different types of systems are sometimes considered:
1. the z_{abs} near z_{em} metal line systems (sometimes referred to as associated systems);
2. the highly displaced metal line systems not exhibiting damped Ly-α;
3. the highly displaced metal line systems exhibiting damped Ly-α; and
4. the weak Ly-α forest systems not normally showing metals.

In the spirit of considering possible controversial issues, I illustrate some data which in my opinion makes one have second thoughts about always adopting a strict intervening interpretation for these four types of systems. The point here is to raise the possibility that some fraction of the narrow line systems have an intrinsic origin.

4.1.1. Z_{abs} Near Z_{em} Systems

As discussed by Foltz *et al.* (1988) in the next paper, these systems tend to preferentially occur in radio-loud QSOs which suggests a dichotomy between them and the BAL systems. However, the strength and velocity extent of some of these systems suggests that they may have a relation to the BAL systems. In Figure 8 three of the strongest of these systems are illustrated. Aside from being generally weaker than the BAL systems and occuring in radio-loud QSOs, these systems tend to have relatively stronger Ly-α absorption and are of lower ionization than the BAL systems. Williams *et al.* (1975) and Morris *et al.* (1986) have derived lower limits on the distance between the central source QSO and the absorbing clouds using CII fine structure line measurements and have concluded that these clouds are at least tens of kiloparsecs from the central source. 3C191 exhibits evidence for absorption-absorption line-locking of 2 CIV doublets.

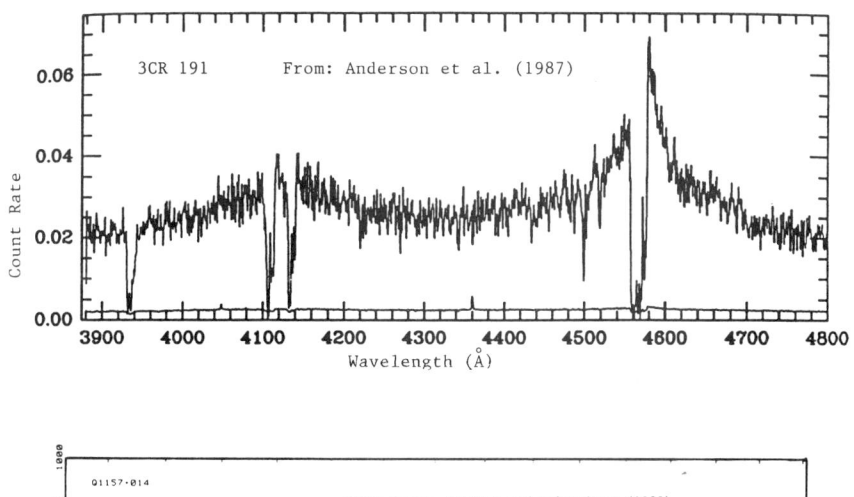

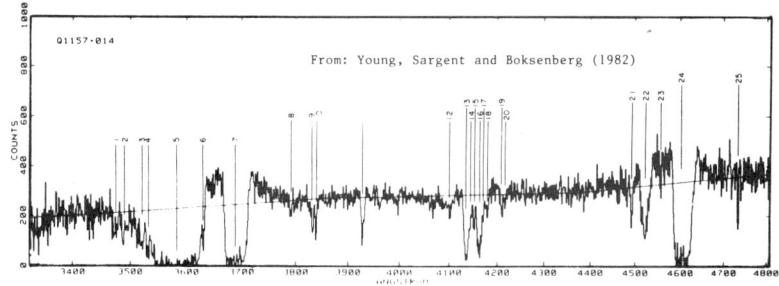

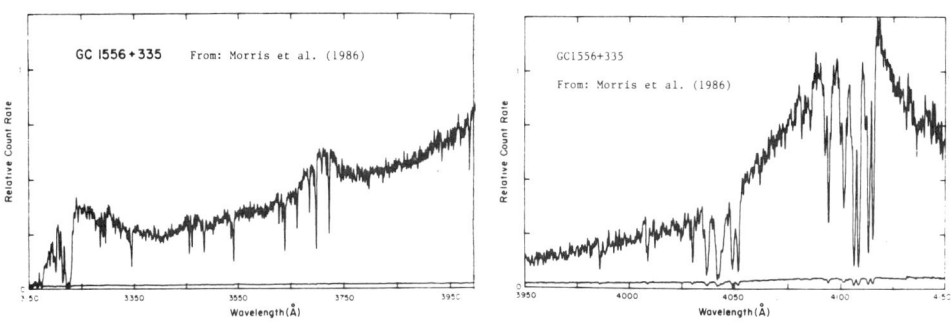

Figure 8. *Three examples of radio-loud QSOs showing strong absorption. The absorption is generally not quite strong enough to qualify for the classification of BALs.*

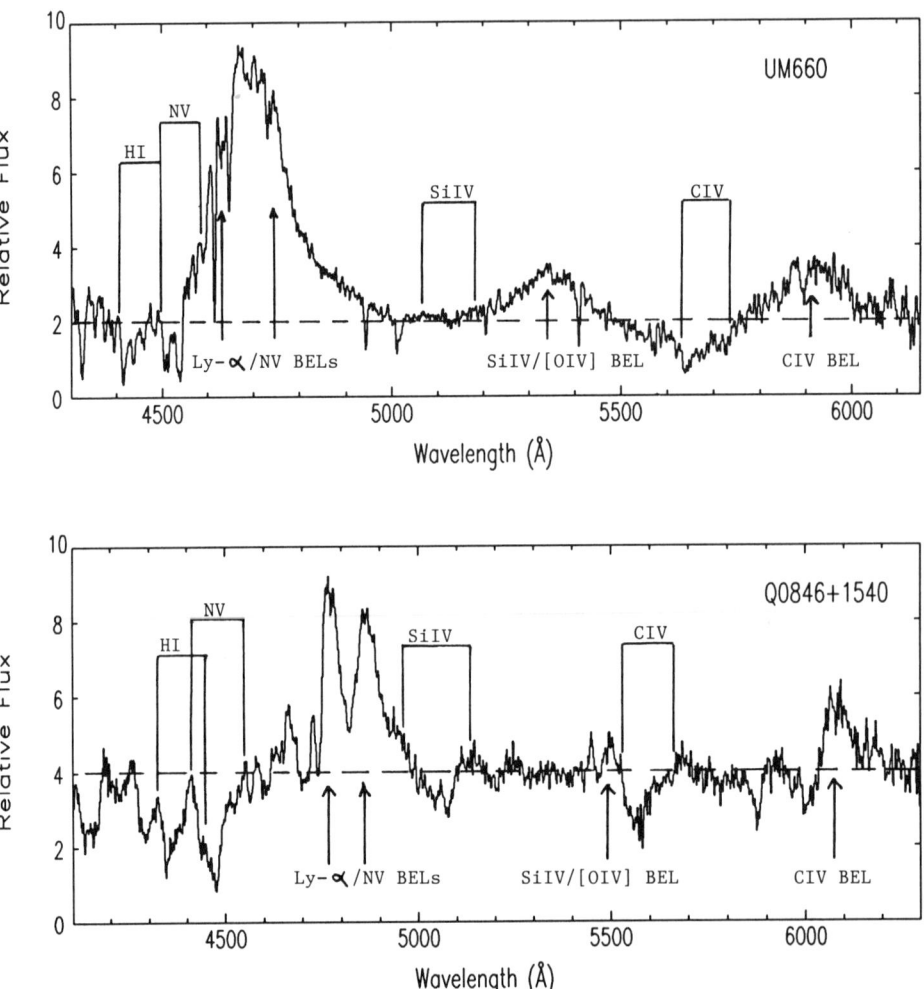

Figure 9. *Two examples of radio quiet QSOs exhibiting absorption which is intermediate in strength between typical narrow absorption lines seen in non-BAL QSOs and the broad absorption lines seen in BAL QSOs.*

4.1.2. Highly Displaced Metal Line Systems

The study of Weymann et al. (1979) suggested that between 0–18,000 km s^{-1} there may be an excess of absorption line systems which cannot be explained by merely excluding the BAL systems and the z_{abs} near z_{em} systems. Of course, the excess of absorption line systems that Weymann et al. talked about could be single isolated intrinsic systems. An important question is: are there any absorption line systems which are not typical BAL systems, but which obviously have a rich absorption line spectrum indicating that spectroscopically, in terms of amount of absorption present, they appear to lie between the BAL type of intrinsic system and any single isolated type of intrinsic system that may exist? Figure 9 shows two spectra which may fit into this category. The property which may distinguish these systems from the majority of other metal line systems is the presence of NV absorption. Of course, a small fraction of the metal line systems show low ions and are inferred to be more than hundreds of kiloparsecs from the central source QSO based on the lack of CII fine structure lines (e.g., Turnshek, Weymann and Williams 1979), leading to energetic arguments which make it difficult to imagine how at least the low ionization systems in this class could be ejected.

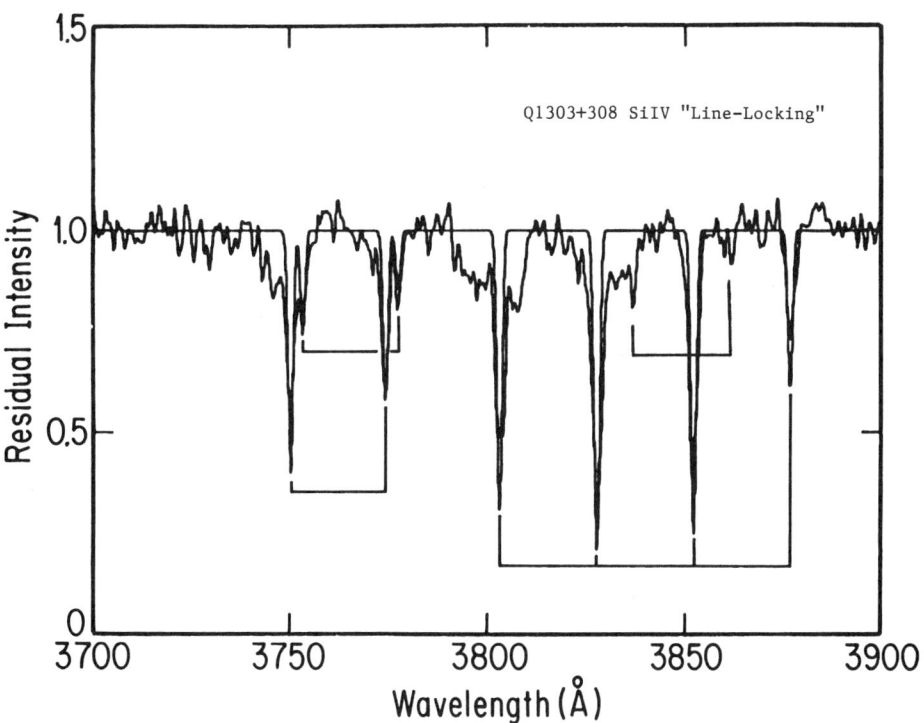

Figure 10. *The spectrum of the BAL QSO Q1303+308 which exhibits evidence for absorption-absorption line-locking in the SiIV absorption lines.*

4.1.3. Damped Ly-α Systems

The damped Ly-α systems are probably the systems which are most secure in terms of the intervening interpretation being correct (see Wolfe 1988). However, it is annoying that three of the spectra of damped systems appear to exhibit evidence for absorption-absorption line-locking. As a reference, Figures 10 and 11 illustrate examples of absorption-absorption line-locking of narrow metal lines in two BAL QSO. Figure 10 shows the most striking case which occurs in the SiIV absorption lines in Q1303+308 (see also the center panel of Figure 5). A more subtle case occurring in Q1413+113 is shown in Figure 11. In this case the $z_{abs} = 1.870$ narrow line

40 D. A. Turnshek

system includes absorption components due to the CIV doublet and the AlII singlet, and so does the $z_{abs} = 1.659$ narrow line system. However, the AlII singlet in the $z_{abs} = 1.659$ system appears line-locked with the blue member of the CIV doublet in the $z_{abs} = 1.870$ system, as can be inferred because the blue member of the CIV doublet has considerably more than twice the strength of the red member of the doublet. In Figure 12 three cases of absorption-absorption line-locking that appear to occur in objects having damped Ly-α absorption are illustrated. These occur in the SiIV lines of PKS0458-020, the CIV lines of Q1244+347 and the MgII lines of MC1331+170. The probablity that this occurs by chance is $\sim$ 10 %, although it is not completely clear how selection bias affects this calculation. In estimating this probability it was assumed that clustering of systems did not occur in the presence of damped Ly-α systems.

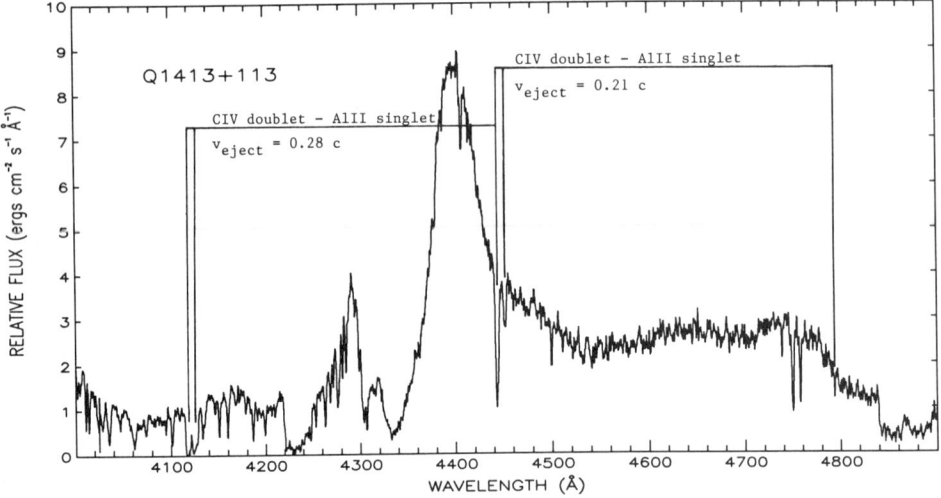

Figure 11. *The spectrum of the BAL QSO Q1413+113 which exhibits evidence for absorption-absorption line-locking between CIVλ1548.2 and AlIIλ1670.8.*

4.1.4. Ly-α Forest Systems

Finally, Figure 13 illustrates some Ly-α forest data in a BAL QSO which at least suggests the possibility that some component of the Ly-α forest may have an intrinsic nature. The BAL QSO Q1413+113 has an identification of resonance line absorption due to FeIII and/or PV in the spectrum. I interpret this as evidence for generally enhanced abundances relative to solar values (see §3.6). However, there is another trough in the Q1413+113 spectrum lying in the Ly-α forest (indicated as unidentified in Figure 12) which I have not been able to identify in terms of resonance line absorption. This unidentified trough cannot be due to NV or SiIV because there is no evidence for corresponding CIV at the appropriate wavelengths. If the unidentified trough is due to Ly-α, it shows no evidence for CIV. Could this apparently smooth absorption be a Ly-α-only ejected BAL trough? If so, its inferred outflow velocity ranges from ~ 10,000 to ~ 22,000 km s^{-1}. The existence of a Ly-α-only ejected BAL trough would raise the question of whether any of the narrow Ly-α-only systems were ejected. I think it is important to keep in mind that because of the frequency of Lyman limit absorption systems and because of the blue cutoff in wavelength produced by the earth's atmosphere, it is normally not possible to explore the Ly-α forest properties in QSO spectra at arbitrarily high inferred ejection velocities.

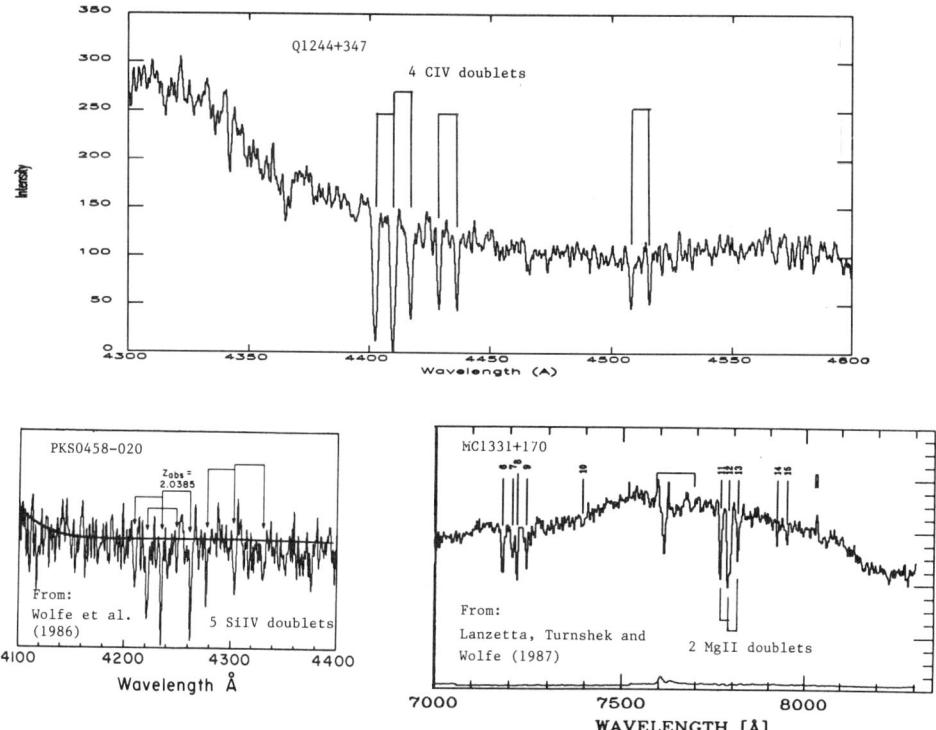

Figure 12. *Three examples of absorption-absorption line-locking in QSOs that have damped Ly-α absorption.*

Hopefully, with the eventual launch of the Hubble Space Telescope, sensitive absorption line surveys at ultraviolet wavelengths will be made. In addition to extending the redshift pathlength that can be surveyed in any one QSO (when Lyman limit absorption is absent), low redshift examples of the presumed intervening objects giving rise to some of these four types of systems may then be imaged, eliminating some of the concerns that I have detailed here.

4.2. The Effect of the BAL Region Outflow

The final question I wish to consider is whether material from the BAL region outflow itself may eventually propagate to large distances from the QSO and give rise to some cosmologically intervening narrow absorption line systems seen in background

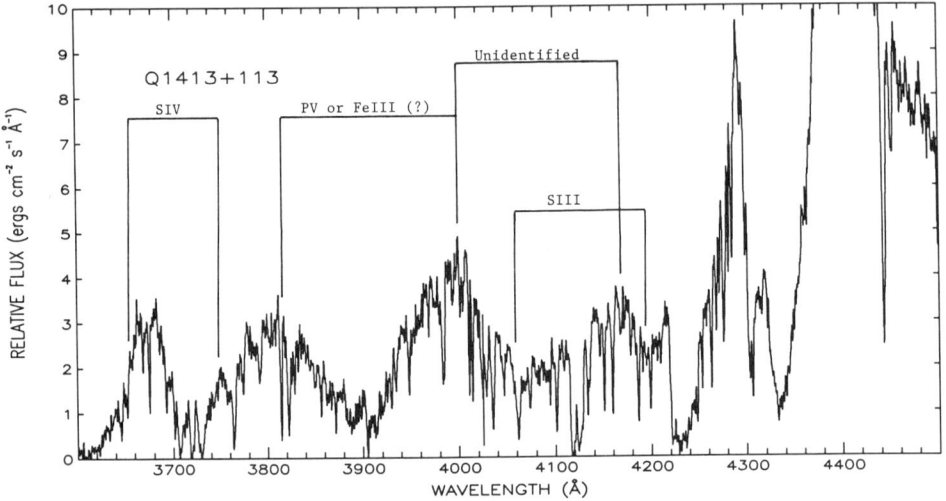

Figure 13. *An unidentified smooth BAL trough in the spectrum of the BAL QSO Q1413+113. If this trough is due to Ly-α, it has no apparent associated metal lines.*

QSO spectra. In particular, the question is whether the (past?) presence of a BAL region in a QSO with redshift z_a could give rise to absorption with a redshift z_a in the spectrum of a background QSO having emission redshift z_b (i.e., $z_a < z_b$ is assumed).

Assuming that the BAL region distance from the central source QSO is $r = 50r_{50}$ pc, that the global BAL region covering factor is $q_c = 0.08q_8$, that the total C^{+3} column density is $N(C^{+3}) = 10^{16}N_{16}$ cm^{-2}, that the typical BAL region outflow velocity is $V = 10^4 V_{10,000}$ km s^{-1}, that the ratio of the mass confining the BAL region to the mass in observable BAL region clouds is $f = 100f_{100}$, that BAL region abundances are enhanced a factor of 10 over solar values by number and that one-third of the carbon is C^{+3}, one finds that:

$$M_{TOT} \simeq 6 \times 10^4 r_{50}^2 q_8 N_{16} f_{100} \; M_\odot \qquad (1)$$

$$M_{metals} \simeq 5 \times 10^3 r_{50}^2 q_8 N_{16} f_{100} \; M_\odot \qquad (2)$$

$$E_{TOT} \simeq 6 \times 10^{55} r_{50}^2 q_8 N_{16} f_{100} V_{10,000}^2 \; \text{ergs} \qquad (3)$$

$$t_{cross} \simeq 5 \times 10^3 r_{50} V_{10,000}^{-1} \; \text{yrs} \qquad (4)$$

where M_{TOT} is the total mass of the BAL region, M_{metals} is the mass in metals of the BAL region, E_{TOT} is the total kinetic energy of the BAL region flow and t_{cross} is the time required for a cloud to cross the BAL region. Note that the assumption $f_{100} = 1$ comes from volume filling factor and cloud evaporation arguments which suggest that the medium confining the BAL clouds is > 100 times more massive than directly observable mass which gives rise to the BALs (e.g., Krolik, McKee and Tarter 1981; Weymann et al. 1982; Schiano 1986). Since t_{cross} is likely to be considerably smaller than a QSO lifetime, BAL region clouds are required to be continuously created or injected. One can then infer the total mass loss rate, dM_{TOT}/dt, and the total power in the outflowing clouds, dE_{TOT}/dt, to be:

$$dM_{TOT}/dt \simeq 12 r_{50} q_8 N_{16} V_{10,000} \ M_\odot \ \mathrm{yr}^{-1} \qquad (5)$$

$$dE_{TOT}/dt \simeq 4 \times 10^{44} r_{50} q_8 N_{16} V_{10,000}^3 \ \mathrm{ergs \ s}^{-1}. \qquad (6)$$

Finally, taking the typical QSO lifetime to be $t_{QSO} = 10^7 t_7$ yrs, one can find M_{DEP} and E_{DEP}, the amount of mass and energy lost or deposited due to the BAL region flow,

$$M_{DEP} \simeq 1.2 \times 10^8 r_{50} q_8 N_{16} f_{100} V_{10,000} t_7 \ M_\odot \qquad (7)$$

$$E_{DEP} \simeq 1.2 \times 10^{59} r_{50} q_8 N_{16} f_{100} V_{10,000}^3 t_7 \ \mathrm{ergs}. \qquad (8)$$

It is clear that in the absence of local forces which would decelerate the BAL flow in the vicinity of the QSO, such a flow will freely expand into the intergalactic medium until it sweeps up its equivalent mass and begins to decelerate. Given the uncertainties, the resulting M_{DEP} and E_{DEP} are not out of line with the Ostriker and Cowie (1981) view of an intergalactic medium dominated by explosions.

Assuming that the BAL material is not locally decelerated in at least some fraction of QSOs (but see §3.5) and adopting the parameters derived above, the material will initially become very highly ionized due to the shock that is generated and, following the calculations of Ostriker and Cowie (1981), eventually will cool on a time scale of several billion years and come to a stop at a distance of ~ 1 Mpc from the central source QSO. The propagation distance, R_{PROP}, depends weakly on E_{DEP}, $R_{PROP} \propto E_{DEP}^{0.2-0.3}$. At moderate to high redshift, R_{PROP} corresponds to a radius on the sky of $\sim 10^2$ arcsec. Considering the number of *observed* QSOs per square degree per unit redshift interval down to faint levels, a significant fraction of the sky may be covered by remnant BAL gas and swept up gas. Moreover, if the number of non-active QSOs (e.g. normal galaxies?) that once had BAL regions were much greater than the number of currently observable QSOs, which is consistent with $t_7 = 1$, then the fraction of the sky per unit redshift effected by BAL 'explosions' could easily exceed unity. Therefore, the remnant ejecta and swept up gas from the BAL regions of QSOs may cause many of the moderate to high redshift narrow absorption line systems seen in QSO spectra. These may be intergalactic clouds, loosely associated with galaxies that were once active. One constraint on this scenario involves the amount of time that it would take the shocked gas to cool. All of this activity would have had to have taken place at an early enough epoch in order for there to be sufficient time for the remnant BAL region gas and the swept up gas to cool.

5. FUTURE WORK

I have discussed how observations of BAL QSOs can be used to constrain models for the BAL region and have speculated on how these constraints can be utilized to fit the BAL region into the framework of existing emission line region models. Additional observational work needs to be done in the areas of understanding selection effects so that BAL QSO and non-BAL QSO properties can be accurately compared. Larger samples are needed. Comparison of properties should not be limited to emission line properties, but should also include optical variability, polarization, continuum, x-ray and radio properties. Results from such work will influence the geometric constraints on models. Work in the area of deriving accurate column densities and extracting emission line profiles also needs to be done so that abundances and the possible correlation between BAL/emission line profile type can be further investigated. Imaging work on the low redshift BAL QSOs is very important as it may eventually provide some clue about aspect angle effects or which types of host galaxies are likely to have BAL regions with the largest BAL region covering factors.

A considerable amount of theoretical work is also necessary. It is important to consider in more detail simple geometric and kinematic models for the BAL flow to see if they can explain the relevant observations. At the same time, it is important to resolve the uncertainties associated with the ionization mechanism and abundances. Only then can the emissivity of the BAL clouds be considered in detail and accurately incorporated into emission line region models. It is also very important to continue to study the details of cloud acceleration, confinement and possible deceleration. This will not only help pin down a detailed model, but will allow a more accurate assessment of the possible pollution of the intergalactic medium by remnant BAL region ejecta.

Finally, in the context of interpreting QSO narrow absorption line work, it is important to continue to explore the possibility that some fraction of the narrow absorption line systems are intrinsic. If this is not properly accounted for, data being used to interpret galaxy halos, galaxy disks and the intergalactic medium at large redshifts may be in error.

I would like to thank Frank Briggs, Paul Coleman, Craig Foltz, Carl Grillmair, Cyril Hazard, Paul Hewett, Allen Schiano, Art Wolfe and especially Ray Weymann for useful collaborations and/or discussions in this work. I would also like to thank Chris Blades, Kostas Routsis and Diane Turnshek for thoroughly reading the manuscript prior to publication.

REFERENCES

Anderson, S. A., Weymann, R. J., Foltz, C. B., and Chaffee, Jr., F.H. 1987, *A. J.*, **94**, 278.

Briggs, F. H., Turnshek, D. A., and Wolfe, A. M. 1984, *Ap. J.*, **287**, 549.

Burbidge, E. M. 1970, *Ap. J. (Letters)*, **160**, L33.

Davidson, K., and Netzer, H. 1979, *Rev. Mod. Phys.*, **51**, 715.

Drew, J., and Gidding, J. 1982, *M.N.R.A.S.*, **201**, 27.

Drew, J., and Boksenberg, A. 1984, *M.N.R.A.S.*, **211**, 813.

Foltz, C. B., Wilkes, B. J., Weymann, R. J., and Turnshek, D. A. 1983, *Pub. A.S.P.*, **95**, 341.

Foltz, C. B., Weymann, R. J., Hazard, C., and Turnshek, D. A. 1984, *B.A.A.S.*, **16**, 1006.

Foltz, C. B., Weymann, R. J., Morris, S. L., and Turnshek, D. A. 1987a, *Ap. J.*, **317**, 450.

Foltz, C. B., Chaffee, Jr., F. H., Hewett, P. C., MacAlpine, G. M., Turnshek, D. A., Weymann, R. J., and Anderson, S. A. 1987b, *A. J.*, **94**, 1423.

Foltz, C. B., Chaffee, Jr., F. H., Weymann, R. J., and Anderson, S. A. 1988, *This volume*.

Giola, I. M., Maccacaro, T., Schild, R. E., Giommi, P., and Stocke, J. T. 1986, *Ap. J.*, **307**, 497.
Grachev, S. I., and Grinn, V. P. 1975, *Astrophysics*, **11**, 20.
Grillmair, C. J., and Turnshek, D. A. 1987, *QSO Absorption Lines: Probing The Universe, A Collection of Poster Papers*. eds. J. C. Blades, C. A. Norman and D. A. Turnshek, (ST ScI Publication), p. 1.
Hartig, G. F., and Baldwin, J. A. 1986, *Ap. J.*, **302**, 64.
Hazard, C., Morton, D., Terlevich, R., and McMahon, R. 1984, *Ap. J.*, **282**, 33.
Hazard, C., Morton, D. C., McMahon, R. G., Sargent, W. L. W., and Terlevich, R. 1986, *M.N.R.A.S.*, **223**, 87.
Hazard, C., McMahon, R. G., Webb, J. K., and Morton, D. C. 1987a, *QSO Absorption Lines: Probing The Universe, A Collection of Poster Papers*. eds. J. C. Blades, C. A. Norman and D. A. Turnshek, (ST ScI Publication), p. 14.
_____. 1987b, *Ap. J.*, **323**, 263.
Hickson, P., and Hutchings, J. B. 1987, *Ap. J.*, **312**, 518.
Junkkarinen, V. T., Burbidge, E. M., and Smith, H. E. 1983, *Ap. J.*, **265**, 51.
_____. 1987, *Ap. J.*, **317**, 460.
Junkkarinen, V. T. 1983, *Ap. J.*, **265**, 73.
Krolik, J. H., McKee, C. F., and Tarter, C. B. 1981, *Ap. J.*, **249**, 422.
Lanzetta, K. L., Turnshek, D. A., and Wolfe, A. M. 1987, *Ap. J.*, **322**, 739.
Lucy, L. 1971, *Ap. J.*, **163**, 95.
Lynds, C. R. 1967, *Ap. J.*, **147**, 396.
Malkan, M., Green, R. F., and Hutchings, J. B. 1987, *Ap. J.*, **322**, 729.
Marti, F., and Noerdlinger, P. D. 1977, *Ap. J.*, **215**, 247.
Moore, R. L., and Stockman, H. S. 1984, *Ap. J.*, **279**, 465.
Morris, S. L., Weymann, R. J., Foltz, C. B., Turnshek, D. A., Schectman, S., Price, C., Boroson, T. A. 1986, *Ap. J.*, **310**, 40.
Netzer, H., and Sheffer, Y. 1983, *M.N.R.A.S.*, **203**, 935.
Osterbrock, D. E., and Mathews, W. G. 1986, *Ann. Rev. Astron. Astrophys.*, **24**, 171.
Ostriker, J. P., and Cowie, L. L. 1981, *Ap. J. (Letters)*, **243**, L127.
Penston, M. V., and Smith, L. 1988, *M.N.R.A.S.*, submitted.
Scargle, J. D., Caroff, L. J., and Noerdlinger, P. D. 1970, *Ap. J.*, **161**, L115.
Schiano, A. V. R. 1986, *Ap. J.*, **302**, 81.
Scott, J. S., Christiansen, W. A., and Weymann, R. J. 1984, unpublished preprint.
Stocke, J. T., Foltz, C. B., Weymann, R. J., and Christiansen, W. A. 1984, *Ap. J.*, **280**, 476.
Stockman, H. S., Angel, J. R. P., and Hier, R. G. 1981, *Ap. J.*, **243**, 404.
Surdej, J., and Swings, J. P. 1981, *Astr. Ap.*, **96**, 242.
Surdej, J., and Hutsemekers, D. 1987, *Astr. Ap.*, **177**, 42.
Turnshek, D. A., Weymann, R. J., and Williams, R. E. 1979, *Ap. J.*, **230**, 330.
Turnshek, D. A., Weymann, R. J., Liebert, J. W., Williams, R. E. and Strittmatter, P. A. 1980, *Ap. J.*, **238**, 488.
Turnshek, D. A., Weymann, R. J., Carswell, R. F., and Smith, M. G. 1984, *Ap. J.*, **277**, 51.
Turnshek, D. A. 1984a, *Ap. J. (Letters)*, **278**, L87.
_____. 1984b, *Ap. J.*, **280**, 51.
Turnshek, D. A., Foltz, C. B., Weymann, R. J., Lupie, O. L., McMahon, R. G., and Peterson, B. M. 1985, *Ap. J. (Letters)*, **294**, L1.

Turnshek, D. A. 1986, *I.A.U. Symposium 119: Quasars*, Reidel, G. Swarup and V. Kapahi, editors, p. 317.

Turnshek, D. A., and Grillmair, C. J. 1986, *Ap. J. (Letters)*, **310**, L1.

Turnshek, D. A., Briggs, F. H., Foltz, C. B., Grillmair, C. J., and Weymann, R. J. 1987, *QSO Absorption Lines: Probing The Universe, A Collection of Poster Papers*. eds. J. C. Blades, C. A. Norman and D. A. Turnshek, (ST ScI Publication). p. 8.

Turnshek, D. A., Foltz, C. B., Grillmair, C. J., and Weymann, R. J. 1988, *Ap. J.*, **325**, in press.

Warren, S., Hewett, P. C., Osmer, P. S., and Irwin, M. J. 1987, *Nature*, **330**, 453.

Weymann, R. J., Williams, R. E., Peterson, B. M., and Turnshek, D. A. 1979, *Ap. J.*, **234**, 33.

Weymann, R. J., Scott, J. S., Schiano, A. V. R., and Christiansen, W. A. 1982, *Ap. J.*, **262**, 497.

Weymann, R. J., and Foltz, C. B. 1983, in *Quasars and Gravitational Lenses* (Proc. 24th Liege Int. Ap. Colloq.), p. 538.

Weymann, R. J., Turnshek, D. A., and Christiansen, W. A. 1985, in *Astrophysics of Active Galaxies and Quasi-Stellar Objects*, J. Miller, editor, p. 333.

Williams, R. E., Strittmatter, P. A., Carswell, R. F., and Craine, E. R. 1975, *Ap. J.*, **202**, 296.

Wolfe, A. M., Briggs, F. H., Turnshek, D. A., Davis, M. M., Smith, H. E., and Cohen, R. D. 1985, *Ap. J. (Letters)*, **294**, L67.

Wolfe, A. M. 1988, *This volume*.

Young, P. J., Sargent, W. L. W., and Boksenberg, A. 1982, *Ap. J.*, **252**, 10.

DISCUSSION

Antonucci: I wonder if you have any thoughts on how BAL quasars got these high abundances?

Turnshek: Well, you would certainly need something like nucleo-synthesis from massive stars.

Antonucci: But as you've quoted the abundances, they are extra-ordinarily high. Does that cause a problem?

Turnshek: Possibly. Part of the problem is that you'd really like to have a good model for the broad absorption line region level of ionization in order to determine the abundances. I skirted over that issue. There is not really a good model for the level of ionization in broad absorption line regions. However, in one of the BAL components up on my poster, you have a case where the C^+ density must be roughly the same as the neutral hydrogen density. Even though we lack a good model for the ionization, its is hard to explain this without significantly enhanced abundances.

J. Bahcall: Dave, I'm just trying to understand your argument that says that the geometry of the BAL quasars is disk-like. What rules out the model where you have exactly the same covering factor, but the geometry is sponge-like. The clouds would cover the entire quasar, but it would have many holes in it.

Turnshek: That's essentially a spherically symmetric geometry, right? The reason why a spherically symmetric geometry has problems has to do with the fact that we observe an enhancement of NV emission in BAL quasars. For the disk-like geometry, I propose that you have individual BAL clouds which are pancake-shaped. Therefore, when a photon from the central source is absorbed and wants to scatter out of an individual BAL cloud, it will preferentially want to leave the cloud in the plane of the large-scale BAL region disk, because the individual pancake-shaped clouds' surfaces are proposed to be oriented perpendicular to the plane of the large-scale BAL region disk. Another way to think about it is to ask the question: how could you have a spherically symmetric geometry and at the same time have NV emission enhanced in BAL quasars (and not in non-BAL quasars), yet still have a small global BAL region covering factor? The two facts are inconsistent.

J. Bahcall: Take your disk, and spread it over the whole solid angle. Now take pieces of the disk and scatter it randomly around the quasar. I don't see how that would differ.

Turnshek: You're saying that we want something like this right here (draws picture). You want clouds all around the central source. In other words, it's not a disk-geometry. There are just clouds everywhere. Therefore, when our line-of-sight misses a BAL region cloud we call the quasar a non-BAL quasar, but when our line-of-sight passes through a cloud, we call the quasar a BAL quasar. It's essentially a spherically symmetric geometry. Well, in that particular case, you would not expect enhancement of NV emission due to resonance line scattering to preferentially take place in BAL quasar samples as compared to non-BAL quasar samples. So only if you can invent a geometry where NV emission is enhanced when BAL region clouds are intersected by our line-of-sight is the geometry consistent with the data. The geometry I've proposed works.

P. Shapiro: Your argument simply says that the cloud size has to be big enough to occult the broad emission line region. They could still be distributed around.

Turnshek: In the particular model I propose, when your line-of-sight passes through the BAL region, you would see BALs, plus the inner photons from the continuum and emission lines would preferentially be scattered such that they leave along the plane of the large-scale BAL region disk. NV emission would be enhanced. However, if we were looking from the pole for example (perpendicular to the plane of the large-scale BAL region disk), a photon would have difficulty escaping the pancake-shaped BAL cloud in that direction. So, you have to postulate two things: (1) that the large scale BAL region geometry is disk-like and (2) that the individual BAL clouds are pancake-shaped. If the individual BAL clouds were spherical clouds, it would not preferentially forward-scatter and back-scatter the photons. That's why we must postulate both a large-scale disk-like geometry and individual pancake-shaped clouds.

Shull: I find that model sort of hard to believe because the column densities are high, and therefore the optical depths would be high. You would then have a much more detailed scattering process, and it would not be preferentially forward-scattering.

Turnshek: In fact, when you assess the observations and do the calculations the line center optical depths for the BAL clouds are typically a few. If we look at some of the profiles that we see (shows viewgraph), the majority of the BAL clouds do not have very high optical depths along our line-of-sight. What I'm postulating, however, is that if we took one individual BAL cloud that it would be pancake-shaped. Therefore, if we could look perpendicular to the large-scale BAL region disk (pole on and a non-BAL quasar), we would be looking into a region where individual BAL clouds have extremely high optical depth but do not cover the source (i.e., a non-BAL quasar), and so photons could not escape the individual BAL clouds in that direction. Therefore, when a Ly-α emission or continuum photon enters an individual pancake-shaped BAL cloud, it would preferentially back-scatter and forward-scatter out of the cloud, enhancing the N V emission via resonance line scattering preferentially in samples of BAL quasars.

Sargent: I have two questions. One is: I think that there is evidence from emission line ratios in quasar spectra, that the emission line regions have fairly normal compositions, Ly-α stronger than C IV and you ionize everything with a power-law continuum. So, how is it then that the broad absorption line material, which as I understand it originates in the broad emission line region, could have a different composition? The second question, which is related, is: How would you produce ordinary-looking sharp absorption line systems by this mechanism which have more or less normal abundances if they started off being highly enriched with heavy elements?

Turnshek: For the first question, as far as the broad emission line region goes, I think that the problems in assessing abundances from broad emission lines are unclear. You have to take an ensemble of clouds that cover a range of temperatures, a range of densities, etc. In other words, you have components that give rise to the broad emission lines. You can't study in detail the properties of one BEL component like you can in the BAL region. So, I would trust abundances determined from BAL profiles much more than I would trust abundances determined from BELs. Also I should say that I don't think it's clear that BAL region clouds have the same properties as the BEL region clouds. I said that the BAL region clouds could give rise to a component of the BEL region, but clearly the broad absorption troughs are capable of covering or occulting an inner region producing the Ly-α BEL. So there are at least two different components.

For the second question, you asked how I might propose that the sharp metal systems that have apparently normal composition, or somewhat lower than normal, might be related to this BAL phenomenon which have high abundances. I assume that this is in the context of the BAL region outflow somehow giving rise to intervening systems along our line-of-sight. If this did occur, there would be a tendency for the BAL flow to sweep up hydrogen that has not been enriched from the intergalactic medium, and the BAL metals would be diluted. Whether that's going on with the ejected systems or not, I don't know.

Wampler: I'd like to enter this conversation. I think that the first thing you said, Wal, that the BEL regions have normal abundances, is not true. I think that in these objects there is evidence for enhanced FeII emission. Now, as Netzer and Wills and so forth pointed out, it looks as if in general in quasars that Fe is up in abundance. However, the models are rather poor yet, and one can have considerable argument about whether Fe is really up in abundance or not, but I think for those BAL quasars that have been studied at the longer wavelengths, there is strong evidence that Fe is even stronger than in the already possibly over-abundant cases of Fe in quasars. For the other lines, carbon and so forth, I think the evidence is not so strong that the abundances are abnormal. However, there may be abnormal abundances for N, but I think that Fe looks very, very strong. So instead of being over-abundant by a factor of ten, it's over-abundant by a factor of 100.

Bregman: Dave, I was wondering, if you have such a model as you've proposed with this disk, you would expect that some fraction of time your line-of-sight to the quasar would be passing through the edge of the disk, and so there should be some distribution that you get for both the shape or the depth of the broad absorption lines and also the intensity of the resonance line scattered emission. I was wondering if you could look at that distribution and use that as evidence for or against particular geometries.

Turnshek: First, I want to make clear that this is the only geometry that I could think of which is consistent with the data. So, if the geometry I've proposed is correct, and your line-of-sight passed through the edge of the large-scale BAL region disk-like structure, you might see more structure in absorption troughs or lower typical optical depths. Therefore, if you were to see more structure in the absorption troughs, you might check NV emission in those types of BAL quasars to see if NV emission is less enhanced. That would be worth trying, but it hasn't been done yet, and samples would have to be very large.

Savage: Back to the abundance issue. You really glossed over a crucial part of it. That is, what assumptions were made in doing the analysis. Ionization corrections and all the rest must play an enormous role in relating one element to another. Can you amplify on that?

Turnshek: The data were sufficiently complicated, that I decided it would be best to present that data in a poster, so the poster showing details is up in the boardroom.

Savage: How about a three-minute summary. How far off can the abundance determinations be depending on assumptions about ionization and so on? What were your assumptions?

Turnshek: Let me show one piece of evidence concerning carbon enhancement relative to hydrogen (shows figure also presented in Poster contribution). Here you see a BAL region component with inferred outflow velocity of 14,000 km s^{-1}. In this figure, you see the SiIV absorption component, the CII absorption component, and the Ly-α absorption component. Now the argument basically goes like this: These lines are probably not saturated, and when you go to a higher resolution (a 4 Å to a 1 Å resolution spectrum) the depths of these lines don't change or get any deeper. So the idea is that you can derive a valid column density on the basis of those observations.

You find that the neutral hydrogen density is just somewhat higher than the singly ionized carbon density. When you put together a photoionization model, and you assume an ionizing continuum as steep as a ν^{-4} power law, you tend to ionize the H° relative to C$^+$ as much as possible because of the ionization potentials. In this case, you find out that to explain the C$^+$ and H° column densities, the carbon abundance has to be enhanced over a factor of 100. If you make the ionizing spectrum flatter and flatter, it gets even worse. Now if you want to put together a photoionization model where you essentially have a power law plus some blackbody spectrum, like Malkan and others proposed, and you do the ionization calculations, you find that by using a lot of blackbody and a little power law, you can make the C enhanced by only a factor of 10 relative to the H. The poster also shows evidence for enhanced Fe and/or P and enhanced S.

Tytler: You didn't comment on the acceleration mechanism. It seems to me that that's a key point in all of this discussion. Does it work? Do you kow what's accelerating the clouds?

Turnshek: Well, the clouds are accelerating. There's no doubt about that.

E. M. Burbidge: How do you look at the energetics?

Turnshek: The energetics have been studied in some detail. I think the only thing I'll say here is that if you look at models for radiation pressure, versus models where the clouds are accelerated by some kind of thermal wind, that generally the thermal wind models are much more capable of producing the kind of accelerations that will give the outflow velocities that you observe. However, one of the results that's always amazed me is this evidence for absorption-absorption line-locking. This evidence would seem to indicate that in terms of playing some minor perturbing role, that radiation pressure has to have some influence. I think observationally that has to be true.

E. M. Burbidge: Yes, I agree. We always use to look at line-locking as a perturbation. We have time for only one more question.

Barthel: Do you know of any non-BAL quasars with strongly enhanced NV emission?

Turnshek: Yes, there are some.

Barthel: Do they have any peculiar properties, or are they just run of the mill quasars?

Turnshek: If you accept the geometry I've proposed, there is nothing saying that the covering is somewhat incomplete even when your line-of-sight passes through the proposed large-scale BAL region disk. You may pass through this region and not intercept any BAL region clouds, but still get enhanced NV emission. You may expect that to happen some fraction of the time. In terms of peculiar properties, what might be worth doing (and I haven't done this, and I don't think anyone else has) is to look for quasars that have enhanced NV emission, and see whether they have the characteristics of other BAL quasars such as enhanced Fe, enhanced AlIII and weaker CIV emission. These are the other emission line properties that seem to occur at this aspect angle.

Barthel: I know of one such object which is variable in the optical.

QSO ABSORPTION SYSTEMS WITH $z_{ABS} \simeq z_{EM}$

Craig B. Foltz, Frederic H. Chaffee, Jr.
Multiple Mirror Telescope Observatory
Smithsonian Institution and the University of Arizona

Ray J. Weymann, and Scott F. Anderson
Mount Wilson and Las Campanas Observatories
Carnegie Institution of Washington

Abstract. We review the controversy regarding the presence of an excess of CIV absorption systems near the emission line redshift and discuss its resolution. We present additional evidence that the presence of such strong, associated absorption may be correlated with the radio properties of the QSO, implying that such absorption could arise in the vicinity of the QSO, either in the host galaxy or in a cluster of galaxies containing the QSO. We are unable to discriminate between these two possibilities at this time. It is also suggested that the presence of strong, narrow HeII $\lambda1640$ emission may be correlated with radio properties of the QSO and/or the presence of associated absorption.

1. 3C 191 – AN HISTORICAL DIGRESSION

The first published reports of the presence of absorption lines in the spectrum of a QSO were presented in a back-to-back pair of letters in the 1966 *Astrophysical Journal* by Burbidge, Lynds and Burbidge (1966) and Stockton and Lynds (1966). The QSO was 3C 191; the system had $z_{abs} \simeq z_{em}$.

In their review of the state of QSO absorption line research fifteen years later, Weymann, Carswell and Smith (1981) refer to 3C 191 as a unique "system posing special problems" for their classification system. A recent spectrum of 3C 191 obtained with the MMT and MMT Spectrograph at 2 Å resolution is presented in Figure 1. Strong absorption is seen over a wide range of ionization (from NV to CII) as is the famous SiII $\lambda1265$ absorption line which arises from the first excited fine-structure level of Si^+.

Six years later, we find ourselves believing that while we still do not understand the nature of the absorber toward 3C 191, and while it is still the only well-established case of SiII fine-structure absorption, we do not think that it is unique in terms of the strength and velocity width of the absorption, or in its velocity displacement from the emission-line redshift. Furthermore, clues in the spectrum of 3C 191 may provide insight into the nature of the absorbing gas in what appears to be a populous class of absorption system.

We will return to considerations of 3C 191 below. First we want to review the controversy which began in the late 1970s and continued through the early 1980s regarding the existence of a population of absorbers which are found preferentially at velocity displacements near the QSO redshift, the systems we have called 'associated' absorption systems. We will then reiterate the apparent resolution of that conflict, and briefly discuss some of the properties of the QSOs demonstrating such absorption.

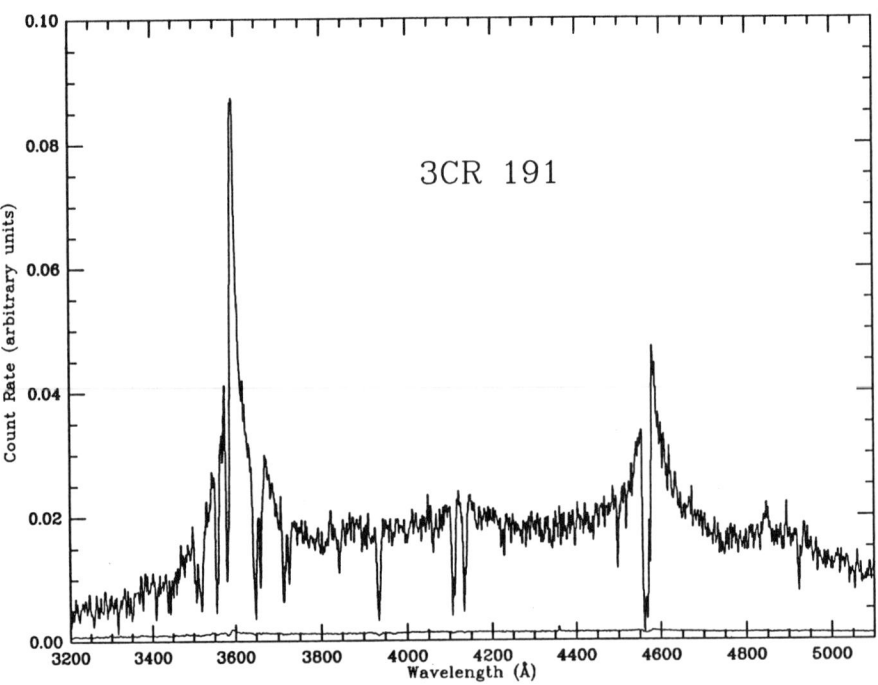

Figure 1. *MMT spectrum of 3C 191 showing the strong associated absorption system at $z = 1.94$. Absorption is present in species covering a wide range of ionization (from NV to CII). The lower spectrum is the 1σ error level as derived from considerations of counting statistics in the object, sky, and dark signals.*

2. IS THERE A STATISTICAL EXCESS OF ABSORPTION SYSTEMS NEAR QSOs?

Weymann et al. 1979 (WWPT) surveyed a sample of 46 intermediate redshift QSOs (35 with $z_{em} > 1.40$) in order to investigate the velocity distribution of CIV $\lambda\lambda 1548, 1550$ absorption lines. [Recall that $v_{ej}/c = (R^2 - 1)/(R^2 + 1)$; $R = (1 + z_{em})/(1 + z_{abs})$.] Since the survey was carried out with uncalibrated photographic plates, quantitative assessment of the signal-to-noise ratio (S/N) was not possible, and accurate absorption equivalent widths were not measured although the approximate strength of the absorption was estimated. Furthermore, because of the relatively high spectral resolution and attendant limited spectral coverage and of the relatively low contrast of the QSO CIV emission line, measurement of z_{em} from the plates was not possible in most cases.

The results of the WWPT study can be summarized as follows:

1. Strong evidence was found for a significant peak in the distribution of the number of CIV doublets per unit velocity per QSO at $z_{abs} \simeq z_{em}$ (within about ± 5000 km s^{-1} of the emission-line redshift);
2. Assuming these 'excess' absorption systems arise in clusters containing the QSO (as suggested by Bahcall 1975 and Weymann and Williams 1978), the distribution of v_{ej} yields a dispersion of $\simeq 750$ km s^{-1} for the putative clusters containing the QSOs;
3. The data provided marginal evidence for an 'ejected' component at velocities $v_{ej} < 18,000$ km s^{-1}.

Subsequently, Young, Sargent and Boksenberg (1982, YSB) surveyed 33 QSOs with $z_{em} \approx 2$. The sample included 3 BAL QSOs, and 1 QSO for which the CIV emission line region was not observed. A photon-counting detector was used resulting in spectra with quantifiable S/N and absorption equivalent width measurements.

The results of the YSB survey were:

1. On large velocity scales ($v_{ej} > 5000$ km s^{-1}), C IV absorption systems are distributed randomly in redshift, consistent with absorption from intervening galaxies;
2. *The observations provided no corroborative evidence for the existence of the peak near z_{em} reported by WWPT;*
3. The two-point correlation function for C IV lines is positive for $500 < \Delta v < 2000$ km s^{-1}, but the excess of systems expected near z_{em} due to the galaxy–galaxy correlation function (assuming QSO hosts are galaxies) was predicted to be below the threshold for detection in this survey and would therefore not account for the excess of systems observed by WWPT.

The apparent discrepancy of these results — WWPT claiming strong evidence for the existence of a population of absorbers near to the QSO (in redshift) and YSB claiming that such a population does not exist — persisted for five years.

In 1986, Foltz et al. (FWPSMC) published their observations of 31 of the WWPT QSOs. These observations were made with the MMT Spectrograph and photon-counting Reticon detector at 1 Å resolution, producing spectra with quantifiable characteristics similar to those presented by YSB. Measurement and analysis of the absorption lines were performed using procedures which were essentially identical to those of YSB. Only those systems in which each member of the CIV doublet had a rest equivalent width (REW) greater than 300 mÅ (the same REW limit used by YSB) were included in the sample. The survey had two major results:

1. The redshift distribution of 'intervening' systems ($v_{ej} > 5000$ km s^{-1}) was identical to that found by YSB;
2. The apparent excess of absorption near z_{em} found by WWPT in essentially the same sample of QSOs persists. Furthermore, it is largely due to ~ 7 strong (REW ≥ 1.5 Å) systems with $|v_{ej}| < 5000$ km s^{-1}. These were denoted "strong", "associated" systems by FWPSMC.

Analysis of the data yielded the following conclusions: First, the associated systems *cannot* be accounted for by a Poisson distribution of absorbers with an equivalent width distribution given by the combination of the YSB and FWPSMC intervening samples. Simulations indicated that the probability of such an occurrence is less than about one part in 10^8. Second, the correlation between the QSOs in this sample and CIV absorbers is *stronger* than the correlation between CIV absorber and CIV absorber. Third, as regards the properties of absorption near the emission line, the FWPSMC and YSB samples are apparently drawn from different parent populations.

FWPSMC speculated that since their sample contains a large fraction of radio-loud (21 of 31), steep-spectrum (10 with $\alpha < -0.7$, where $S_\nu \propto \nu^\alpha$) QSOs while the YSB sample does not (13 radio-loud, 2 steep-spectra), and since in the FWPSMC sample, six of the associated systems occur in radio-loud QSOs, perhaps the occurrence of strong associated absorption is related to the radio properties of the QSO. This leads us to ask the question we consider next.

3. DO STRONG ASSOCIATED SYSTEMS OCCUR PREFERENTIALLY IN RADIO LOUD QSOs?

Since three of the objects known to have complex absorption near z_{em} were 3CR sources (3CR 191, was known from early studies; 3CR 205 and 3CR 298 were FWPSMC QSOs), we decided to observe all of the 3C and 3CR QSOs satisfying the criteria $z_{em} \geq 1.25$; $m_V < 19.0$ The details of this study have been presented by Anderson et al. (1987).

By inclusion in the 3C and 3CR catalogs, the objects in this sample satisfy the criterion: $S(179\ MHz) > 9$ Jy for the 3CR sources and a similar criterion at 158 MHz for the 3C QSOs. By selection, most objects have steep radio spectra. Indeed, all but 3C 446 have 6–11 cm spectral indices (Veron-Cetty and Veron 1985) steeper than -0.7. The sample includes the objects: 3CR 9, 3CR 181, 3CR 191, 3CR 205, 3C 243, 3CR 268.4, 3CR 270.1, 3CR 298, 3CR 432, 3C 446, and 3CR 454. Because of its nomenclature, 3CR 48/54 (0141+339) was also included but, despite its nomenclature, it is a member of neither the 3C, nor 3CR catalogs.

Spectral observations with 1 Å resolution were obtained with the MMT Spectrograph. The S/N was high enough such that in all cases the 3σ upper limit on the REW of an unresolved CIV doublet at a redshift near the CIV emission line was > 0.6 Å. The spectra are all reproduced in Figure 2.

The results of this '3C Mini-Survey' can be summarized as follows.
1. Six objects (191, 205, 243, 268.4, 270.1, and 298) show associated absorption with REW ≥ 1.5 Å. Four other objects (48/54, 181, 432, and 454) show weaker absorption (REW $\simeq 0.5$ Å) within 5000 km s^{-1} of z_{em}.
2. Excluding 3CR 48/54 from consideration, we estimate from the combined intervening (i.e. displaced from the emission redshift) sample of YSB and FWPSMC, that only ~ 0.1 intervening system with REW ≥ 1.5 Å is expected within all 11 of the ± 5000 km s^{-1} velocity windows centered on the CIV emission line of the individual QSOs.

These results were construed as evidence for *some* correlation of the presence of associated absorption with radio properties. However, since the 3C 'mini-survey' contains mostly steep-spectrum sources, it cannot be used to investigate the possible correlation of the occurrence of associated absorption with radio spectral index. Furthermore, the 3C sample contains powerful sources so it cannot be used to investigate correlations with radio power. We are currently surveying a larger, homogeneous sample of QSOs which we denote the 'Radio-Loud Survey'. The sample being considered is selected from the catalog of Veron-Cetty and Veron (1985) according to the following criteria: Declination $> -20°$, $S(11\ cm) > 100$ mJy, $V < 18.5$, $1.25 \leq z_{em} \leq 2.4$.

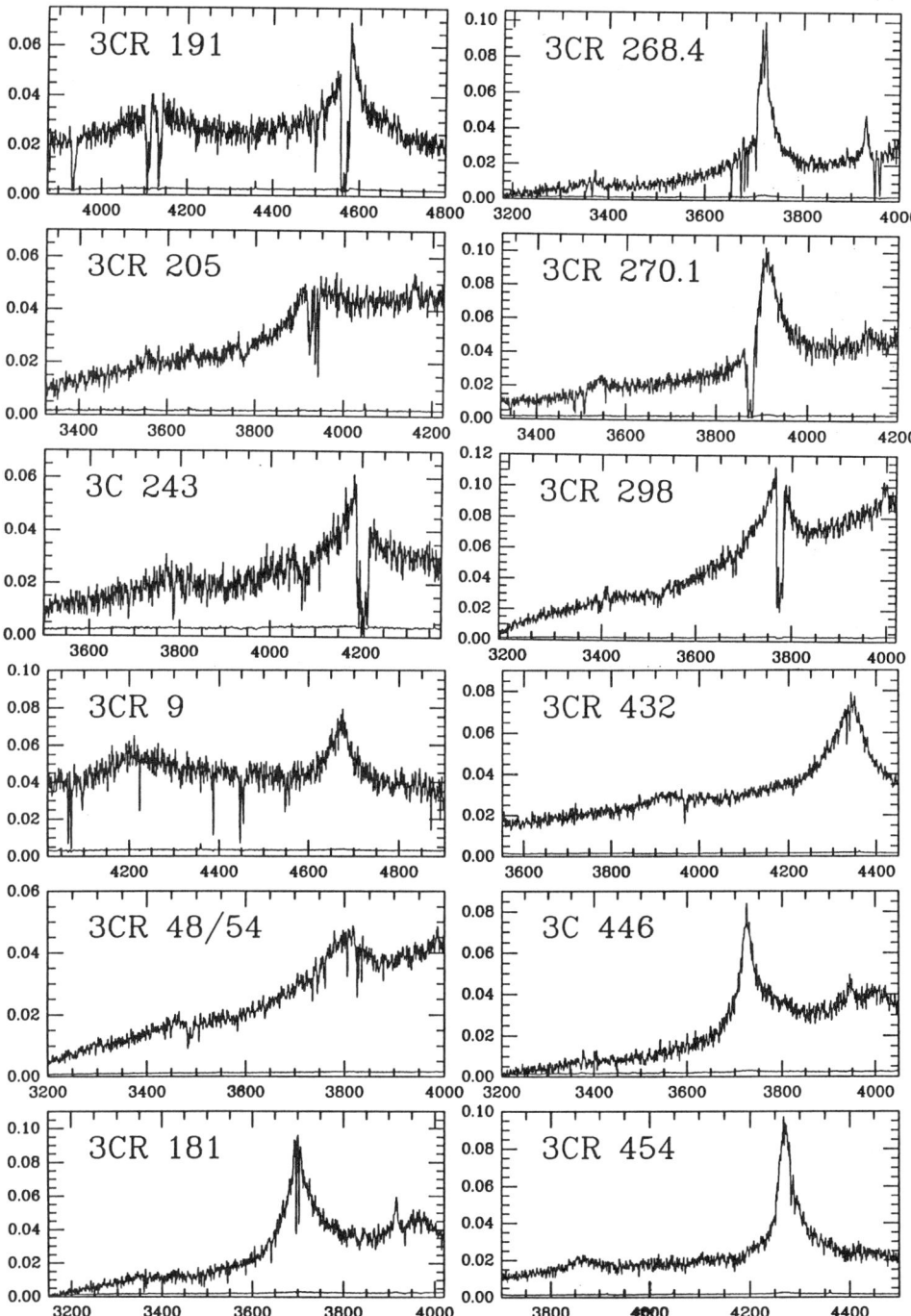

Figure 2. Spectra of QSOs in the 3C mini-survey. In all cases, the strong emission line near the right edge is CIV λ1550. Note the strong associated absorption in the top six spectra.

The full sample consists of ~ 160 objects. Some of these may be excluded due to optical faintness. At this time we are > 50 % complete with data available on 88 objects. The spectroscopy is similar to that of the 3C mini–survey, although some of the observations are being carried out at 2 Å resolution. Although the sample is magnitude-limited, it spans 4 magnitudes in absolute magnitude.

The preliminary results can be summarized as follows.

1. Of the 88 objects in the current sample, 22 show associated CIV absorption with REW ≥ 1.5 Å.
2. There is a tendency for associated absorption to occur preferentially in steep-spectrum sources. There is an apparent correlation with luminosity which may be the result of a paucity of steep-spectrum QSOs with $M_V < -27$ (see discussion below).
3. Associated systems exist in QSOs with bent, straight, and unresolved radio morphologies.
4. There is currently no statistical evidence for outflow ($\langle v \rangle = 200$ km s^{-1} ± 1600 km s^{-1}). However, in a poster presented at this workshop, Junkkarinen (1987) has shown that some apparently real cases of 'infall' still exist (including 3C 298, $\Delta v \simeq -300$ km s^{-1}). Junkkarinen's sample is basically a sub–sample of the Radio Loud Survey but, by intentional selection, is biased in favor of infall systems. Junkkarinen derives $\sigma \simeq 400$ km s^{-1} for $z_{abs} > z_{em}$ systems.

In another poster presented at this conference, Møller and Jakobsen (1987a,b) suggest that the data available in the literature are consistent with the presence of associated absorption's being anti–correlated with QSO optical luminosity and that this correlation may be more fundamental than any correlation between absorption properties and radio properties. Their claim is that associated systems do not occur in the most luminous QSOs ($M_V < -27$). In order to address this question, we have made a preliminary analysis of the radio-loud sample. We stress the preliminary nature of these results. We will consider correlations between the strength of any associated absorption and the luminosity and radio spectral index of the QSOs. The uncertainties in all these quantities are fairly large and difficult to quantify at this time. Specifically we note the following.

1. In general, the upper limit on the REW of any associated absorption is calculated assuming that the individual members of the doublet are unresolved. We underestimate the upper limit in cases where the absorption is weak and broad. Furthermore, in some cases we have not yet actually measured the upper limit on the absorption REW; in what follows, we estimate all of these to be of order 0.4 Å.
2. The absolute magnitudes of the QSOs were determined from the apparent magnitudes tabulated in the Veron catalog. The uncertainty due to variability and inaccurate magnitude estimates is unknown.
3. The radio spectral indices are heterogeneous, ranging from accurate, nearly-simultaneous, broad wavelength baseline measurements to values derived from the flux densities at 6 and 11 cm tabulated in the Veron catalog. Again, the uncertainties introduced by source variability and beam size effects are unknown.

In Figure 3 we plot the REW (or limit thereon) of any CIV absorption complex within ± 5000 km s^{-1} of the CIV emission line against the radio spectral index α. Strong associated absorption appears to arise in objects with steep radio spectra as suggested by FWPSMC. In Figure 4, we plot REW(CIV) versus M_V. The effect noted by Møller and Jakobsen is evident in the figure – all but one of the QSOs with

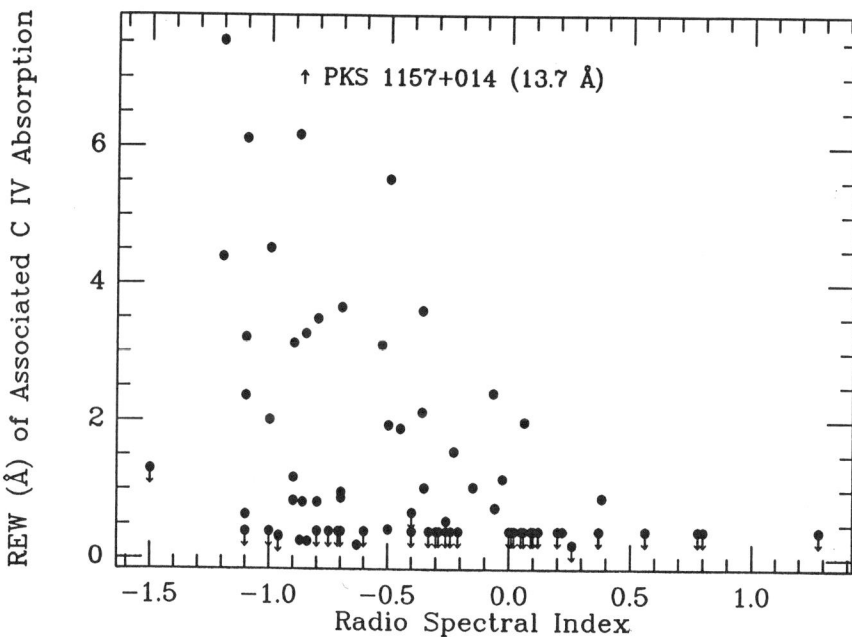

Figure 3. *Plot of the REW of CIV associated absorption versus radio spectral index for 88 QSOs in the radio-loud Survey.*

REW(CIV)> 1 Å are fainter than $M_V = -27$. We believe that this apparent correlation is an artifact resulting from a paucity of steep-spectrum, optically-luminous, radio-loud QSOs. This is demonstrated in Figure 5 where we plot M_V versus the radio spectral index. Note that there are few QSOs in the sample with $M_V < -27$ and only one with $\alpha < -0.7$ and $M_V < -27$. In their study, Møller and Jakobsen (1987a,b) consider a sample of QSOs including a number of optically-selected objects. High resolution observations of optically-selected QSOs have been biased toward optically luminous objects since, of course, such objects are easier to observe than low luminosity objects at the same redshift. Whereas such a bias does exist for the radio-loud Survey as a result of our imposing a faint magnitude cutoff in the sample selection, the sample does include relatively low luminosity objects. If radio-loud objects are combined with an optically-selected sample biased toward high luminosity objects, then any correlation of the properties of the associated absorption with radio properties will be manifested as an apparent correlation with optical luminosity.

Given what we consider to be strong evidence that associated absorption tends to occur in radio-loud QSOs and Broad Absorption Lines (BALs) tend to occur in radio-quiet QSOs (Turnshek 1988), we now pose the following question:

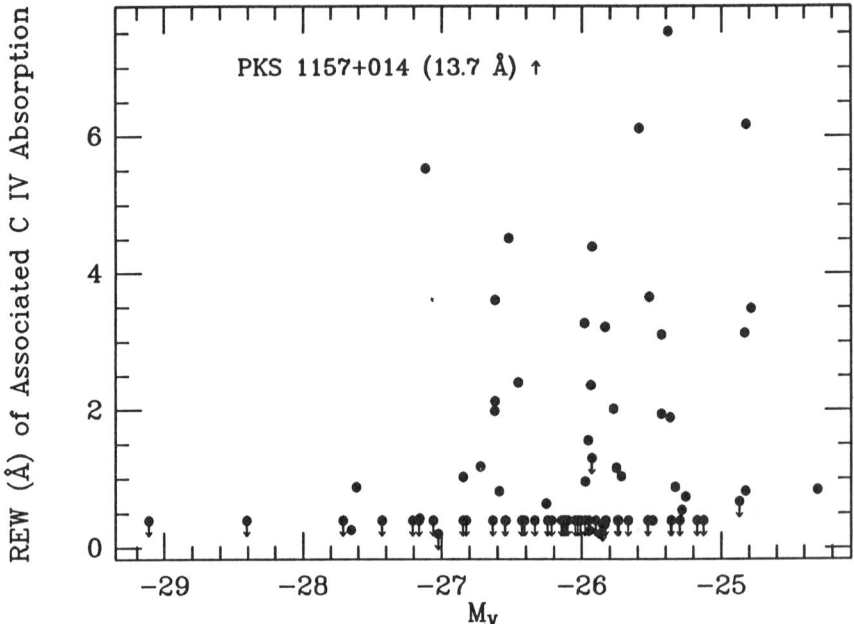

Figure 4. *Plot of the REW of CIV associated absorption versus absolute visual magnitude for 88 QSOs in the radio-loud Survey.*

4. IS THERE A DICHOTOMY BETWEEN THE OCCURRENCE OF BAL ABSORPTION IN RADIO-QUIET QSOs AND ASSOCIATED ABSORPTION IN RADIO-LOUD QSOs?

First we consider the question: Does associated absorption tend to be found in radio-loud QSOs? From the radio-loud CIV survey, 22 QSOs out of the sample of 88 observed to date show strong (REW $\geq$ 1.5 Å) absorption within 5000 km s^{-1} of the emission-line redshift. Combining the radio-quiet members of the YSB and FWPSMC samples, 1 of 29 show such absorption (and that object, 1256+357, is a weak radio emitter with $S_{11cm} \simeq 11$ mJy). If the true incidence of associated absorption, p, is approximated by the union of these two samples, then $p \simeq (1+22)/(29+88) \simeq 0.20$. Using this approximation, the joint probability of choosing one or zero objects without associated absorption in a sample of 29 and 22 or more with such absorption in a sample of 88 would be: $P(\leq 1, 29, p) \cap P(\geq 22, 88, p) \simeq 0.00134$.

Now we consider the question: Is BAL absorption found preferentially in radio-quiet QSOs? First, Stocke *et al.* (1984) carried out a survey of 30 BAL QSOs (including PKS 1157+014; see Figure 6) and found that their radio properties differed from a sample of > 400 optically-selected QSOs at the 99.9 % confidence level in the sense that the BAL QSO sample has a smaller fraction of strong radio sources. Second, from the combination of the Radio-Loud CIV survey and 13 published digital spectra from the same sample, only one QSO (PKS 1157+014) shows BAL characteristics, and it only marginally. We adopt 3 %, 5 %, and 10 % as reasonable guesses for the true incidence of BAL QSOs among *optically-selected* QSOs. If these apply to

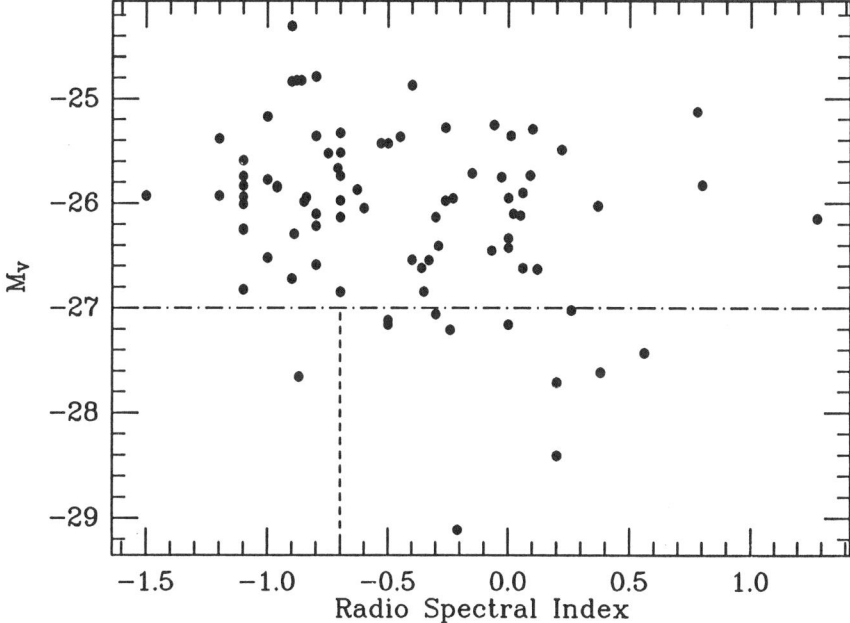

Figure 5. *Plot of the absolute visual magnitude versus radio spectral index of 88 QSOs in the radio-loud Survey. Note the paucity of luminous, steep-spectrum sources.*

radio-loud QSOs as well, then:

$P(\leq 1, 101, .03) = .19$ 'wishful thinking'
$P(\leq 1, 101, .05) = .036$ 'suggestive'
$P(\leq 1, 101, .10) = .0003$ 'conclusive'

5. 3C 191 REVISITED

As noted above, 3C 191 is still unique in that it contains the only absorption line system with well-documented SiII fine-structure absorption. A detailed analysis of this system assuming that the QSO photoionizes the absorbing cloud was carried out by Williams et al. (1975). The results of their analysis still apply:
1. In order to understand the excitation of the SiII fine-structure line, $n_e \geq 10^3$ cm^{-3};
2. The distance between the continuum source and the absorbing cloud is ≥ 10 kpc.
3. If the absorbing cloud covering factor $(\Omega/4\pi)$ is greater than about 0.1, emission from the absorbing clouds will contribute substantially to the observed emission lines.

In this context we note that in a poster presented at this workshop, Barthel (1987) suggests that the presence of strong CIV absorption near z_{abs} might be expected to be found preferentially in radio sources showing distorted or 'bent' structure if the same material responsible for the absorption also produces the distortions in the radio

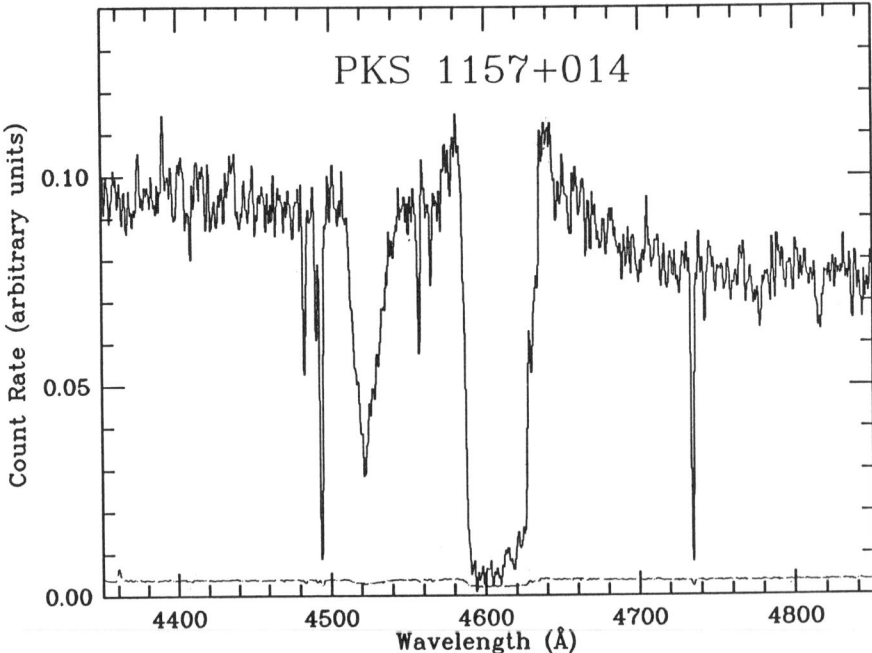

Figure 6. *Spectrum of the region around the CIV emission line in PKS 1157+014, an object with very strong associated absorption near 4610 Å as well as a BAL-like feature near 4520 Å. This is the only BAL-like feature among 101 radio-loud QSOs whose spectra we have examined.*

plasma. If this hypothesis is correct, the characteristic scales agree with the distance between 3C 191 and the absorber, i.e. a few to tens of kiloparsecs.

Note also that the electron density inferred by Williams *et al.* (1975) is higher than in typical interstellar clouds. Perhaps they are being confined by ram pressure connected somehow to the radio plasma.

Finally, if $\Omega/4\pi$ is large, we expect substantial HeII $\lambda1640$ emission (as well as [OIII] $\lambda5007$) from the absorbing clouds. Note the relatively strong and narrow HeII emission in 3C 191 in Figure 1 and particularly in 3CR 181 and 3CR 268.4 which are re-plotted in Figure 7. In the 3C mini-survey we find six objects with measurable HeII $\lambda1640$ emission. The mean FWHM is ~ 800 km s^{-1}, much less than the FWHM of the CIV emission lines.

6. SPECULATIONS

We find that the apparent general correlations between the presence of strong associated absorption with steep radio spectra on the one hand and BAL absorption with weak radio emission on the other reminiscent of the general correlations between the optical spectra, radio properties, and properties of the QSO fuzz which have been investigated for low-redshift QSOs by Stockton and Boroson and their collaborators. Also compelling is the presence of strong narrow HeII $\lambda1640$ emis-

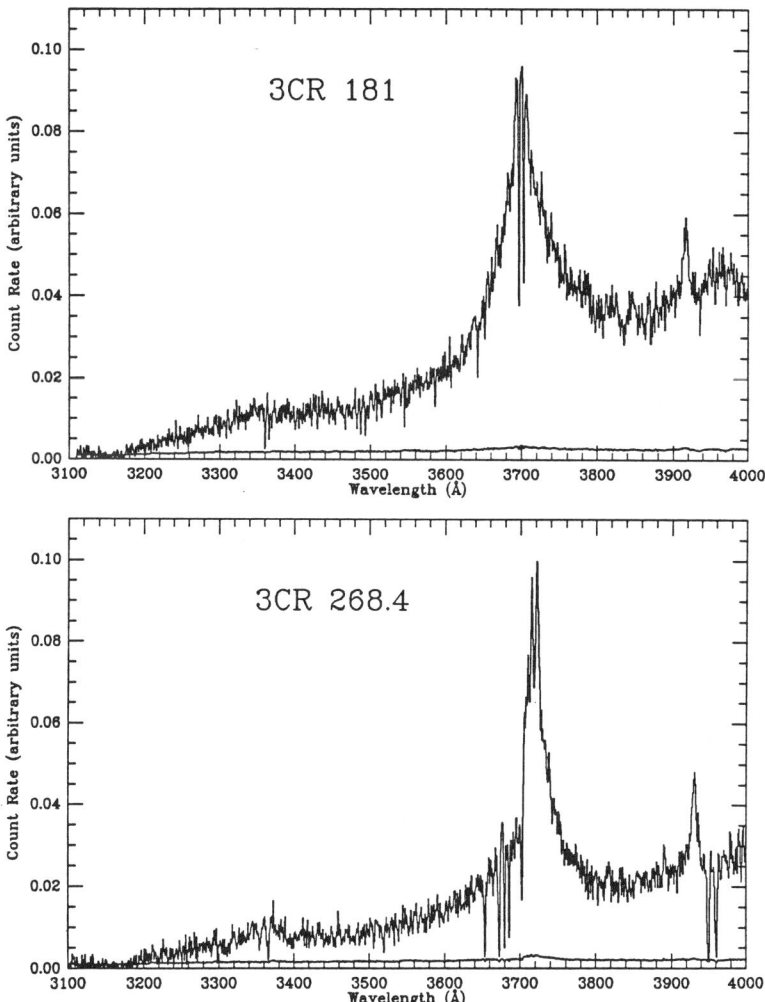

Figure 7. *Enlarged plots of the spectra of 3CR 181 and 3CR 268.4 from Figure 2. Note the strong and narrow HeII λ1640 emission lines longward of the CIV emission.*

sion in steep-spectrum radio QSOs, just the type which at low redshift demonstrate strong forbidden lines in their nuclear spectra and whose nebulosity is dominated by forbidden-line emission. This may be circumstantial evidence supporting an intrinsic origin for the associated absorption (i.e., the absorption arises in the gas responsible for the nebular [OIII] and HeII emission). We summarize how the absorption properties may fit into this general scheme in Table 1.

One possible correlation which is suggested by the table is between the presence of strong FeII emission and BAL absorption. During the discussion of the previous talk, it was emphasized by J. Wampler that BAL QSOs may have unusually strong FeII emission in comparison to other QSOs. This is a point worth pursuing, as is the

Table 1. Are There Two Types of QSOs?

Property	First Type	Second Type
A. Confirmed for low redshift QSOs[1]:		
Nebulosity dominated by:	Forbidden emission lines	Continuum
Nuclear forbidden lines:	Strong (e.g. [OIII])	Weak
Nuclear permitted lines:	Very broad and irregular	Less broad and smooth
Nuclear FeII emission:	Weak	Strong
Radio spectrum:	Steep	Flat or quiet
Radio morphology:	Extended	Compact or quiet
B. Speculation for high redshift QSOs:		
Absorption at $z_{abs} \sim z_{em}$:	Strong, associated CIV complexes	BALs
He II $\lambda1640$ emission:	Strong and narrow[2]	?
Distance to absorbing material:	Tens of kiloparsecs.	Tens of parsecs.

[1] Boroson, Persson, and Oke (1985); Stockton and MacKenty (1987).
[2] From the same region as the [OIII] emission and/or CIV associated absorption clouds?

correlation of HeII emission with radio properties.

We can not yet discriminate conclusively between an intrinsic origin for the associated absorption and a situation in which the absorption arises in clusters containing the QSOs. In the case of a cluster origin, it may be as simple as: BAL QSOs have spiral galaxy hosts; spirals eschew rich clusters. On the other hand, Yee and Green (1987) have shown that (at least at low redshift) radio-loud QSOs are found preferentially in clusters. However, this does not explain the apparent HeII correlation.

In support of an intrinsic origin for the associated systems, Weymann et al. (1987) argue that under a specific, but plausible set of assumptions (BAL clouds are embedded in a sub-relativistic wind, clouds are photoionized, covering factor $\simeq 10\%$, $V(flow) \simeq 15,000$ km s^{-1}), the kinetic power in a BAL flow is of order a few percent of the optical luminosity, an amount similar to that required to power extended radio structure. Perhaps the basic mechanism that powers the BAL flows (at early epochs?) is related to that later diverted into powering the radio lobes. This is compatible with the evolutionary scenario sketched by Briggs, Turnshek and Wolfe (1984) in which a BAL QSO evolves into a radio-loud QSO. In this scheme, PKS 1157+014 represents the presumably rare transition object.

In either case, certain critical observations are required to discriminate between the two scenarios and investigate the properties of either the extra-nuclear absorbing gas or the clusters: First, it is essential that we obtain better systemic redshifts for those objects with associated absorption. This work has been spearheaded by Junkkarinen in his observations of [OII] $\lambda3727$ and other forbidden lines. Second, it is essential to have a large sample of radio-quiet objects for comparison with the radio-loud survey to understand correlations of associated absorption with, for example, optical luminosity. In order to discriminate between our suggestions and those of Møller and Jakobsen (1987a,b), it may be necessary to observe the associated absorption properties of a sample of intrinsically faint radio-quiet objects. This is a fine program for any of our competitors to carry out. Finally, observations of the HeII $\lambda1640$ emission strength and width in a sample of radio-quiet or flat-spectrum sources and FeII emission in a sample of BAL QSOs would confirm or refute some of the correlations suggested in Table 1.

REFERENCES

Anderson, S. A., Weymann, R. J., Foltz, C. B., and Chaffee, Jr., F. H. 1987, *A. J.*, **94**, 278.
Bahcall, J. N. 1975, *Ap. J. (Letters)*, **200**, L1.
Barthel, P. 1987, *QSO Absorption Lines: Probing the Universe, A Collection of Poster Papers*. eds. J. C. Blades, C. A. Norman, D. A. Turnshek, (ST ScI Publication), p. 19.
Boroson, T. A., Persson, S. E., and Oke, J. B. 1985, *Ap. J.*, **293**, 120.
Briggs, F. H., Turnshek, D. A, and Wolfe, A. M. 1984, *Ap. J.*, **287**, 549.
Burbidge, E. M., Lynds, C. R., and Burbidge, G. R. 1966, *Ap. J.*, **144**, 447.
Foltz, C. B., Weymann, R. J., Peterson, B. M., Sun, L., Malkan, M. A., and Chaffee, Jr., F. H. 1986, *Ap. J.*, **307**, 504 (FWPSMC).
Junkkarinan, V. 1987, *Absorption Lines: Probing the Universe, A Collection of Poster Papers*. eds. J. C. Blades, C. A. Norman, D. A. Turnshek, (ST ScI Publication), p. 21.
Møller, P., and Jakobson, P. 1987a, *Absorption Lines: Probing the Universe, A Collection of Poster Papers*. eds. J. C. Blades, C. A. Norman, D. A. Turnshek, (ST ScI Publication), p. 28.
_____. 1987b, *Ap. J. (Letters)*, **320**, L75.
Stocke, J. T., Foltz, C. B., Weymann, R. J., and Christiansen, W. A. 1984, *Ap. J.*, **280**, 476.
Stockton, A., and Lynds, C. R. 1966, *Ap. J.*, **144**, 451.
Stockton, A., and MacKenty, J. W. 1987, *Ap. J.*, **316**, 584.
Turnshek, D. A. 1988, *This volume*.
Veron-Cetty, M.-P., and Veron, P. 1985, *ESO Sci. Rep*, No. 4.
Weymann, R. J., Anderson, S. A., Foltz, C. B., and Chaffee, Jr., F. H. 1987, in *Proceedings of the Thirteenth Texas Symposium on Relativistic Astrophysics*, ed. M. Ulmer, in press.
Weymann, R. J., Carswell, R. F., and Smith, M. G. 1981, *Ann. Rev. A. A.*, **19**, 41.
Weymann, R. J., and Williams, R. E. 1978, *Phys. Scripta*, **17**, 217.
Weymann, R. J., Williams, R. E., Peterson, B. M., and Turnshek, D. A. 1979, *Ap. J.*, **234**, 33 (WWPT).
Williams, R. E., Strittmatter, P. A., Carswell, R. F., and Craine, E. R. 1975, *Ap. J.*, **202**, 296.
Yee, H. K. C., and Green, R. F. 1987, *Ap. J.*, **319**, 28.
Young, P., Sargent, W. L. W., and Boksenberg, A. 1982, *Ap. J. Suppl.*, **48**, 455 (YSB).

DISCUSSION

Wolfe: It's an interesting result that you have obtained. The associated absorption correlating with the steep spectrum sources is just the opposite of what you'd expect, because when you have a compact radio source, you have a small source putting out energy. Has anybody done a VLBI study of your sample of steep spectrum radio sources, because very often objects that are steep spectrum at low frequency turn out to have compact cores at high frequencies.

E. M. Burbidge: That's what I wanted to ask too.

Foltz: Not that I know of. Peter, do you know if anyone has done that?

Barthel: No.

Wampler: I have a comment and a question. The comment is that in looking at flat spectrum radio sources, it seems to me that very often you do see HeII emission, but I don't think I've ever seen narrow HeII emission like you've shown in your spectra. If that is characteristic of these objects, then that's important. The question I have is: it seems to me that of the 10 or 15 or so Bright Quasar Survey (PG) objects, there are a fair number, say 5 or 6, that have associated absorption. What absorption equivalent width do you consider strong?

Foltz: We've arbitrarily adopted 1.5 Å (in the rest frame).

Wampler: At the level of 1.5 Å in that survey there must be 4 or 5 objects, maybe more, that have strong associated absorption, and those objects are radio quiet and bright.

Foltz: Well, that certainly doesn't fit in the picture I've given.

Barthel: I would like to make some comments. From straight and distorted radio sources one can also infer sometimes the presence of very dense clouds in an environment. In the presence of very compact hot spots that we map with VLBI, it's typical to find that you need a very dense medium around them. Also, this is true from rotation measures of radio knots in jets. A beautiful example is 3C205 on which we published a paper a year ago, and another example is 3C191 mentioned in the posters.

Shull: I have a question about x-rays. Your correlation with the radio suggests that maybe it's a jet or something that's powering this. Has anyone looked at the correlation with the x-ray spectrum or steepness?

Foltz: We have not.

Shull: Were they bright enough to have been done?

Foltz: They are quite faint. The 3C objects were between 18.2 and 19 mag.

Miley: Are the statistics too sparse to say anything about redshift?

Foltz: They are too sparse.

Miley: I think that's a very important point. The work that Peter Barthel and I have done really indicates that right at about redshifts of 1.5 the sources become very bent.

Foltz: That's probably about the mean redshift for our sample, because at much lower redshift than that we can't observe them very well, and at much higher redshift they typically become too faint for us to observe. Most of the objects that we have observed are between redshifts of 1.3 and 1.6. At least the 3C objects are right in that regime.

Miley: In the radio, the bending really comes in right at about that redshift, and so I really think it's important to explore the redshift dependence.

Foltz: Well, it certainly is important to continue the observations to lower redshifts, but we simply can't do that.

Turnshek: I have two questions. What is the typical width of that narrow HeII emission feature?

Foltz: The mean value is 800 km s^{-1} for the 6 objects that we could measure in our 3C survey.

Turnshek: For objects with associated absorption systems, were there any data on NV emission? What I would be interested in is any evidence that NV emission is enhanced in these objects, like it is in BAL QSOs.

Foltz: Again, typically there is not information on NV emission because the redshifts are too low. Typically we're observing around redshift 1.5.

Wolfe: Is anything known about Faraday rotation or Faraday depolarization in these objects compared to the objects with very displaced absorption systems?

Kronberg: A few years ago Judith Perry and I did a study to look for a correlation between Faraday rotation and absorption line strength (1982, *Ap. J.*, **263**, 518; 1984, *Ap. J.*, **279**, 19). We found indeed that there is a correlation which has since become clearer. The tentative conclusion was that the systems with strong FeII and MgII absorption certainly have an excess Faraday rotation. Maybe some of the higher excitation line systems didn't, but that's very tentative.

Foltz: However, those weren't limited to ejected or intrinsic systems, right? Those were intervening systems as well as systems near the emission line redshift.

Kronberg: There was no separation between different types of systems, except that there was some hint, just to go further on speculation, that some of the intrinsic systems maybe did have a higher Faraday rotation. We're looking at that at the moment. The other thing we've done is rotation measure maps of 2 or 3 objects. The results are not entirely conclusive, but I can show them to you afterward if you're interested.

Barthel: I'd like to point out that for VLA studies the resolution is very critical, because it seems that the rotations are very confined to knots in jets.

Ozernoy: In principal, it is not excluded that there could be objects intermediate between spirals and ellipticals. In this connection, I wonder if you see any quasars with characteristics intermediate between type 2 and type 1 as you mentioned?

Foltz: Well, there are certainly QSOs with weaker absorption. We find them down to several hundred milliangstroms where basically our data run out. Some of those are expected purely to be extensions of the intervening population. I don't think that I can discriminate at this time between what you suggest and an intervening population.

Norman: We had some discussion of the geometry of BAL regions. I just wonder if someone could comment on whether that could be jet-like?

Foltz: Do you want to comment on that, Dave?

Norman: They may be rather like the small inner jets you see in Seyfert Galaxies.

Turnshek: Using jets has the advantage that there is still a small covering factor. The problem is, with a single jet-like geometry, you can never hope to produce a symmetric NV BEL by resonance line scattering.

Norman: Let's say you have the same covering factor in a jet-like geometry as in a disk-like geometry.

Turnshek: A covering factor of 5–10 % in a single outflowing jet would produce emission from resonance line scattering, but it would all be blue scattered emission if the jet were directed toward you. You need to have some kind of cylindrical symmetry, like a disk, to get the symmetric NV BEL from resonance line scattering. Oppositely directed jets cannot be ruled out.

Foltz: Are you thinking about entraining material?

Norman: Yes, right. An extended jet.

Foltz: I think that the associated systems do not have very high velocities, except for perhaps in 1556+335 that Simon Morris and others have studied (1986, *Ap. J.*, **310**, 40) where we see a system that is very complex at 5000 km s^{-1}. Perhaps another such system is in 0215+015 (see Blades et al. 1985, *Ap. J.*, **288**, 580).

Bregman: I was wondering if in the radio sources that have associated absorption, if there's some correlation with the radio structure and whether or not you have absorption and where it is? For example, just in the sense that a lot of the radio structures (e.g., some radio lobes) are known to have gas around them causing emission, so one might also expect them to be causing absorption. If you saw them end-on, you would expect to see more absorption. Is that the case?

Foltz: The numbers are too small to make detailed correlations. I will point out that there are examples of strong associated absorption in unresolved straight sources, bent sources, etc. So we see examples in basically all morphologies. One good example of a bent structure is 3C270.1 which Stocke, Burns and Christiansen (1985, *Ap. J.*, **299**, 799) have recently studied.

Kronberg: It might just be of interest to mention that we have just completed a radio study at 6 wavelengths of the Molonglo sample of quasars identified by Cyril Hazard with the attempt of looking at the structure and the Faraday rotation mapping of the whole set. That should be quite interesting.

Junkkarinen: In that new radio sample that you've been surveying, did you find anything with excited CII fine structure lines?

Foltz: We typically don't observe that far down into the blue and I can't say that the data are in presentable enough form yet to answer that.

Junkkarinen: How many SiII $\lambda1526$ absorption lines did you find without SiII $\lambda1533$ find structure?

Foltz: I can't answer that at this time. That's very preliminary.

Junkkarinen: So you don't have any distance estimates?

Foltz: Not really. The three absorption systems that do have distance estimates are 1556+335, 1157+014 and 3C191. Derived distances all turn out to be greater than 50 kpc or so.

Miley: One thing I should point out in answer to Joel Bregman's question is that you wouldn't necessarily expect a correlation (between type of radio structure and presence of absorption). This is because, of course, when you see a bent source the jet is coming out on the plane of the sky, whereas for absorption, the absorption clouds are between you and the optical object. So whether a jet just happens to bump into a cloud may determine whether it's bent or not.

Foltz: You're just using a covering factor argument.

Dahari: Is there any evidence for infall like $z_{abs} > z_{em}$?

Foltz: Vesa, I think, can answer that better, but there certainly are cases like that. In fact 3C298, one of the quasars I showed you, has absorption that's apparently infalling at 400 km s^{-1}, and that evidence has persisted based on observation of [OII] 3727 emission. Is that true, Vesa?

Junkkarinen: Yes, that's true, and there is one where the redshift from [OII] 3727 emission suggests an infall of 1200 km s^{-1}. So there is good evidence.

Barthel: We've found a number of red wings on CIV emission profiles in objects with distorted sources which would also argue for infall.

THE Ly-α FOREST: OBSERVATIONAL EVIDENCE FOR LINE EVOLUTION

Richard W. Hunstead
School of Physics
University of Sydney

Abstract. The distribution of Ly-α forest clouds is examined as a function of redshift, rest equivalent width and HI column density. From a maximum likelihood analysis of line samples spanning a wide range in z ($1.5 - 3.78$), there is compelling evidence for strong evolution in line density with redshift. Within individual QSOs, however, there is a countervailing trend (the inverse effect) which appears to be restricted to a wavelength interval close to Ly-α emission and is most plausibly explained by enhanced ionization of the Ly-α clouds in the vicinity of the QSO. The distribution of Ly-α equivalent widths is surprisingly uniform with redshift and spectral resolution and this leads to the prediction of a steep dependence of HI column density on z. A detailed study of the redshift dependence of the column density distribution is not possible at present but is potentially of great importance in understanding the evolution of Ly-α clouds.

1. INTRODUCTION

One of the most striking features of the spectra of high-redshift QSOs is the plethora of narrow absorption lines which occur at wavelengths shorter than Ly-α emission. In this region (commonly known as the Ly-α forest) the majority of absorption lines are Ly-α and the HI clouds giving rise to them are referred to colloquially as Ly-α clouds. A small proportion of the lines can be identified with heavy elements in well-defined systems which are thought to arise in intervening galaxies; Ly-α is usually very strong in these systems. The Ly-α clouds, on the other hand, are believed to form an intergalactic population not associated with galaxies. In a given redshift interval, the number density of Ly-α lines from QSO to QSO is found to be consistent with Poisson statistics and this argues strongly for a cosmological origin and distribution of the clouds (see also Sargent 1988, this volume). The detection of a significant trend in cloud population or cloud properties with redshift could therefore be of vital importance in elucidating the physical conditions in the early universe.

Since the Ly-α forest lines are so numerous and are accessible to ground-based telescopes over a wide range of redshift, $z = 1.5$ to > 4, their properties can be examined statistically and trends with redshift can be investigated. In analysing the observational evidence for evolution, I will be discussing three Ly-α distribution

functions:
1. the number density of Ly-α clouds as a function of redshift (dN/dz);
2. the distribution of Ly-α equivalent widths (dN/dW); and
3. the distribution of H I column densities (dN/dN_c).

Inevitably, there are shortcomings in the observations: inadequate resolution, insufficient signal-to-noise ratio, incomplete knowledge of detector characteristics and so on. Despite these limitations, it has been possible to make considerable progress in understanding the *global* properties of the Ly-α clouds. Specific information concerning the physical conditions and chemical abundances in individual clouds is still rather sparse and the observations are difficult. For this reason, a theoretical picture for cloud formation and evolution can only be painted with the broadest of brushes and the picture is bound to change as the observational data are improved.

Over the past decade there has been considerable controversy as to whether the cloud distribution is uniform in co-moving space or whether the number density changes with epoch (i.e. z). An important factor in resolving the controversy was the extension of the redshift baseline afforded by the $z = 3.78$ QSO 2000-330 (Peterson et al. 1982; Hunstead et al. 1986a). A high-resolution spectrum of 2000-330 in the region of Ly-α emission is shown in Figure 1 together with a corresponding spectrum of the $z = 2.06$ QSO 1256-175. This figure reveals clear differences between the two spectra and effectively sets the scene for this review. As discussed in §2, for a non-evolving population we expect a modest increase in Ly-α line density in going from $z = 2.06$ to 3.78 but nothing as dramatic as seen in these two spectra. Furthermore, ignoring the strong metal-system Ly-α in 1256-175, there is also a striking difference in the average line strength at the two redshifts. These straightforward comparisons obviously point to strong evolution in the Ly-α cloud population with redshift. There are other important features in these spectra, which will be covered later.

A brief historical summary of the conflicting results is given in §2, followed by an account in §3 of the most recent analyses of line evolution by Murdoch et al. 1986 (hereafter MHPB) and Tytler 1987 (hereafter T87). The inverse effect, a weakening of the global trend within individual QSOs, is discussed in §4. The distribution of Ly-α equivalent widths is examined in §5 and shown to be essentially independent of resolution. In §6 the results of §3 and §5 are combined to give rough predictions for the distribution of column densities N_c and the evolution of N_c with redshift; the properties of the measured N_c distributions for three QSOs are discussed briefly. Future areas for observation are examined in §7 and the conclusions are given in §8.

2. HISTORICAL BACKGROUND

The first evidence for evolution in Ly-α line density was presented by Peterson (1978), based on a preliminary study of four QSOs. For clouds of cross-section σ and number density ρ_o per unit co-moving volume, he showed that the corresponding number density of absorption lines per unit redshift is given by

$$\frac{dN}{dz} = \sigma \rho_o \frac{c}{H_o}(1+z)(1+2q_o z)^{-1/2}. \tag{1}$$

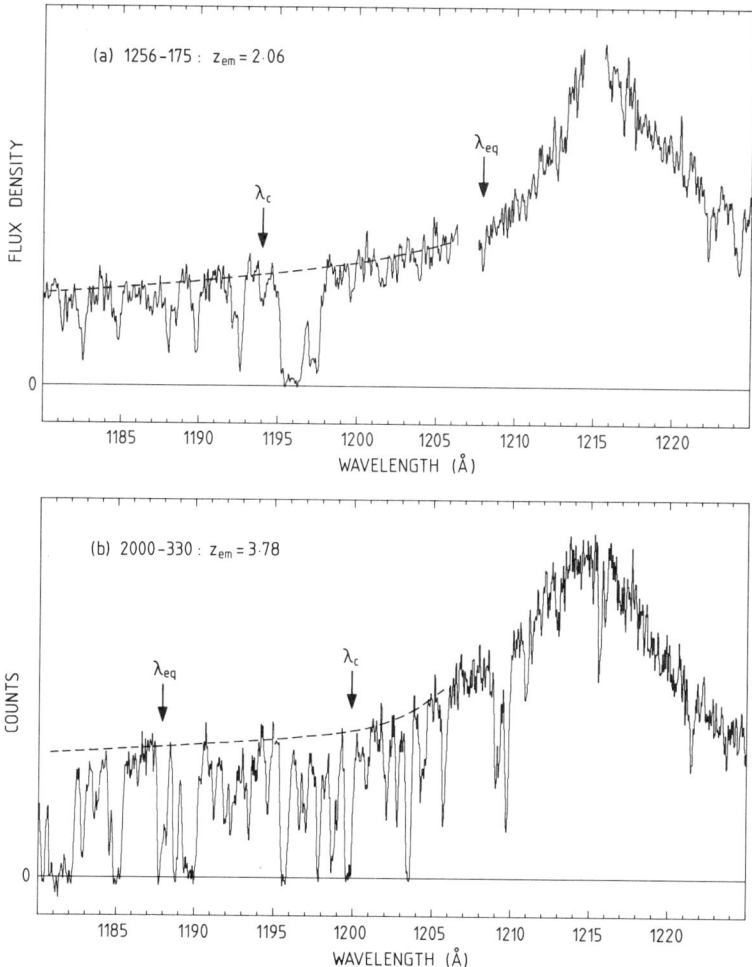

Figure 1. *AAT spectra showing Ly-α emission and part of the Ly-α forest in two QSOs: (a) 1256-175 at z = 2.06, and (b) 2000-330 at z = 3.78. Each spectrum has been shifted to the QSO rest frame and an approximate continuum level is shown. The original resolution was 0.6 Å FWHM. The spectrum of 1256-175 has been edited to remove a strong SiIV λλ1393,1402 doublet (at $z_{abs} = 1.649$) in the Ly-α emission profile; the strong absorption line near 1196 Å is Ly-α in a heavy element system at $z_{abs} = 2.009$. λ_c marks the approximate point where the Ly-α emission line meets the continuum and λ_{eq} corresponds to the distance at which the Lyman limit flux from the QSO is equal to the estimate of the diffuse UV background by Tytler (1987).*

For a non-evolving cloud population, $\sigma\rho_o$ is constant and dN/dz then depends weakly on redshift. Peterson found that the average value of dN/dz for each QSO increased strongly with z, suggesting that $\sigma\rho_o$ was also increasing with redshift. However, his line sample was not clearly defined and it is likely that the line count at high z was artificially boosted by including intrinsically weaker lines than were counted at lower z.

Ellis (1978) used a different approach and applied it to an inhomogeneous sample of absorption lines drawn from the literature. He found no evidence for evolution and, moreover, found no tendency for an increase in dN/dz with z in either of the two QSOs from Peterson (1978) with published spectra.

Sargent et al. (1980; hereafter SYBT) pointed to the need to define a uniform sample by imposing a lower limit to the *intrinsic* equivalent width $W[\equiv W(obs)/1+z]$ of the Ly-α lines. For their sample of five QSOs and $W \geq 0.32$ Å, SYBT used the method of maximum likelihood (ML) to fit the individual lines to an expression of the form

$$\frac{dN}{dz} = \left(\frac{dN}{dz}\right)_o (1+z)^\gamma \qquad (2)$$

and found $\gamma = 0.48 \pm 0.54$. Note that the typical results for no evolution from eqn. (1) are $\gamma = 1$ for $q_o = 0$ and $\gamma = 1/2$ for $q_o = 1/2$. The SYBT result was therefore in direct conflict with Peterson's (1978).

Subsequent authors adopted the functional form of eqn. (2) and the values of γ reported prior to 1986 are listed in Table 1. Brief details of most of these estimates and the samples on which they are based have been summarised by MHPB. The range of answers is confusing, especially when it is realised that the data sets are not independent. The results of Carswell et al. (1982) are particularly puzzling: using Peterson's method (least-squares fit to the coarsely-binned data for each QSO) they find positive evolution, yet when γ is evaluated for each QSO individually and then averaged, the implied evolution is in the opposite sense. They concluded that "in individual QSOs the line density tends to decrease with increasing redshift, while for the sample of QSOs as a whole there is a small net increase with redshift". Carswell et al. pointed out that their results were difficult to reconcile with cosmologically distributed clouds, unconnected with the QSO, and sought an explanation in terms of a hidden selection effect (e.g. due to line blending, uncertain continuum estimation) in their line sample.

Table 1. Estimates of γ prior to 1986

Reference	γ
Peterson (1978)	Evolution ($\gamma > 1$)
Ellis (1978)	No evolution
Sargent et al. (1980; SYBT)	0.48 ± 0.54
Weymann et al. (1981)	2.5
Zou et al. (1982)	No evolution
Young et al. (1982)	1.81 ± 0.48
Carswell et al. (1982)	1.4 ± 0.7 (Peterson)
	0.6 ± 0.6 (ML)
	-2.1 ± 1.5 ($\bar{\gamma}$)
Peterson (1983a)	2.2 ± 0.4
Phillipps and Ellis (1983)	-0.4 ± 0.3 ($q_o = 0$)
Sargent and Boksenberg (1983)	2.1
Peterson (1983b)	2.36 ± 0.36
Atwood et al. (1985; ABC)	1.7 ± 1.0

3. RECENT DETERMINATIONS OF γ

3.1. The MHPB Analysis

The sample of Ly-α lines used in the comprehensive study by MHPB differed from most of the earlier samples in two important respects.

1. The range in absorption redshift (1.5 - 3.78) was the largest accessible from the ground at that time. Extension of the redshift interval was achieved by adding 2000-330 (z_{em} = 3.78, Hunstead et al. 1986a) and the bright BL Lac object 0215+015 (z_{em} = 1.715, Blades et al. 1985) to the existing sample of nine QSOs used by Young, Sargent and Boksenberg (1982).
2. Ly-α lines associated with definite heavy-element systems were excluded on the grounds that they form a separate population. Investigations of heavy-element and Lyman limit systems (summarised by Sargent 1987) suggest that their evolution is weak or non-existent.

Attention was also given to ensuring that the line sample was homogeneous by restricting the sample to lines observed at resolutions in the range 0.8 - 1.5 Å FWHM with the same type of detector (IPCS, Boksenberg 1978).

MHPB used the method of maximum likelihood (ML) to determine γ and its error. The advantages of ML are that it gives the minimum variance and allows considerable flexibility in the selection or exclusion of data. In addition, it does not require arbitrary grouping of the data which has been an undesirable feature of many of the analyses in Table 1. For their full sample of 277 lines with $W \geq 0.32$ Å from 11 QSOs, MHPB obtained $\gamma = 2.17 \pm 0.36$, a value significantly greater than expected for non-evolving clouds. For comparison, the value obtained when Ly-α lines in heavy-element systems were included (298 lines) was 1.79 ± 0.35, supporting the contention that such systems evolve much more weakly than the Ly-α clouds.

Although ML gives the best estimate of γ for the assumed form of eqn. (2), various statistical tests are needed to ascertain how well that form fits the data. MHPB used a Kolmogorov test to show that the *global* fit was excellent and a χ^2 test to establish that the number of lines in each QSO was consistent with a Poisson distribution. However, the key test was to investigate the compliance of the trend of dN/dz in individual QSOs with the global trend. In this non-parametric test (referred to by MHPB as the Q test), the expected value of Q is 0.5 if the lines are distributed within QSOs according to the global distribution. They found a mean Q of 0.454 ± 0.017 which indicated a significant (2.7σ) tendency for relatively fewer lines of the requisite strength to lie close to the emission redshift i.e. there is a trend within QSOs running counter to the global trend. This was dubbed by MHPB the inverse effect.

The Q test was unable to distinguish between an inverse trend distributed throughout the redshift range of each QSO or an effect concentrated to one extreme. When MHPB imposed an upper cutoff wavelength, λ_c, to each QSO to exclude lines occurring within the Ly-α emission profile, the significance of the Q test fell from 2.7σ to a negligible 1.2σ and γ increased slightly to 2.31 ± 0.40. This implied that the inverse effect was localised to the vicinity of the QSO; further discussion of the inverse effect is given in §4. For graphical presentation the data from this latter sample have been grouped coarsely into redshift bins and plotted in Figure 2 together with the ML fit. The point at lower left is from IUE observations of the $z = 1.33$ QSO 1634+706 (Jenkins et al. 1987); it is clearly consistent with the extrapolation from higher redshifts.

3.2. The T87 Analysis

Tytler (T87) used an expanded sample of 473 Ly-α lines with $W \geq 0.36$ Å formed from 19 QSOs; ten of these are in common with the MHPB sample. The additional objects covered a wider range of resolution than MHPB (0.5-2.4 Å) making the T87 sample less homogeneous. In order to quantify the inverse effect (referred to in T87 as the "system anomaly"), an extra term was introduced which depends on the wavelength of each absorption line in a given QSO relative to the wavelength of Ly-α emission. Equation (2) then becomes

$$\frac{dN}{dz} \propto (1+z_a)^\gamma \left(\frac{1+z_a}{1+z_e}\right)^{-\delta} \propto (1+z_a)^{\gamma-\delta}(1+z_e)^\delta \tag{3}$$

where z_a is the redshift of an absorption line in a QSO with emission redshift z_e. The T87 estimate for γ was 2.30±0.36 and for δ was 3.4±1.0. However, Tytler argued that because of the existence of the inverse effect (as revealed now by a significant value for δ), "all claims that published γ values are indicative of Ly-α system evolution should be considered very suspect". This was despite stating that the result for γ was insensitive to the value of δ.

Hunstead et al. (1987b) have reanalysed the MHPB sample using the two-parameter approach of eqn. (3). For comparison, both one- and two-parameter fits have also been applied to the T87 sample. A summary of the results is presented in Table 2. There is clearly little to choose between the δ of T87 and the Q parameter of MHPB in establishing the inverse effect. The most striking feature of the table is the consistency among the various estimates of γ, all of which are significantly greater than expected for non-evolving clouds in a universe with $q_o \geq 0$. Hunstead et al. have explored the effects of changing the equivalent width limits and cutoff wavelengths in each sample but there was no noticeable effect on γ. In all cases a Kolmogorov test indicates a good fit to the assumed functional form for dN/dz.

Although T87 takes the view that the mere existence of the inverse effect casts doubt on the reality of Ly-α cloud evolution, this can not be supported on statistical (or any other) grounds. The significance of γ is determined by its standard deviation (properly evaluated to allow for correlation between the estimates of γ and δ), regardless of the value or significance of δ. Furthermore, when the λ_c cutoffs are imposed, δ becomes effectively zero for the MHPB sample and insignificant (1.2σ) for the T87 sample.

Table 2. ML estimates of γ and δ for the MHPB and T87 samples

Ly-α line sample	1-parameter fit		2-parameter fit		
	γ	$Q - 0.5$	γ	δ	N
MHPB:					
Full sample	2.17 ± 0.36	−0.046 ± 0.017	2.30 ± 0.38	2.1 ± 1.1	277
With λ_c em cutoffs	2.31 ± 0.40	−0.020 ± 0.019	2.30 ± 0.42	−0.1 ± 1.5	222
T87:					
Full sample	2.28 ± 0.33	−0.044 ± 0.013	2.43 ± 0.34	3.1 ± 1.0	473
With λ_c em cutoffs	2.62 ± 0.38	−0.018 ± 0.015	2.74 ± 0.40	1.5 ± 1.3	373

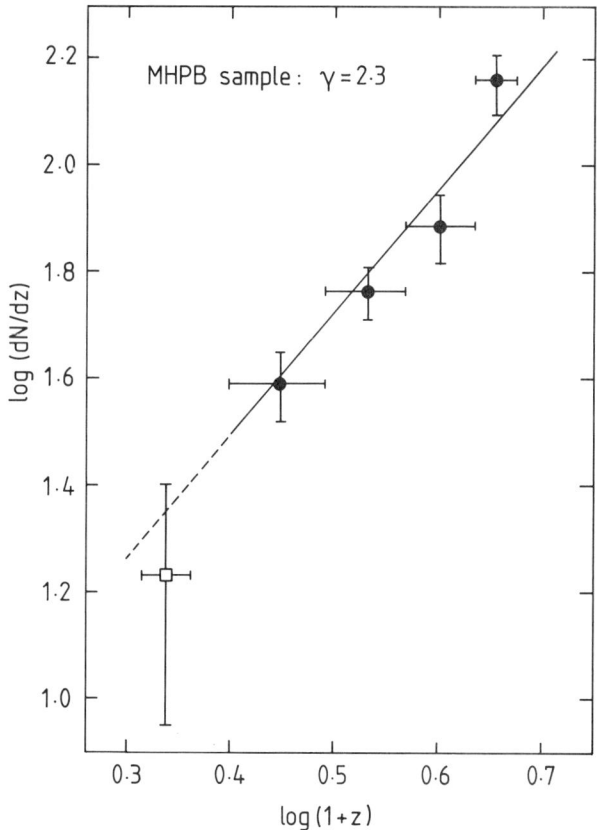

Figure 2. *Log (dN/dz) plotted as a function of log (1 + z) for the MHPB sample with λ_c emission cutoffs. The line drawn is the ML fit to the 222 lines and corresponds to $\gamma = 2.3$. Grouping of the data is solely for presentation: vertical bars are 1σ (i.e. $\sqrt{N}$) errors and horizontal bars define the redshift interval. The point at lower left (□) is from IUE data reported by Jenkins et al. (1987) at this meeting.*

3.3. Is Line Blending Important?

Concern has often been expressed about the effects of blending on the line counts (SYBT, Carswell et al. 1982, T87), to the extent that T87 has claimed that line blending at resolutions ~ 1 Å seriously compromises any attempt to examine evolution in line density. As described by SYBT, in a sample defined with $W > W_o$, the observed line density will be boosted by blends of weaker lines (i.e. those with $W < W_o$) and depressed by blends among members of the sample (with $W > W_o$). Due to the exponential distribution of equivalent widths (§5), the blending of weak lines is *expected* to dominate, although it is not obvious how the contamination will change with redshift and, therefore, how it is likely to affect an estimate of γ. These competing effects can be assessed indirectly by looking for changes in the equivalent width distribution with spectral resolution. In fact, MHPB found that the normalized W distribution did not change significantly over a wide range in resolution (refer to §5), suggesting that blending at resolutions ~ 1 Å does not lead to serious *systematic* errors in line density as a function of redshift.

Parnell and Carswell (1987) have addressed this question directly by using simulated data matching the redshift intervals of the MHPB sample with a line distribution according to eqn. (2). The line parameters were chosen to mimic the distributions found by Carswell *et al.* (1984, hereafter C84) and Atwood, Baldwin and Carswell (1985, hereafter ABC) from high-resolution data. The simulated data were then degraded in resolution and signal-to-noise ratio to match the real data. The derived values of γ were a surprisingly close match to the input values, for all $\gamma \leq 3$. This important study confirms that intermediate resolution spectra are adequate for studying line density evolution.

3.4. Extension to Low Redshift

Bechtold *et al.* (1984) and O'Brien, Gondhalekar and Wilson (1986) have attempted to extend the measurement of γ to low redshifts by using low resolution IUE and ground-based spectra. Instead of counting lines, they measure the depression of the spectrum below an assumed continuum, extrapolated from the region redward of Ly-α emission. The average depression between Ly-β and Ly-α (excluding the emission line wings) is proportional to the total equivalent width of *all* absorption lines and is approximately proportional to dN/dz for Ly-α clouds. For z between 1.0 and 3.5, Bechtold *et al.* obtained $\gamma = 1.29 \pm 0.21$ which is consistent with no evolution for $q_o = 0$. On the other hand, for a sample extending to lower redshift, O'Brien *et al.* found $\gamma = 1.67 \pm 0.23$ which confirms evolution at the 3σ level for $q_o = 0$.

This approach relies heavily on the continuum extrapolation being valid. For seven high-redshift QSOs ($\langle z_{em} \rangle = 2.98$), Steidel and Sargent (1987) established clearly, from carefully calibrated spectra at ~ 5 Å resolution, that power-law continua defined redward of Ly-α emission also fitted the Ly-α forest region down to Ly-β emission. In order to test the validity of the γ estimates given above, the MHPB sample has been adjusted to emulate the situation at low resolution: heavy element lines have been added in, Ly-α lines in heavy-element systems have been retained and the λ_c cutoffs applied. An ML fit gives $\gamma = 1.66 \pm 0.34$, in excellent agreement with the result of O'Brien *et al.* (1986); the added lines are clearly responsible for lowering the value of γ. The close correspondence between the γ estimates suggests further that there is no significant *intrinsic* steepening of the continua in low-z QSOs below 1200 Å, at least down to Ly-β (Kinney *et al.* 1985).

4. THE INVERSE EFFECT

4.1. Its Wavelength Range

The results summarized in Table 2 present a consistent picture, in which there is a strong global trend toward greater line density at high redshift but a weakening of this trend within individual QSOs. The first step in trying to understand this inverse effect is to determine the wavelength interval over which it applies. The *empirical* result from the exclusion of lines falling within the Ly-α emission profile is that the inverse effect arises from a deficiency of Ly-α lines in the immediate vicinity of the QSO (within ∼4 Mpc; MHPB) and that λ_c defines its *approximate* range. This can be seen in Figure 1 from the marked difference in line density between the portions of spectrum to the red and blue of λ_c. The claim in T87 that the inverse effect applies over a much greater wavelength interval (extending down to 1125 Å in the QSO rest frame) is based on very weak evidence and is inconsistent with the results in Table 2.

4.2. Possible Physical Explanations

Two possible explanations for the inverse effect were considered by MHPB:
1. increased ionization of Ly-α clouds in the vicinity of the QSOs; and
2. some clouds being too small to occult fully the QSO broad emission line region (BELR).

The latter possibility was tested directly by SYBT and by Zou et al. (1982) who compared the properties of lines close to Ly-α emission with those having $z_{em} - z_{abs} \geq$ 0.1 and found no significant (>2σ) differences. This lack of evidence for partial occultation of the BELR is consistent with the detection of common Ly-α absorption in multiple QSO images (Weymann and Foltz 1983, Foltz et al. 1984) which implies that at least some clouds extend over several kpc, more than three orders of magnitude greater than the presumed size of the BELR. It has been argued (Barcons and Fabian 1987) that common absorption could arise through clustering of much smaller clouds, in which case clusters of closely-spaced Ly-α lines should be detectable at high spectral resolution along a single line of sight. There is some evidence for weak clustering of Ly-α lines in the velocity range 50 - 290 km s^{-1} (Webb 1987).

Turning to the first possibility, Weymann, Carswell and Smith (1981) and Sargent, Young and Boksenberg (1982) have already proposed an enhancement of the Lyman continuum flux to explain the incidence of high-ionization metal systems with $z_{abs} \sim z_{em}$. However, a problem in pursuing this approach has been the uncertainty in the diffuse UV background radiation at high redshift. Recent estimates (Ostriker and Ikeuchi 1983, Gondhalekar 1983, ABC, Bechtold et al. 1987, T87, Bajtlik, Duncan and Ostriker 1988) show significant differences which arise from the assumed luminosity distribution of QSOs at $z \geq 3$, the degree of absorption by HI clouds which are optically thick in the Lyman continuum and the contribution from young galaxies or possibly pregalactic stars. If we assume a lower limit for the intensity of the background at the Lyman limit of $J_\nu = 10^{-22}$ erg cm^{-2} s^{-1} Hz^{-1} sr^{-1} (at $z \sim 2.5$) then for most of the QSOs in the MHPB and T87 samples, the individual QSO flux is 10-50% of this value at the distances corresponding to λ_c. Provided the diffuse radiation field is not substantially greater than this lower limit, it is therefore plausible to attribute the inverse effect mainly to the increased ionization of Ly-α clouds in the neighbourhood of luminous QSOs.

The situation is somewhat different for the highest redshift QSO in the sample, 2000-330, since at a distance corresponding to λ_c its flux is several times greater than the above lower limit to the background. This is illustrated in Figure 1 using the T87 parameter λ_{eq}; this wavelength corresponds to the radial distance at which the QSO Lyman continuum flux is equal to the ionizing background adopted by T87. Since there is only weak evidence in 2000-330 for a reduction in line density in the vicinity of Ly-α emission, a substantially higher value of J_ν is implied and may indicate a significant contribution from young galaxies at $z \sim 3.5$ (Carswell et al. 1987). This raises the interesting possibility of using the inverse effect to trace the redshift evolution of the UV background.

Several further possible explanations of the inverse effect have been put forward, including inhomogeneous line lists, unidentified H_2 lines, QSO ejecta and very large scale inhomogeneities at $z \sim 2.5$. All of these proposals present serious difficulties and must remain speculative until the wavelength range of the inverse effect is better defined.

4.3. Can it Explain the Inconsistencies in γ?

As pointed out by MHPB, the effect on γ of a countervailing trend in dN/dz within QSOs will depend on the Δz of the sample and the method of analysis used. When the redshift range is small, or if QSOs are treated individually and then averaged (Carswell et al. 1982), the inverse effect will tend to cancel or even swamp the global evolution. The larger the range in z_{em}, the less will be the influence of the inverse trend within individual QSOs. The method used by Peterson is the least affected, although it is less precise due to the heavy grouping of the data. There is also an expected bias in this method which will tend to exaggerate the steepness of the evolution; this is because of the limited wavelength coverage of the low-z objects (close to Ly-α emission), which will have their line count depressed due to the inverse effect.

5. DISTRIBUTION OF EQUIVALENT WIDTHS

5.1. Analysis of the Distribution

For their sample of lines from five QSOs (including Ly-α in heavy element systems), SYBT found that for $W \geq 0.16$ Å the distribution of Ly-α equivalent widths was well fitted by an exponential function of the form

$$n(W) = \partial^2 N/\partial W\, \partial z \approx (N^*/W^*)\, e^{-W/W^*} \qquad (4)$$

with $W^* = 0.36 \pm 0.02$ Å. This form ignores any z-dependence but was consistent with their finding of no evolution in dN/dz (Table 1). The parameter W^* was determined from an ML fit to the grouped data; the assumption that W^* is independent of z was supported by a rank correlation test between W and z for the whole sample.

MHPB reanalysed the SYBT sample by applying ML to the individual lines but excluding Ly-α lines in known heavy element systems. They obtained $W^* = 0.28 \pm 0.02$ Å (the corresponding value given in the Appendix of SYBT is 0.29). For the full MHPB sample with $W \geq 0.32$ Å the result was $W^* = 0.31 \pm 0.02$ Å. When weak lines were included in the sample, MHPB found significant deviations from an exponential form which were confirmed by higher resolution data from C84 for 1101-264 at 20 km s^{-1} FWHM and from ABC for 0420-388 at 33 km s^{-1}. For graphical presentation it is convenient to group the data into intervals ΔW but normalization is necessary to take account of the dependence of dN/dz on z. For a reference redshift of 2.44, MHPB defined the number density of lines per unit redshift per unit equivalent width, $n^*(W)$, as

$$n^*(W) = \sum_i \left(\frac{3.44}{1+z_i}\right)^\gamma \left(\Delta W \sum_k \Delta z_k\right)^{-1} \qquad (5)$$

where the sum is over lines with W_i in the range ΔW and Δz_k is the redshift interval covered by QSO k.

The equivalent width distribution $n^*(W)$ is plotted in Figure 3 over the range $W = 0.06 - 0.7$ Å. The line drawn is the ML fit to the MHPB sample for $W \geq 0.32$ Å. In addition to the data points shown previously by MHPB (their Figure 3), this figure includes new high resolution data for 2000-330 (Hunstead et al. 1986b; 1987a,c) and the high S/N but much lower resolution data of Boulade et al. (1986) for the luminous QSO 1225+317 ($z_{em} = 2.2$). This expanded data set confirms the conclusions of MHPB that:

1. a single exponential is not a good fit over a wide range in W; and
2. there is no obvious change in $n^*(W)$ with resolution over the range 20-300 km s^{-1} FWHM.

For $W \geq 0.20$Å the ML estimate of W^* for the high resolution 2000-330 data is 0.26 ± 0.02 Å, in close agreement with the corresponding value of 0.26 ± 0.03 Å from ABC; the appropriate MHPB value is 0.30 ± 0.02 Å, while for the T87 sample with $W \geq 0.36$ Å, I obtain $W^* = 0.305 \pm 0.014$ Å. Therefore, while there may be a marginal increase in the slope of the $n^*(W)$ distribution with increasing resolution, it is far less than one might intuitively expect or than was predicted by the blending corrections of SYBT. This may simply reflect an intrinsic difficulty in deblending the Ly-α forest region fully, even at 20 km s^{-1} resolution.

The steep rise in $n^*(W)$ reported by MHPB for $W < 0.2$ Å is confirmed by the more recent data, even those at much lower resolution from Boulade et al. (1986), and presumably marks the transition from the logarithmic to the linear region of the curve-of-growth. It helps to explain the much smaller value of W^*, 0.16 ± 0.02 Å, found by C84 for a small sample of lines in 1101-264 with $W > 0.03$ Å. In this case the higher resolution enabled many additional weak lines to be detected without making a significant change to the line distribution for $W > 0.2$ Å. However, C84 attributed the difference in slope to a fundamental change in $n(W)$ with resolution and later papers (e.g. ABC, T87) have unfortunately reasserted this claim without checking its validity.

5.2. Correlations Between W and z

For both the SYBT and MHPB samples there is no significant rank correlation between W and z. However, MHPB found that when the Spearman rank correlation coefficient was evaluated separately for each QSO, nine out of 11 estimates were negative with a weighted mean of $\bar{\rho} = -0.172 \pm 0.061$. The single-sided probability of finding such a negative correlation by chance is 0.2 %. Being a non-parametric test, the rank correlation can only determine the overall sense of a trend. However, this result can be understood in terms of the inverse effect if the relatively fewer lines occurring near Ly-α emission tend also, on average, to be weaker (refer Fig. 1). This would be expected if the inverse effect arises from the enhanced ionization of Ly-α clouds near luminous QSOs.

A similar correlation (at the 2.3σ level) was found by SYBT in testing the ejection hypothesis for the Ly-α clouds. They dismissed this result because much of the effect seemed to be due to one object, 2126-158, and they attributed this to blending. (MHPB also noted that 2126-158 gave the largest contribution to $\bar{\rho}$ but that it was not significantly different from the mean.) The SYBT test was repeated by Zou et al. (1982) on a sample of nine QSOs (including the five from SYBT) and they obtained a

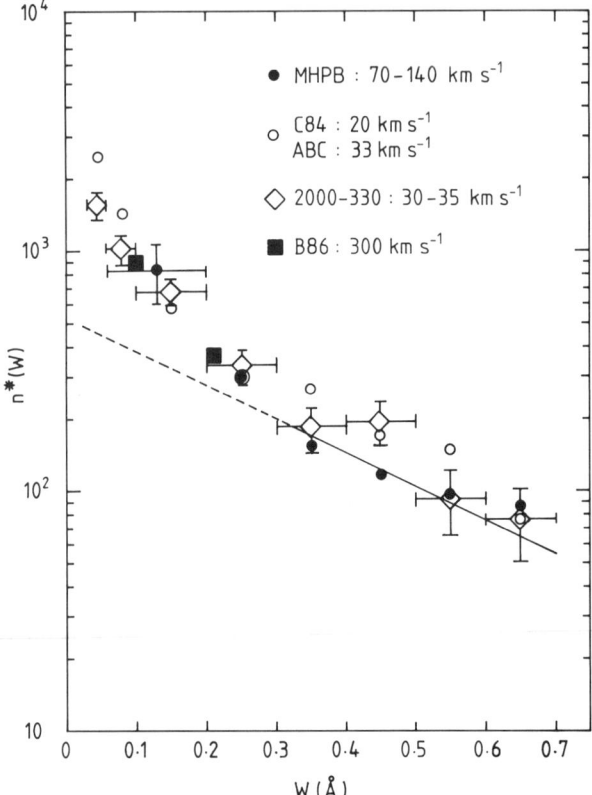

Figure 3. *Distribution of intrinsic equivalent widths W of Ly-α lines for data of different spectral resolution, as indicated (B86 = Boulade et al. 1986). This figure is adapted from Figure 3 of MHPB: $n^*(W)$ is the density of lines in each W bin, normalized to $z = 2.44$, and the straight line drawn is the ML fit for $W \geq 0.32$ Å from MHPB. For clarity, the bin widths (horizontal bars) and 1σ errors (vertical bars) are shown only for the most recent data for 2000-330 (Hunstead et al. 1987c).*

2.6σ correlation; inexplicably, this result was also dismissed as being consistent with a uniform W distribution within each QSO. A further test carried out by T87 compared the cumulative distribution functions for Ly-α lines falling above and below a QSO rest-frame wavelength of 1125 Å. These subsamples were compared *globally* rather than *within* QSOs and predictably yielded a null result.

6. EVOLUTION OF INDIVIDUAL CLOUDS

The most surprising feature of the Ly-α equivalent width distribution is its apparent uniformity with spectral resolution and with redshift. MHPB reported a marginal (2σ) trend between W^* and z in the sense of larger W^* (and, therefore, larger average W) at higher redshift. However, for the purposes of an initial analysis, W^* will be assumed to be constant.

Regardless of the detailed physical processes responsible for the evolution of the neutral hydrogen component of the clouds, the observed increase of dN/dz with z implies that the Ly-α equivalent width of individual clouds must decrease with time, so that for many clouds, W falls below the sample limit (0.32 Å for the MHPB sample) at lower redshift. The combined distribution of N with both z and W may then be written

$$\frac{\partial^2 N}{\partial z \partial W} = \frac{A_o}{W^*} e^{-W/W^*}(1+z)^\gamma \qquad (6)$$

where A_o is constant. This implies that the evolution of W with z has the form

$$W = \gamma W^* \ln(1+z) + \text{constant}, \qquad (7)$$

which implies in turn that, for $W \geq 0.32$ Å, ΔW is independent of W for a given Δz. For realistic values of the velocity dispersion parameter b and for most of the range $W = 0.3 - 1.0$ Å, the curve of growth may be approximated by

$$W = k \ln N_c + \text{constant}, \qquad (8)$$

where k depends on b. From eqns. (7) and (8) it follows that

$$N_c \propto (1+z)^\eta \qquad (9)$$

where $\eta = \gamma W^*/k$. For a typical Ly-α cloud with $N_c \sim 10^{14-15}$ cm^{-2}, the data from ABC suggest a b-value of around 30-40 km s^{-1}. Considering the equivalent width limit of 0.32 Å, MHPB adopted $b = 40$ km s^{-1} and used a weighted average value of k to obtain an indicative value of $\eta \approx 6$. If the weak trend between W^* and z is taken into account, this leads to a higher estimate of $\eta \sim 8$.

On a purely phenomenological basis, these results point to a strong evolution with redshift in the H I column density of individual clouds. The empirically determined values of η are consistent with theoretical models in which the clouds are highly ionized and exist in pressure equilibrium with a hot intergalactic medium (Ostriker and Ikeuchi 1983, ABC, Ikeuchi and Ostriker 1986). It is important to note, however, that eqn. (8) holds only for a narrow range of W. In this region an exponential distribution in W corresponds to a power-law distribution in N_c (for constant b) which has the form

$$dn(N_c) \propto N_c^{-\beta} dN_c, \qquad (10)$$

with $\beta = k/W^* + 1$. For $\eta \approx 6$ this gives an indicative value of $\beta \approx 1.4$.

The redshift dependence of the column density distribution is potentially the most valuable tool of all for exploring the evolution of Ly-α clouds. There are, of course, inherent difficulties in determining HI column densities (C84, ABC), and their distribution function will be much less certain than that for the equivalent widths. If the distribution shows deviations from a smooth power law (e.g. a change in slope at high or low N_c) this will allow clouds of a particular N_c to be tracked as a function of z. If the simple evolution law of eqn. (9) holds, then for $\eta \approx 6$, $\Delta(\log N_c) \approx 1.0$ between $z = 3.5$ and 2.0.

The column density distribution has been determined in several QSOs from Voigt profile fitting to high resolution data. In each case, a power law of the form of eqn. (10) was adopted and an ML fit used to determine β. The results are summarized in Table 3. The β estimates are broadly similar and somewhat greater than the rough prediction from eqn. (10). As pointed out by all the authors, the reliability of column density estimates depends on many factors, particularly the S/N ratio of the spectrum and the availability of unblended higher-order Lyman lines. In the case of 2000-330, which was observed independently at different telescopes, there are significant differences in interpretation. Carswell et al. (1987) find a poor fit to a single power law and propose instead a knee in the distribution near $\log N_c = 14.35$. On the other hand, Hunstead et al. (1987a), from higher S/N data, find that a single power law is a good fit for $\log N_c > 13.25$.

The preliminary column density distribution for 2000-330 from Hunstead et al. (1987a) is shown in Figure 4. There is some apparent flattening of the distribution in the region of $\log N_c = 14.5$ but Kolmogorov goodness-of-fit tests applied to a range of samples with differing lower cutoffs show acceptable fits to a power law. There is also no significant difference between the values of β found for two subsamples split at $\log N_c = 14.35$, although the ML slope in each region is slightly steeper than the overall slope (Table 3). Therefore, while this independent analysis of the column density distribution in 2000-330 does not confirm the abrupt change in slope claimed by Carswell et al. (1987), there are indications that the distribution may not be a smooth power law. This is encouraging for the prospect of using the shape of the distribution to determine the differential evolution of Ly-α clouds with redshift.

Table 3. Details of the column density distribution in three QSOs

Object	Range in z_{abs}	Range in $\log N_c$	β	N	Reference
1101-264	1.89-2.13	> 13.0	1.68 ± 0.10	45	C84
0420-388	2.72-3.12	14.0-16.7	1.89 ± 0.14	38	ABC
2000-330	3.3 -3.75	> 13.75	1.71*	60	Carswell et al. (1987)
		13.75-14.35	1*		
		≥ 14.35	2*		
2000-330	3.43-3.78	> 13.25	1.57 ± 0.05	149	Hunstead et al. (1987a)
		13.25-14.35	1.73 ± 0.14	109	
		> 14.35	1.65 ± 0.10	40	

* These estimates refer to the combined 2000-330 and 0420-388 line lists.

7. ISSUES REQUIRING FURTHER OBSERVATIONS

7.1. Does the Inverse Effect Depend on QSO Luminosity?

There is persuasive evidence from §4 and from a recent analysis by Bajtlik et al. (1988) that the inverse effect can be accounted for entirely by enhanced ionization of HI clouds in the vicinity of luminous QSOs. A crucial test of this hypothesis will be to observe less luminous QSOs over the full range of accessible redshifts. Provided accurate spectrophotometry is available, it should then be possible to explore in detail the luminosity dependence of the inverse effect and, thereby, the redshift evolution of the UV background.

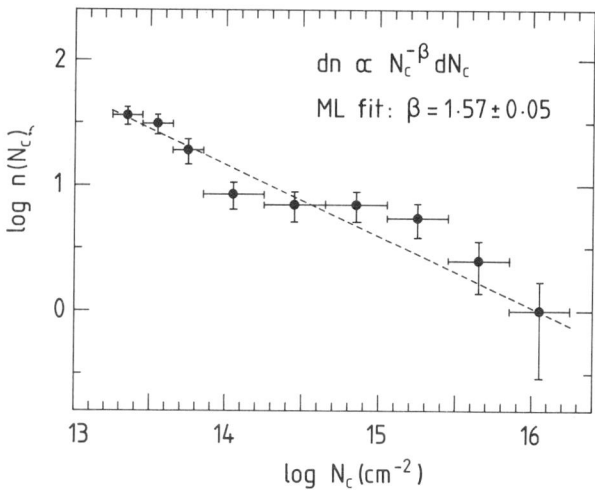

Figure 4. *Distribution of H I column densities $n(N_c)$ in 2000-330 for $\log N_c > 13.25$ and $\langle z \rangle = 3.6$. The dashed line is the ML fit to the 149 individual clouds. The grouping is for presentation only: horizontal bars define the bin widths and vertical bars are $1\sigma(\sqrt{N})$ errors.*

7.2. Do the Ly-α Clouds Contain Metals?

It has generally been assumed that, if the Ly-α clouds constitute a genuine intergalactic population, they must be of primordial composition. This assumption has been reinforced by the lack of success in attempts to detect associated heavy elements, notwithstanding the fundamental difficulty in detecting metals in clouds with such small average HI column densities. Two methods have been employed. The first involves stacking the QSO spectra in the rest frame of each Ly-α line and then looking for significant absorption in the position of metal lines (Norris, Hartwick and Peterson 1983, Sargent and Boksenberg 1983). Apart from a marginal detection of OVI by Norris et al., results from this approach have been negative. The second method involves a direct search for specific metal lines at the well-determined redshift of individual high-column clouds ($N_c > 10^{16}$ cm^{-2}). The latter approach has been pursued by Chaffee et al. (1985, 1986) for SiIII $\lambda 1206$ and CIII $\lambda 977$, again with negative results. Both methods are subject to a high level of line confusion from the Ly-α forest.

As spectra of superior resolution or S/N ratio are obtained, additional weak metal lines will undoubtedly be discovered in association with the stronger Ly-α forest lines (e.g. Blades et al. 1985, Meyer and York 1987). For this reason, it is important to continue searching in high-column clouds for metal lines, both in the forest and redward of Ly-α emission. As a result of the possible detection by Meyer and York of SiIII $\lambda 1206$ in a Lyman-only system at $z_{abs} = 3.047$ in 2000-330, we recently carried out a preliminary search for SiIII and CIII in other strong Ly-α systems in this object (see Blades 1988, this volume). The initial detection rate is encouraging, although a detailed assessment of the chance rate still needs to be carried out. Of course, any additional unrecognised metal systems constitute a contamination of the

Ly-α samples; however, this does not weaken the case for evolution, since the value of γ is already highly significant with these lines included. If these putative metal systems belong to the same static or slowly evolving population as the systems already identified (refer §3.1), then their removal from the Ly-α samples will tend to increase still further (beyond 2.3) the estimate of γ for the remainder.

7.3. What are the Properties of the Ly-α Clouds at Low N_c?

Any model for the Ly-α clouds will be firmly constrained by the precise shape of the N_c distribution at low column densities (e.g. Barcons and Fabian 1987) and the evolution of this distribution with redshift. As pointed out by C84, a power-law distribution of H I column densities must turn over or cut off at low N_c for the cloud numbers to be finite. In the case of 1101-264, C84 believe that the cutoff is near their detection limit ($\sim 8 \times 10^{12}$ cm^{-2}) "since the continuum regions of the spectrum show no signs of the population of weak lines predicted by extrapolating the $p(N, z)$ relation to $N(\mathrm{HI}) = 10^{12}$ cm^{-2}". The distributions in 0420-388 and 2000-330 presented by Carswell et al. (1987) show a flattening below $\log N_c \sim 14.3$ although this is not seen in 2000-330 by Hunstead et al. (1987a) (refer Figure 4). In addition, the "best" data for 0420-388 (ABC, Figure 2) show a steep increase in numbers down to $\log N_c = 13.5$.

The true form of the distribution at low column density remains highly uncertain. This is not surprising, considering the difficulty in obtaining data of sufficiently high S/N ratio ($\geq 15:1$) at the high resolutions required. Whereas stronger lines require higher-order members of the Lyman series to give reliable N_c and b values, at $\log N_c \sim 13.5$ the Ly-α lines move onto the linear portion of the curve of growth and the line parameters are determined adequately by a profile fit to Ly-α alone (e.g. Chaffee et al. 1983). Observers (and time assignment committees!) are therefore encouraged to aim for much higher S/N ratios than has been typical up till now; there may well be a few surprises in store.

8. CONCLUSIONS

(1) *Evolution of number density dN/dz.* The number density dN/dz of Ly-α clouds (with rest equivalent width $W > W_o$) is usually assumed to vary as $(1+z)^\gamma$. The literature up to 1986 records a bewildering range of estimates for γ, with contradictory results being obtained when different methods were applied to the same data. By extending the redshift baseline to $z = 1.5 - 3.78$ and excluding Ly-α lines in heavy-element systems, MHPB obtained $\gamma = 2.17 \pm 0.36$ which differs from the non-evolving exponent by $3.25\sigma (q_o = 0)$ or $4.6\sigma (q_o = 1/2)$. This demonstrates beyond reasonable doubt that the hydrogen clouds were more numerous at earlier epochs.

(2) *Trend of dN/dz within QSOs—the inverse effect.* Both the MHPB analysis and a subsequent analysis by Tytler (1987) revealed a significant countervailing trend (or inverse effect) within individual QSOs. When lines close to Ly-α emission are excluded, the effect is no longer significant and the estimate of γ is increased slightly; the best current estimate is $\gamma = 2.3 \pm 0.4$. The most plausible explanation is that the inverse effect arises from the enhanced ionization of Ly-α clouds in the vicinity of luminous QSOs which would lead naturally to a deficit of lines ($> W_o$) at redshifts close to Ly-α emission and help to explain the inconsistencies among the earlier estimates of γ. This hypothesis can be tested readily by observing less luminous QSOs

over the full range of accessible redshifts. It is likely that such a study will also yield valuable information on the redshift evolution of the intergalactic radiation field.

(3) *Distribution of equivalent widths.* The equivalent width distribution shows surprising uniformity with redshift and no obvious change with spectral resolution over the range 20-300 km s^{-1} FWHM. However, an exponential distribution function is *not* a good fit over a wide range in W, with a steepening of the distribution for $W < 0.2$ Å as the clouds move from the saturated to the linear region of the curve of growth. There is no significant overall rank correlation between W and z but a significant negative correlation is found when the results for individual QSOs are averaged. This is consistent with the QSO-ionization interpretation for the inverse effect.

(4) *Evolution in HI column density of individual clouds.* The observed increase of dN/dz with z implies that the equivalent width of individual clouds must decrease gradually with time and eventually fall below the sample limit. The composite $n(W, z)$ distribution, together with an approximation to the HI curve of growth, leads to the prediction of a steep dependence of HI column density on redshift; it also leads to a power-law distribution of column densities. Direct measurements are available for the column density distributions in three QSOs. A power law is a good fit for $\log N_c > 14.5$ but the form of the distribution at lower column densities is still open to question. Despite difficulties in observation and data reduction, there are encouraging prospects for using the column density distribution at different redshifts to trace the detailed evolution of the Ly-α clouds.

It is a pleasure to thank Hugh Murdoch, Max Pettini and Chris Blades for many illuminating and stimulating discussions on these topics.

REFERENCES

Atwood, B., Baldwin, J. A. and Carswell, R. F. 1985, *Ap. J.*, **292**, 58 (ABC).
Barcons, X. and Fabian, A. C. 1987, *M.N.R.A.S.*, **224**, 675.
Bajtlik, S., Duncan, R. C. and Ostriker, J. P. 1988, *Ap. J.*, in press.
Bechtold, J., Green, R. F., Weymann, R. J., Schmidt, M., Estabrook, F. B., Sherman, R. D., Wahlquist, H. D. and Heckman, T. M. 1984, *Ap. J.*, **281**, 76.
Bechtold, J., Weymann, R. J., Lin, Z. and Malkan, M. A. 1987, *Ap. J.*, **315**, 180.
Blades, J. C., Hunstead, R. W., Murdoch, H. S. and Pettini, M. 1985, *Ap. J.*, **288**, 580.
Boulade, O., Kunth, D., Sargent, W. L. W. and Vigroux, L. 1986, *Pub. A.S.P.*, **98**, 1140.
Boksenberg, A. 1978, In Optical Telescopes of the Future: Proc. ESO Conference, ed. F. Pacini, W. Richter and R. N. Wilson, p. 497.
Carswell, R. F., Morton, D. C., Smith, M. G., Stockton, A. N., Turnshek, D. A. and Weymann, R. 1984, *Ap. J.*, **278**, 486 (C84).
Carswell, R. F., Webb, J. K., Baldwin, J. A. and Atwood, B. 1987, *Ap. J.*, **319**, 709.
Carswell, R. F., Whelan, J. A. J., Smith, M. G., Boksenberg, A. and Tytler, D. 1982, *M.N.R.A.S.*, **198**, 91.
Chaffee, F. H., Foltz, C. B., Bechtold, J. and Weymann, R. J. 1986, *Ap. J.*, **301**, 116.
Chaffee, F. H., Foltz, C. B., Röser, H-J., Weymann, R. J. and Latham, D. W. 1985, *Ap. J.*, **292**, 362.

Chaffee, F. H., Weymann, R. J., Latham, D. W. and Strittmatter, P. A. 1983, *Ap. J.*, **267**, 12.
Ellis, R. S. 1978, *M.N.R.A.S.*, **185**, 613.
Foltz, C. B., Weymann, R. J., Röser, H-J. and Chaffee, F. H. 1984, *Ap. J. (Letters)*, **281**, L1.
Gondhalekar, P. M. 1983, *M.N.R.A.S.*, **204**, 997.
Hunstead, R. W., Murdoch, H. S., Blades, J. C. and Pettini, M. 1986b, *Proceedings of IAU Symposium 119, Quasars*, ed. G. Swarup and V. K. Kapahi (Dordrecht: Reidel), p.569.
Hunstead, R. W., Murdoch, H. S., Peterson, B. A., Blades, J. C., Jauncey, D. L., Wright, A. E., Pettini, M. and Savage, A. 1986a, *Ap. J.*, **305**, 496.
Hunstead, R. W., Murdoch, H. S., Pettini, M. and Blades, J. C. 1987b, *Ap. J.*, submitted.
Hunstead, R. W., Pettini, M., Blades, J. C. and Murdoch, H. S. 1987a, *Proceedings of IAU Symposium 124, Observational Cosmology*, ed. A. Hewitt, G. Burbidge and L. Z. Fang (Dordrecht: Reidel), p. 799.
Hunstead, R. W., Pettini, M., Blades, J. C. and Murdoch, H. S. 1987c, In preparation.
Ikeuchi, S. and Ostriker, J. P. 1986, *Ap. J.*, **301**, 522.
Jenkins, E. B., Caulet, A., Wamsteker, W., Blades, J. C., Morton, D. C. and York, D. G. 1987, *QSO Absorption Lines: Probing the Universe, A Collection of Poster Papers.* eds. J. C. Blades, C. A. Norman and D. A. Turnshek, (ST ScI Publication), p. 34.
Kinney, A. L., Huggins, P. J., Bregman, J. N. and Glassgold, A. E. 1985, *Ap. J.*, **291**, 128.
Meyer, D. M. and York, D. G. 1987, *Ap. J. (Letters)*, **315**, L5.
Murdoch, H. S., Hunstead, R. W., Pettini, M. and Blades, J. C. 1986, *Ap. J.*, **309**, 19 (MHPB).
Norris, J., Hartwick, F. D. A. and Peterson, B. A. 1983, *Ap. J.*, **273**, 450.
O'Brien, P. T., Gondhalekar, P. M. and Wilson, R. 1986, In New Insights in Astrophysics: 8 Years Of UV Astronomy with IUE, ESA SP-263, p. 435.
Ostriker, J. P. and Ikeuchi, S. 1983, *Ap. J. (Letters)*, **286**, L63.
Parnell, H. C. and Carswell, R. F. 1987, *M.N.R.A.S.*, in press.
Peterson, B. A. 1978, *Proceedings of IAU Symposium 79, The Large Scale Structure of the Universe*, ed. M. S. Longair and J. Einasto (Dordrecht: Reidel), p. 309.
_____. 1983a, *Proceedings of IAU Symposium 104, Early Evolution of the Universe and its Present Structure*, eds. G. O. Abell and G. Chincarini (Dordrecht: Reidel), p. 349.
_____. 1983b, in *Proc. 24th Liège Internat. Colloquium, Quasars and Gravitational Lenses*, ed. J. P. Swings (Liège: Institut d'Astrophysique, Université de Liège), p. 563.
Peterson, B. A., Savage, A., Jauncey, D. L. and Wright, A. E. 1982, *Ap. J. (Letters)*, **260**, L27.
Phillipps, S. and Ellis, R. S. 1983, *M.N.R.A.S.*, **204**, 493.
Sargent, W. L. W. 1987, *Proceedings of IAU Symposium 124, Observational Cosmology*, ed. A. Hewitt, G. Burbidge and L. Z. Fang (Dordrecht: Reidel), p. 777.
Sargent, W. L. W. and Boksenberg, A. 1983, in *Proc. 24th Liège Internat. Colloquium, Quasars and Gravitational Lenses*, ed. J. P. Swings (Liège: Institut d'Astrophysique, Université de Liège), p. 518.

Sargent, W. L. W., Young, P. and Boksenberg, A. 1982, *Ap. J.*, **252**, 54.

Sargent, W. L. W., Young, P. J., Boksenberg, A. and Tytler, D. 1980, *Ap. J. Suppl.*, **42**, 41 (SYBT).

Steidel, C. C. and Sargent, W. L. W. 1987, *Ap. J.*, **313**, 171.

Tytler, D. 1987, *Ap. J.*, **321**, 69 (T87).

Webb, J. K. 1987, *Proceedings of IAU Symposium 124, Observational Cosmology*, ed. A. Hewitt, G. Burbidge and L. Z. Fang (Dordrecht: Reidel), p. 803.

Weymann, R. J., Carswell, R. F. and Smith, M. G. 1981, *Ann. Rev. Astr. Ap.*, **19**, 41.

Weymann, R. J. and Foltz, C. B. 1983, *Ap. J. (Letters)*, **272**, L1.

Young, P., Sargent, W. L. W. and Boksenberg, A. 1982, *Ap. J.*, **252**, 10.

Zou, Z.-L., Chen, J.-S., Bian, Y.-L., Tang, X.-Y. and Cui, Z.-X. 1982, *Acta. Astrophys. Sin.*, **2**, 253 [translated in *Chin. Astr. Ap.*, **7**, 31 (1983)].

DISCUSSION

Miley: Is there any evidence that the inverse effect could be related to the radio characteristics? I ask this because of what we heard concerning (i) the prevalence of metal line systems with $z_{abs} \simeq z_{em}$ in radio loud quasars, and (ii) the possible unification between the Ly-α and metal systems.

Hunstead: That's an interesting question. I can't answer it off-hand, but I suspect that we would have noticed such a trend if it had been obvious.

Morton: There was one thing that wasn't clear to me in your discussion of dN/dz. Does your equivalent width limit refer to the rest frame or the observed frame? Also, as I remember, don't you still have a range of something like a factor of two in your equivalent width limits?

Hunstead: They are all rest frame equivalent widths. The dN/dz analysis used a cut-off at 0.32 Å in the rest frame and it was a uniform sample.

Spitzer: Would you indicate again how you determined hydrogen column densities from the observed equivalent widths?

Hunstead: For the lower resolution (1 Å) data the column density distribution is derived analytically from the equivalent width distribution, making use of the approximate relationship between W and N_{col} for the relevant part of the curve-of-growth. The analysis predicts a power-law distribution in N_{col} and this is just what is found when the columns are determined properly, i.e., by profile-fitting to high resolution data.

Shapiro: Is there anything known about the possible correlation between the inverse effect and the X-ray luminosity of the background QSO?

Hunstead: I'm not sure what is known about the X-ray luminosity of these objects. A point that needs to be stressed, of course, is that this is a sample of QSOs which is amongst the most luminous that can be put together, for obvious reasons. So, I suspect that many are indeed X-ray sources.

Duncan: I would like to make two brief comments. The first has to do with the equivalent width distribution of Ly-α clouds. If the distribution of clouds in *column density* is an unbroken power law, and if clouds have velocity width of order 35 km s^{-1} with little intrinsic variation (which is consistent with measurements), then you *expect* the equivalent width distribution to turn upward near $W = 0.2$ Å, in agreement with your figure, because the clouds become unsaturated there.

Hunstead: That's absolutely right. In fact, we pointed this out in Murdoch *et al.* (1986, *Ap. J.*, **309**, 19).

Duncan: My second comment has to do with the inverse effect. Stanislaw Bajtlik, Jeremiah Ostriker, and I have just completed a study of this based on the data sample of Tytler. We find a positive correlation with luminosity that is consistent with the ionization model for this effect, although the statistical significance is not high. Furthermore, we find a distribution of clouds near quasars which fits the ionization explanation for the effect, for certain assumed models of the background UV radiation, $J_\nu(z)$. If the ionization explanation is correct, the $J_\nu(z)$ models which lead to good fits must be close to the *true* UV background. For more details we have a poster upstairs.

Tytler: I'd like to hear more about your detection of metals in high column density Ly-α clouds. How many of them?

Hunstead: This is very recent work and I would rather defer answering your question quantitatively until we have had a chance to examine the statistics properly. Basically, we looked for SiIII λ1206 in 2000–330 at the precise redshifts (obtained from profile fitting) of Ly-α clouds with $N(HI) \geq 10^{15}$ cm^{-2}. There appears to be an excess of detections over chance.

Wolfe: I want to ask you about finding weak lines at low redshifts. Isn't it true that, in general, the data has higher signal-to-noise as you go towards Ly-α emission and so, on the average, your threshold equivalent width will be lower as you get near the emission redshift? Has it been taken into account that the signal-to-noise varies significantly, even within one spectrum? You wouldn't expect to find any weak lines in the blue.

Hunstead: I agree that the signal-to-noise will be non-uniform, but I think the cut-off that has been used in all of these samples is well above the detection limit, even for the noisiest data.

PROPERTIES OF THE Ly-α CLOUDS

R. F. Carswell
Institute of Astronomy
Cambridge

Abstract. Spectroscopy of QSOs with resolution $R \gtrsim 10^4$ is providing a powerful tool for the study of intervening absorbing clouds. The redshifted Lyman lines are almost invariably resolved, and suggest a Doppler parameter $b = \sqrt{2}\sigma \sim 35$ km s^{-1} on average. The HI column densities and clustering properties may also be inferred, subject to the uncertainty over the number of closely spaced components. Ionization and density estimates depend quite critically on the sizes of the clouds, and the only useful size estimates come from observations of the gravitationally lensed images of 2345+007 by Foltz et al. (1984). The heavy-element abundances in the clouds are generally inferred to be low, but depend strongly on uncertain ionization levels.

1. INTRODUCTION

There are now an increasing number of investigations into QSO absorbing clouds using high resolution spectra. Many of these have dealt with a range of lines covering a few systems which show heavy-element absorption, while some cover the numerous systems for which only Lyman lines are usually detectable. The heavy-element system studies usually have the aims of determining the scale for velocity structure, a quantity that we might expect to be similar to that for a galaxy or cluster of galaxies, and element abundances. Since there are only a few such systems in each QSO, and few QSOs have been studied in this way because of the large amounts of telescope time involved, there is as yet little statistical information available. On the other hand, the Lyman line systems are very numerous, with some hundreds of detectable Ly-α lines in the spectra of the highest redshift objects. Thus, while there is less information on the properties in individual cases, questions concerning the distributions of some of the determinable quantities can be addressed.

We have to infer almost all we know about the Ly-α absorbing clouds from spectroscopic observations of QSOs, linking this with what we may think about the environment containing them at high redshifts. For any particular case, it has to be along the line of sight to a bright high redshift QSO, and the raw information consists of an absorption line profile (which at intermediate resolution is usually a remarkably good approximation to the instrument profile, or a few instrument profiles blended) about some mean wavelength. From this we derive the redshift, which is usually straightforward enough, and then, with increasing uncertainty, the equivalent width of the line, component structure, HI column density and velocity width Doppler parameter for each component. There is some additional information which comes from extending the observations to longer wavelengths than the Ly-α lines are found, and that concerns the presence (or, more strictly, usual absence) of heavy-element lines. Also, close pairs or gravitationally lensed QSOs can provide invaluable information on length scales in directions perpendicular to the lines of sight to the background QSOs. In this contribution we consider what we have learned about the properties of the Lyman forest clouds from such observational material.

2. DERIVING DISTRIBUTIONS

If we are to measure absorption lines then the first thing we have to determine is the continuum level against which these lines are seen. As is evident from the intermediate dispersion spectrum of any QSO with emission redshift $z_{em} \sim 3$ or more, this is a non-trivial task. The continuum estimate shown for a region in the Lyman forest of the spectrum of the $z_{em} = 3.12$ QSO 0420–388 (Figure 1) illustrates the difficulties quite well—there are large regions of spectrum, occasionally hundreds of Angstroms in the observed frame, where there is little confidence that the continuum level is reached at all. It is not at all clear that the continuum level should not be somewhat higher in the region 4600–4800Å for example. More extreme examples are provided by higher redshift QSOs, such as the $z_{em} = 3.78$ QSO PKS2000–330 (Hunstead et al. 1986). At lower redshifts, where the Ly-α line density is lower, and at higher resolution, the continua are better defined. In the $z = 2.14$ QSO Q1101 − 264, for example, there were no difficulties in establishing reliable continuum levels at intermediate (1.5Å, Carswell et al. 1982) or high (0.25Å, Carswell et al. 1984) dispersion. Indeed, the two are consistent with each other.

The continuum level problem is not unique to QSO spectroscopy, of course, but I know of no satisfactory way of allowing for continuum uncertainties other than to perform an absorption-line analysis for several different continuum determinations and examining the scatter in the results. Given the large amount of work involved it is not surprising that this has not been done except in a few test cases (e.g. Carswell et al. 1987). An alternative, rather unsatisfactory, approach is to consider only systems for which the central line intensities are close to zero, when continuum uncertainties will have less effect on the equivalent widths of the lines. For most of the work described below the results are based on best estimates of the continuum level, with little (or no) knowledge of how errors in this level affect the results.

Standard QSO absorption-line studies have been at a resolutions of about 75 - 100 km s^{-1}, and while they have provided us with a number of useful insights into the nature of the line absorbing regions, detailed investigations seem to require somewhat higher resolution than is normally attainable with a standard intermediate dispersion spectrograph. Information relevant for the redshift distribution of the numbers of lines is available from such lower resolution work, with the most recent compilations described by Murdoch et al. (1986), Hunstead (1988), and Tytler (1987a). There are uncertainties about the underlying redshift dependence of the number of absorbers obtained in this way, since blending effects could be important, but it is clear that the data are not consistent with an unevolving co-moving population of clouds. If the number of systems per unit redshift is $\propto (1+z)^\gamma$, then, for Ly-α systems with rest equivalent widths > 0.32Å, $\gamma \sim 2$ - 2.5 (see Hunstead, 1988). From higher resolution ($R \sim 10^4$) data $\gamma = 1.76 \pm 0.46$ for systems with $\log N > 13.75$ (Carswell et al.

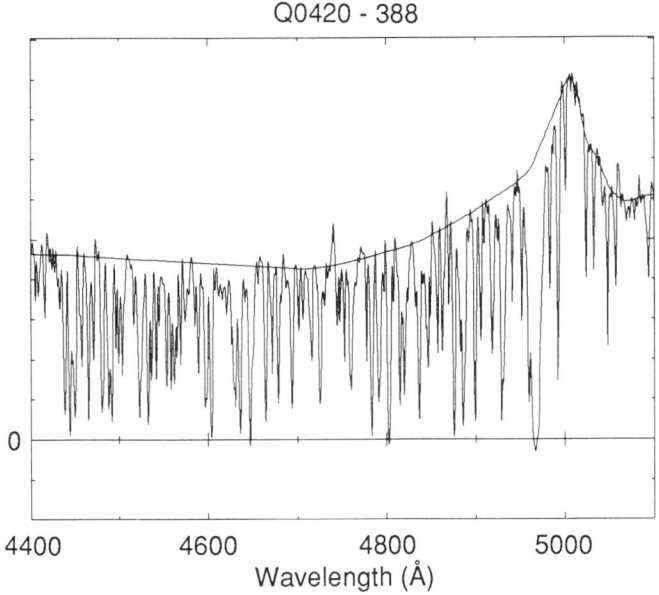

Figure 1. *Part of the spectrum of the $z = 3.12$ QSO Q0420-388 at a resolution of 1.5Å, and an estimate of the continuum level (based on the spectrum of Atwood, Baldwin and Carswell 1985).*

1987). This result is not significantly different from that determined from the lower resolution spectra, but since the line selection criteria differ it is not obvious that there must be detailed agreement.

Higher resolution studies have been undertaken recently by a number of groups, following a pioneering study of small regions of the spectrum of PHL957 by Chaffee et al. (1983). As is usual in interstellar line work, the higher the resolution the more the components you see. However, there is a natural limit to this, set by the thermal widths of the lines at the temperatures in the gas clouds. Thus it would be helpful if we knew what temperatures to expect! Hydrogen is particularly useful in this regard, since it is the lightest element so has the highest thermal velocity dispersion at any temperature. For 10^4 °K, the Doppler parameter $b = \sqrt{2}\sigma = 13$ km s^{-1}, corresponding to a FWHM of 21.5 km s^{-1}. Thus, with adequate S/N and a resolution of 20 km s^{-1} or so, one should be able to resolve all Ly-α lines for those systems with temperatures in excess of 10^4 °K. Chaffee et al. 1983 report on one line where the Doppler parameter b in the range 12 - 17 km s^{-1}, but it is clear from Figure 2 that most, if not all, of the Ly-α lines are resolved at $\sim$ 20 km s^{-1} resolution.

All we can hope for from a single absorption line, or a series of lines from the same lower level in an ion, is a measure of how well it is approximated by a Voigt profile (convolved, of course, with the instrument profile), and some determination of the velocity spread and ion column density. In general we first try to fit single Voigt profiles, partly because the formalism is well developed, but also from the belief that a normal distribution in velocity should be a reasonable approximation under a range of circumstances. This gives a redshift, z, a Doppler parameter, b, and ion

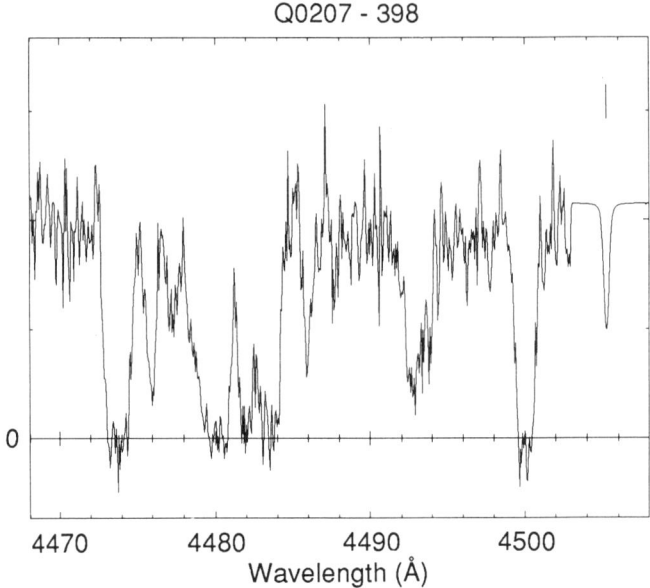

Figure 2. *The Ly-α absorption lines in part of the spectrum of the $z = 2.81$ QSO 0207-398 (from Webb 1987a). The resolution (FWHM) is 21 km s^{-1}, and the instrument resolution is shown by a simulated unresolved line in a short section of continuum at the long wavelength end (under the tick mark).*

column density, N in each case. Of course, if a single Voigt profile is a poor fit, then we tend to try blends of two or more until the fit is acceptable. In this way we are able to determine values for the parameters for a number of systems, as the minimum number of components required by the data. If we adopt this minimum component approach systematically, then we may overestimate the intrinsic line widths, perhaps underestimate the degree of clustering on small velocity scales, but, except under fairly contrived circumstances, obtain reasonable total ion column densities. This last point has been investigated in some detail by Jenkins (1986).

Two datasets have been studied intensively to determine the basic parameters of the Lyman forest clouds. Two high redshift objects, 0420-388 at $z = 3.12$ (Atwood, Baldwin and Carswell 1985) and PKS2000-330 at $z = 3.78$ (Carswell et al. 1987) were observed using the echelle spectrograph at the CTIO 4-m telescope, yielding a spectral resolution of 30-35 km s^{-1} (FWHM) over a useful, but not necessarily continuous, wavelength range from about 4000-6000Å. A number of QSOs with redshifts ranging from $z = 2.1$ to 2.8 have also been observed using the long focal length camera on the Cassegrain spectrograph at the AAT, with resolution in the range 20-25 km s^{-1}. Since these higher resolution spectra are from a conventional spectrograph, the wavelength coverage is limited to 140Å, and so it has taken a long time to build up an adequate sample of lines. Observations of one object have been published (1100-264, Carswell et al. 1984); the rest are available but not widely so (Webb 1987a; Webb, Irwin and Carswell, in preparation).

A major difference between the two samples, apart from the resolution, is the redshift range observed. For the lower resolution CTIO data, profile fitting was undertaken only for those systems for which at least one higher order Lyman line was in the observed wavelength range as well as Ly-α, and no systems were considered where the Ly-α lines could be affected by higher order Lyman lines at higher redshifts. For the AAT material the redshifts are usually so low that Ly-β is often not observable, so the results come from fitting Ly-α only in all but a few cases. We have, of course, compared the results of fitting Ly-α alone with those from fitting Ly-α and Ly-β, and undertaken trials with simulated data, to make sure that the Lyα fitting procedure gives reliable results.

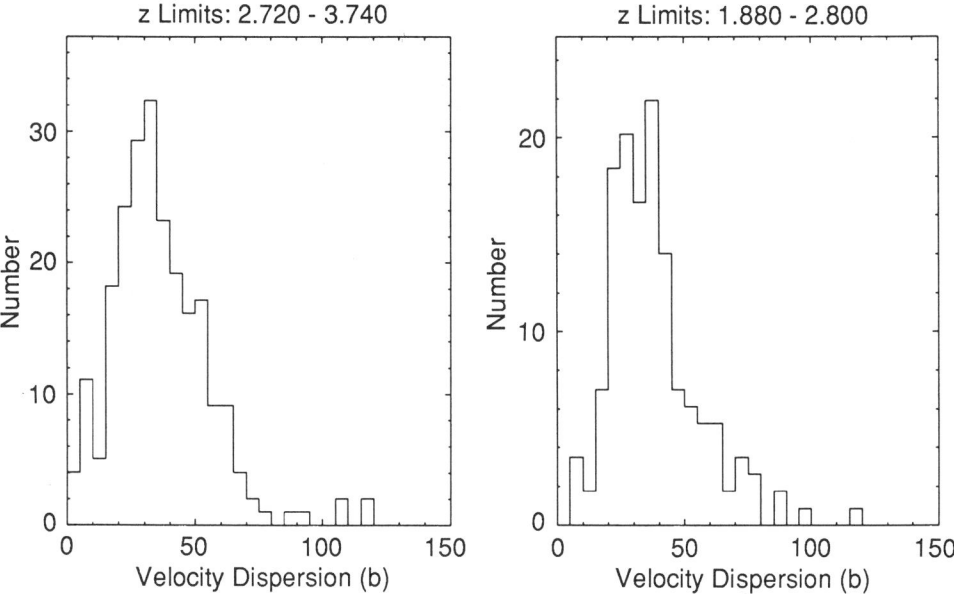

Figure 3. *The inferred HI Doppler parameter distributions derived from CTIO spectra at high $z \sim 3.4$, and AAT data at low $z \sim 2.4$, redshifts. Note that the spread in values in each case contains a significant component due to measurement errors.*

From these data the observed Doppler parameter distribution peaks at about 30–35 km s^{-1} with no strong redshift dependence (Figure 3, see also Carswell *et al.* 1987). The first moment about the mean is about 10–15 km s^{-1}, with the precise value depending on the resolution and S/N in the spectra. There may be some tendency for the single lines to have lower Doppler parameters (see Atwood, Baldwin and Carswell 1985) but at present such differences are significant at just the 2σ level. Webb (private communication) has shown that lines in complexes have not been under- or over-fitted relative to single features, so any differences are not likely to be due to simple variations in analysis technique as applied in the two cases.

At least some of the dispersion in the b-values must arise from errors in the measurements, but it is not clear how much. In principle we may compare deviations from the mean value with the computed error estimate. However the latter quantity is based on an ellipsoidal approximation to the probability contours (see Webb 1987a), and for systems where the line center is optically thick the actual error contours are asymmetric (see Carswell *et al.* 1984), so it is not obvious how this comparison should be made. Trials with artificial data designed to mimic the real data at 0.25Å resolution from the AAT as far as possible, with an imposed intrinsic Gaussian spread of Doppler parameters, suggest that the distribution may have a $\sigma \sim$ 5–10 km s^{-1} over the redshift range $2 \lesssim z \lesssim 3$. There is no discernible trend for the mean to change with redshift, so these trials suggest that there is an intrinsic spread in values.

If there is an intrinsic spread in Doppler parameters, and the line widths have a purely thermal origin, then the simple picture in which the clouds are pressure confined by a hot intergalactic medium and in thermal balance with the ultraviolet background cannot be correct. In this picture, the pressure and thermal balance require, for a given pressure and ionizing flux, a unique density and temperature (Carswell et al. 1984). Such a simple model is unlikely to be correct, even if the basic picture is right. One obvious complication is the presence of fluctuations in the background radiation. Another is the interpretation of the Doppler widths as thermal widths. A Doppler parameter b in the range 30–35 km s^{-1} corresponds to a temperature in the range 5–7×10^4 °K, which is rather too hot for models without some carefully contrived conditions. More natural temperatures of around 2×10^4 °K are consistent with the data only if most of the apparent width arises through bulk motions of some sort. This is a (weak) argument that gravity may play a role.

Along with estimates of the Doppler parameter one derives an HI column density for each absorbing system. A considerable amount of effort has gone into estimating the HI column density distribution function, partially as a result of the realization that the redshift evolution of this distribution can provide some quite strong constraints on the properties of the clouds (see Carswell et al. 1987). The position of any break in the distribution as a function of redshift, and the reasonable assumption that we might identify clouds at the same positions in the distribution relative to the break at all redshifts, would allow us to follow their evolution in some detail.

This hope is not realized, at least yet. Generally, the number of systems Ψ with HI column density N is adequately described by a power law $d\Psi \propto N^{-\beta} dN$, where $\beta \sim 1.75$, at least over the range $13.5 < \log N < 15.5$. There has been a suggestion that the distribution is flatter at low column densities, or, equivalently, steeper at high column densities, for high redshift ($z > 2.7$) systems derived from resolution $R \sim 10^4$ spectra (Carswell et al. 1987). However, this is somewhat tentative and is not seen in a lower redshift, higher spectral resolution ($R \sim 18000$) sample (Webb 1987a). The two distributions are shown for comparison in Figure 4. There are a number of dangers in interpreting the HI distribution directly, related mainly to incompleteness at the low column density end and uncertainties at the high column density end, so such differences are difficult to assess. A description of these is provided by Carswell et al. 1987. Perhaps the most important new result to emerge from the high resolution Lyman line studies recently is the discovery that these systems are weakly clustered (Webb 1987a,b; see also Webb, Carswell and Irwin 1984). A sample of a little over 200 systems in 6 objects yields an excess in the two-point correlation function in the range up to 200–300 km s^{-1}, with a peak amplitude in $\xi(r)$ of about 0.5 ± 0.15 at the lowest velocity separations detectable (~ 70 km s^{-1}) due to the finite S/N and intrinsic line widths. The two-point correlation function is shown in detail by Webb (1987a, b). The presence of this peak runs counter to the claim by Sargent et al. (1980) that the Lyman line systems are not clustered (their lowest velocity scale was about 300 km s^{-1}, so there is no real conflict), though it further supports their general conclusion that the correlation function is much flatter than we would expect if the Lyman systems were associated with galaxies in clusters. However, it is as well to bear

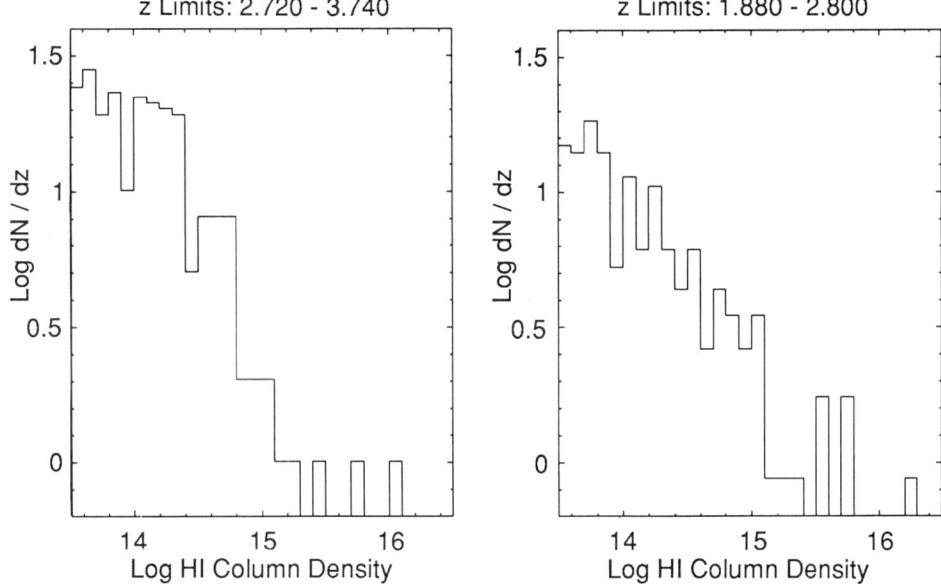

Figure 4. *The inferred HI column density distributions at high $z \sim 3.4$ and low $z \sim 2.4$ redshifts. The lowest plotted point corresponds to a single system lying in that bin.*

in mind that the Webb result may provide us with only a lower limit to the degree of clustering of the Lyman systems, since the analysis is based on the minimum number of components required to fit each line profile or complex at the available S/N and resolution.

3. CLOUD PROPERTIES

All these considerations provide some clues as to the nature of the clouds in the Ly-α forest, but provide us with little to investigate their physical properties. We can determine the redshifts, Doppler parameters, and HI column densities, but still have little idea of their densities, ionization, abundances and geometry. A single key observation provides the means to help unravel the first three, but at the cost of assuming something about the last item, the geometry. Observations of the spectra of two images of the gravitationally lensed QSO 2345+007 by Foltz et al. (1984) showed that the two lines of sight contained a number of common Ly-α absorbers. From this they were able to infer a lower limit to the size somewhere in the range ~ 5 - 25 kpc for systems with high ($\gtrsim 1$Å) rest equivalent widths. They note that a few lines were not common to both spectra, so this scale could be a rough characteristic size rather than a limit. If we assume that they were measuring single systems, with Doppler parameters of about 30 - 35 km s^{-1}, then the corresponding HI column densities are about 10^{15} cm^{-2}. If we further assume that the clouds are roughly isotropic, then for this estimated size we find that the neutral hydrogen density is somewhere in the region of 10^{-8} - 10^{-9} cm^{-3}. If we know the ionizing background flux from QSOs, or young galaxies, at high redshifts (Bechtold et al. 1987), then we may infer the hydrogen density and ionization from the HI density and this background flux. Such calculations yield densities of order 10^{-3} cm^{-3} and neutral hydrogen fractions of order 10^{-5} or so, with uncertainties of about an order of magnitude. Total masses of order $10^7 M_\odot$ would then be required.

Knowledge of the density and ionizing flux also allows estimates of the heavy-element ionization levels, and hence, from the absence (usually) of heavy-element lines, upper limits to their abundances. Generally, CIII is likely have the highest column density of the heavy-elements, and from its absence it is possible to set a heavy-element/hydrogen ratio upper limit of $10^{-3.5}$ solar in a $z = 3.321$ system towards S5 0014+81 (Chaffee et al. 1986).

If, however, the fundamental assumption of near isotropy is incorrect, and we are seeing highly flattened systems, then the picture may be different. The ionization may be quite low, the density high, and the heavy-element abundances as high as $\sim 10^{-1.5}$ solar in the S5 0014+81 case, depending on how extreme a flattened system exists. Chaffee et al. (1986) point out that the clouds would then have to be only about 1 pc thick. Some of the points applicable in these circumstanceas have been discussed by Barcons and Fabian (1987). Another crucial point is that the Foltz et al. (1984) observations may not apply to single clouds, but instead correlated agglomerations of clouds as in e.g. a galactic disk, and so the size constraint is misleading. This is difficult to rule out, but it is noteworthy that the velocity spread is lower, at a few tens of km s^{-1}, than we might expect if we were looking along two sight-lines through a rotating galaxy.

The ionization levels in the clouds could be determined, in principle, if we know the ionizing background flux. At redshifts near that of the QSO the ultraviolet flux from the QSO itself will further ionize the material in the clouds and reduce their apparent numbers, a point first noted by Weymann, Carswell and Smith (1981), and subsequently explored by Murdoch et al. (1986). See also the reviews by Hunstead (1988), Sargent (1988) and Tytler (1988) in this volume. If the clouds are predominantly neutral, then the falloff in the numbers of detectable systems would occur nearer the emission redshift than if they were highly ionized. The difficulty is that a larger ionizing background flux than predicted would have a similar effect, at least to the level we can expect to determine it, and this quantity is largely unknown.

Two further pieces of evidence suggest that the Ly-α systems may share a number of properties with the heavy-element systems. Tytler (1987b) has pointed out that there is continuity in the N(HI) distribution between the two classes, suggestive of a single population. There are some difficulties in determining the HI column densities in the range $10^{15} \lesssim N(HI) \lesssim 10^{17}$ cm^{-2}, in particular, so this topic merits further examination. A similar suggestion has come from the work of Meyer and York (1987), who also have found weak CIV lines in some cases, calling into question the common assumption that the Lyman systems generally have low heavy-element abundances. However, it remains clear from the Chaffee et al. (1986) result that at least some systems must have heavy-element abundances considerably below solar values.

4. CONCLUSIONS

The large amount of observing time which has gone into the study of the Ly-α forest systems has provided us with some useful results, but there is still a considerable amount of work to be done. Most of the high resolution spectra obtained so far have a S/N (per resolution element) ~ 20, and an improvement of a factor of two or more is needed to address some of the questions now raised.

At present it is not clear if the Lyman system clouds are pressure confined by a hot intergalactic medium, or are predominantly gravitationally bound. The hot IGM picture, particularly, carries the consequence that all clouds have the same temperature and density at a given redshift where the ionizing (and heating) background and the confining pressures are fixed. Thus if the measured Doppler parameters are similar at a given redshift, this would argue that the Doppler parameter $b = \sqrt{2}\sigma \sim 30$ km s^{-1} arise from thermal broadening (so $T \sim 6 \times 10^4$ K), or clusters of systems with lower temperatures. Higher S/N spectra should enable us to establish whether or not there is an intrinsic Doppler velocity spread, and check if there is clustering of the clouds on velocity scales ≤ 100 km s^{-1}.

The HI column density distribution is determined roughly in the range $13.5 < \log N(HI) < 15.0$ to be a power law with numbers $\propto N^{-1.75}$. The lower limit is set by S/N considerations, and the higher one because of lack of systems. There are hints that the distribution may not be a pure power law, and that there is a flattening towards smaller values of N. If this is true, then it offers the possibility of following the redshift evolution of the cloud ionization and sizes in reasonable detail, since the position of any 'knee' as a function of redshift tells us which clouds we may identify with each other of the redshift range. This then provides powerful constraints for any Lyman system cloud models. Knowledge of the shape of the distribution at low HI column densities may also serve as a constraint for evaporation rates of the material in the clouds.

Estimates of the sizes of absorbers are crucial for estimating the densities, ionization levels and heavy-element abundances in the clouds. So far these have to come from searches for common absorption in gravitationally lensed or accidentally close sight-lines. These are rare, and with one notable exception, the objects are so faint or so far apart that only the crudest size limits can be derived. Further observations of high redshift lensed QSOs would be a very important contribution.

REFERENCES

Atwood, B., Baldwin, J. A., and Carswell, R. F., 1985. *Ap. J.*, **292**, 58.

Barcons, X., and Fabian, A. C., 1987. *M.N.R.A.S.*, **224**, 675.

Bechtold, J., Weymann, R. J., Lin, Z., and Malkan, M., 1987. *Ap. J.*, **315**, 180.

Carswell, R. F., Whelan, J. A. J., Smith, M. G., Boksenberg, A., and Tytler, D., 1982. *M.N.R.A.S.*, **198**, 91.

Carswell, R. F., Morton, D. C., Smith, M. G., Stockton, A. N., Turnshek, D. A., and Weymann, R. J., 1984. *Ap. J.*, **278**, 486.

Carswell, R. F., Webb, J. K., Baldwin, J. A., and Atwood, B., 1987. *Ap. J.*, **319**, 709.

Chaffee, F. H., Jr., Weymann, R. J., Latham, D. W., and Strittmatter, P. A., 1983. *Ap. J.*, **267**, 12.

Chaffee, F. H., Jr., Foltz, C. B., Bechtold, J., and Weymann, R. J., 1986. *Ap. J.*, **301**, 116.

Foltz, C. B., Weymann, R. J., Röser, H.-J., and Chaffee, F. H., Jr., 1984. *Ap. J.*, **281**, L1.

Hunstead, R. W., 1988, *This volume*.

Hunstead, R. W., Murdoch, H. S., Peterson, B. A., Blades, J. C., Jauncey, D. L., Wright, A. E., Pettini, M., and Savage, A., 1986. *Ap. J.*, **305**, 496.

Jenkins, E.B., 1986. *Ap. J.*, **304**, 739.
Meyer, D. M., and York, D. G., 1987. *Ap. J.*, **315**, L5.
Murdoch, H. S., Hunstead, R. W., Pettini, M., and Blades, J. C., 1986. *Ap. J.*, **309**, 19.
Sargent, W. L. W., Young, P. J., Boksenberg, A., and Tytler, D., 1980. *Ap. J. Suppl.*, **42**, 41.
Sargent, W. L. W., 1988, this volume.
Tytler, D., 1987a. *Ap. J.*, **321**, 69.
_____. 1987b. *Ap. J.*, **321**, 49.
_____. 1988, *This volume*.
Webb, J. K. 1987a. *Ph.D. thesis, Cambridge University.*
_____. 1987b. In *I.A.U. Symposium 124*, "*Observational Cosmology*", ed. A. Hewitt, G. R. Burbidge and L. Z. Fang, p. 803.
Webb, J. K., Carswell, R. F., and Irwin, M. J., 1984. *Bull. Am. Astron. Soc.*, **16**, 733.
Weymann, R. J., Carswell, R. F., and Smith, M. G., 1981. *Ann. Rev. Astron. Astrophys.*, **19**, 41.

DISCUSSION

Tytler: Have all Lyman lines in heavy-element sytems been excluded?

Carswell: We have omitted those we know about but we haven't checked nearly as carefully as Meyer and York have for example. If there's a Ly-α with weak metals it will have been left in.

Webb: Could I just make a comment on that? What I did is I made an estimate of the contamination that the heavy-elements would give. I just assumed that the interlopers could be CIV and made an estimate of the integrated numbers under the equivalent width distribution, estimated by Jacqueline Bergeron, and said okay, if you have all those contaminants in there, can it account for all the clustering? And of course, it can account for some but not all of it. Also, another point is that you don't see a peak in the correlation function at 500 km s^{-1} roughly, which corresponds to the $\lambda 1548$, $\lambda 1550$ CIV splitting. So, if it were CIV contamination, then it's most likely that you would know.

Tytler: Yes, I was thinking more that they'd be Ly-α lines with metal lines because you have or haven't observed the CIV region. That was something I was asking actually. Have you observed CIV?

Carswell: No.

Webb: And what you're suggesting is a possibility.

Tytler: I had another question. It would be interesting to know your best guesses for, say, the minimum and maximum b-values that the Ly-α clouds have.

Carswell: Aaaah!

Tytler: Well, you've got six hundred!

Carswell: (Laughs) Yes, but I don't know if the tails of the distribution are dominated by measurement error or reality.

J. Bahcall: Let me make a comment, because I hear you hesitate. I have a comment which has the opposite side to your remark. I'm enormously impressed by the *constancy* of the b-value, not by the maximum. I'm not impressed that you may marginally have measured the dispersion. I'm impressed by the fact that it could have been 10 and it could have been 50, but in fact it settles at about 35 km s^{-1}, and I wonder if Richard Hunstead has a comment on that. Is that common; do you agree that that shows up in your data, also?

Hunstead: We find somewhat smaller b-values than Bob has found in the object that we have in common.

J. Bahcall: I see. How small?

Hunstead: We find an average b-value of about 20 km s^{-1}. We also find a correlation between b and N that's a little bit stronger than the one that Bob finds.

J. Bahcall: 20 km s^{-1}. And what is your resolution?

Hunstead: It's the same as his: 30–35 km s^{-1} FWHM or an instrumental b-value of around 20 km s^{-1}.

Carswell: Hold on here. It may be better to bring this up in more detail later, but we certainly differ. We have the approach that we will fit as few components as possible, consistent with the signal-to-noise, and then we go back, after having done all the fitting, over the distribution of errors, actually the distribution of chi square for the resultant fits to make sure that's consistent with, again, our hypothesis that we haven't underfitted or overfitted the data. As far as we can tell—it is consistent. It is very difficult, in fact, to be particularly certain that our analyses reflect reality. These are uncertainties due to observational or analysis details, but all one can do is try and fit as few components as possible to the data and then try to infer things from that. We may, in fact, have missed components, but if we have, it's because of the finite signal-to-noise.

J. Bahcall: Let's see, Richard, you have one object in common with Bob?

Hunstead: Yes.

J. Bahcall: In that case, you got 20 km s^{-1}; he got 35 km s^{-1}?

Carswell: That's right, yes.

Hunstead: We have higher signal-to-noise data obtained with a conventional spectrograph. We're fairly confident about the level of scattered light in the spectrograph.

Bob's observation was made with an echelle spectrograph where I think it is questionable. There may well be scattered light which certainly will affect fitting if you are using all the lines in the Lyman series.

Carswell: I doubt that scattered light is serious when so many of our lines have zero central intensity. Also, there is negligible scattered light outside the stellar spectrum.

J. Bahcall: Does anybody else have a common object?

Wolfe: Yes. I want to comment on this. Could you take each other's data and analyze it?

Hunstead: We have done this already. Bob has had a go at analysing our data.

Carswell: I have had a go at analysing three small regions chosen from your data where the disagreements were greatest. I offered Dick our data.

Wolfe: What happened?

Carswell: I find fewer components than they do.

Hunstead: But you did find larger columns than you found on your own data.

Carswell: Oh, sure. I mean, that's the sort of thing that happens when you're near the flat part of the curve of growth for several Lyman lines. The uncertainties are large, and when dealing with complex blends we may both be wrong, that is another point.

Shull: I have a question and comment on the same subject. Do you get a b-value by fitting Ly-α, Ly-β, Ly-γ, or by fitting the width?

Carswell: We get the b-value if we have Ly-α, Ly-β, Ly-γ, etc., then we fit Ly-α, Ly-β, Ly-γ, etc., etc. But there are cases where that's impossible. We have worried that we may be biasing things by fitting only the Ly-α lines under these circumstances. So what we did was we had one unfortunate summer student go off and fit 600 artificially created lines, and see whether he got out the distribution that went in. In the case of velocity dispersions we tried this and sure enough the mean value agreed with the simulation but of course the dispersion about the mean was higher.

Shull: My question, what I'm worried about is, even if you did fit the b-value well, when you go to that column density, you get a number like 20 or 35 km s^{-1}. That can't be thermal, it must be many components. Therefore the column density you get is going to depend on the distribution and velocity of those components and you are just going to get a lot of uncertainty there.

Carswell: Sure. There's no denying that in all cases we could have a lot of components making up the lines we see. The only thing we can say is that it's not a few. So you are asking for fairly large clustering effects on scales of order 30–40 km s^{-1}. I cannot deny this for a second.

Webb: I just have another comment concerning the apparent difference in velocity dispersion that the two groups seem to find. Although the higher resolution AAT sample that Bob described is not in common with both groups, one thing that you can do with data like that is to select sub-samples. One thing that we have done is to take lines which come in apparently clumps, be they clumps due to random distribution, or whatever, and take those apart, and just use more reliable lines which are apparently single, and then look at the distribution functions for those. What is found in fact in the sample that was described, is that for the apparently single lines, the mean velocity dispersion parameter is the same as for the apparently blended lines. But as you would expect for the apparently blended lines, the values are broader, so in that respect our profile fitting procedures are consistent.

Lewis: Does the functional form and parameter value of the equivalent width distribution tell you anything about the physical processes that affect the Ly-α clouds or about their origin, stability, or evolution? Yes, or no?

Carswell: Moving right along — (Laughter).

Rees: I would like to ask about the reality of the one or two systems that are claimed to have low b-values. In particular, it was a claim in a paper by Chaffee et al. (1983, Ap. J., **267**, 12) a few years ago of one system which had a line which was anomalously low for any photoionization model. Some theorists have made a lot of this, and I think, Bob, you had one even lower case. So I'd like to know if you feel that these should be taken seriously as being a problem for photoionization models?

Carswell: I have seen nothing that would, with my hand on my heart, have to be less than say 20 km s^{-1}.

Hunstead: This is our less than perfect analysis of the Ly-α forest lines in 2000-330 (shows viewgraph). The open circle is the Chaffee et al. point. I believe we are seeing a lot of low-b/low-N clouds. These points are just very coarsely grouped means.

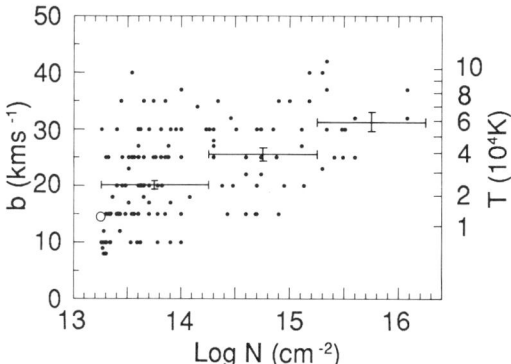

Figure Caption. *b versus log N for Ly-α clouds in 2000-330, together with coarsely-grouped means and standard errors. The temperature scale assumes that the b-values arise solely from thermal motions in the HI gas. (From Hunstead et al. 1987, IAU Symposium #124, "Observational Cosmology" eds. A. Hewitt et al., p. 799.)*

J. Bahcall: What are the error bars?

Hunstead: At the low end I would say probably ± 5 km s^{-1}. However, at higher columns, it's often difficult to determine a b-value and a column density independently of one another.

Carswell: Can I ask you how you get ± 5 km s^{-1} when your instrumental resolution is around 35 km s^{-1}?

Hunstead: No, it's 20 km s^{-1}. Our instrumental b-values are around 18–25 km s^{-1}.

Carswell: Still ...

Hunstead: It's difficult. (Laughter) It's clearly a function of signal-to-noise ratio.

Rees: What you're telling us is a very important point, because if they are really well below 15 km s^{-1}, then that does rule out any model with high photoionization.

J. Bahcall: I'd like to make an expression of hope here, and a question to both of you. Everybody has shown this viewgraph so far; all the speakers; and I hope everybody will continue to show it because it contains a lot of information here. But ... I wonder if you guys would comment on the reliability, the certainty, of the measurements of the column densities in the region in which they're big and with these big b-values. How can you do that with any reliablity?

Hunstead: For a damped line you can't determine a b-value.

J. Bahcall: How do you get the column density, is that a reliable thing you can do in this region?

Carswell: Yes. It is very easy to fit to the broad damping wings, even at quite low resolution.

J. Bahcall: What happens if the line goes down to zero?

Carswell: It has to. It is very important to go down to zero for anything above HI = 10^{18} cm^{-2}, even at resolutions of ~ 2Å or more.

J. Bahcall: But when it's down to zero? Couldn't you have an infinite column density?

Carswell: No, because what you're now measuring is the damping wings.

J. Bahcall: Suppose you had a very small velocity width.

Carswell: It's dominated completely by the damping wings.

Wolfe: If you have a 21 cm line, then that's the way to get a velocity dispersion. In most cases, the velocity dispersion σ is less than 10 km s^{-1}.

J. Bahcall: It is very small.

Wolfe: Well, there is one very well studied case. The effective velocity dispersion is 10 km s^{-1}. In fact there are two features, where each feature has a sigma, (I'm not using b now, I'm using sigma), of 5 km s^{-1}. But those are fairly different than the things you guys have been talking about.

Carswell: Basically, anywhere above here (at high column density), the doppler parameter is 35 km s^{-1} or less, then you don't care what it is in determining the column density.

Wolfe: Also the damped systems lie above that curve? Isn't that correct? The damped systems lie above the extrapolation?

Shull: Since $b = 20 - 40$ km s^{-1} is too large for the HI to be thermally broadened, it is likely that the Ly-α clouds contain many velocity components. In this case, the relation between W and N becomes quite uncertain [$W \propto \sqrt{\ln N}$ for a Maxwellian distribution, but $W \propto \ln N$ for an exponential distribution]. I worry that you have underestimated the errors in determining N(HI) and that the break in the distribution with N at $N \sim 10^{14}$ cm^{-2} could be an artifact of curve-of-growth saturation.

Carswell: We worried about this, but I think the fact that we fitted all Lyman lines available, most of which are not saturated, means that we should not get an artificial break at 10^{14} cm^{-2}. It is also interesting that the break is less pronounced in the AAT data where only Ly-α was available.

ABSORPTION LINES AND GALAXY FORMATION

Martin J. Rees
Institute of Astronomy
Cambridge

Abstract. Some implications of absorption lines are assessed in the light of evidence that galaxies have larger radii, and formed more recently, than seemed the case when the phenomenon was first discovered. The high column density systems could be due to gas still falling into galaxies at $z \simeq 2 - 3$. The interpretation of Ly-α systems, as is well known, remains controversial. If the 'cold dark matter' cosmogony were correct, these systems would be attributable to clouds of gas gravitationally confined in 'minihalos' whose dimensions and z-dependence can be calculated. Other issues relating to cloud confinement and ionization are briefly discussed.

1. THE SYSTEMS OF HIGH COLUMN DENSITY

1.1. Dimensions and Gaseous Content of Galaxies at $z > 2$

Among the earliest conjectures about multiple absorption line systems was Bahcall and Spitzer's (1969) proposal that the absorption arose in galactic halos along the line-of-sight. This idea has survived, in modified form, as a popular interpretation of metal-line systems, but the Ly-α only systems soon came to seem embarrassingly numerous.

But if the Ly-α lines had only recently been discovered, and Bahcall and Spitzer were theorising now rather than nearly 20 years ago, their perspective might be different. There is now firm evidence that the luminous parts of galaxies are embedded in dark halos extending out to $\gtrsim 100$ kpc. The free-fall time for protogalactic material is of order $(2GM/R_{turn}^3)^{-1/2}$, where R_{turn} is its radius at turnaround. The collapse factor before a non-dissipative system virialises is < 2 (though it can, of course, be larger when radiative cooling permits dissipation). If galaxies were only 10 kpc in size, they could therefore have formed on a timescale of 10^8 years, from material that 'turned around' at a redshift $z \simeq 20$. But the collapse phase of galaxies whose diffuse halos extend out beyond 100 kpc would have been greatly prolonged.

The mean density of the Universe at the turnaround time of the halo material must be lower than the density of the proto-halo itself (by a factor 5.5 in the simplest model for a spherical protogalaxy), so a system whose mean density at turnaround was very low cannot have collapsed until a correspondingly late epoch. The low densities and long dynamical timescales in the outer parts of halos therefore have the crucial implication—irrespective of what these halos actually consist of—that galaxy formation (and, more specifically, the formation of the outer parts of disc galaxies) was not completed until the Universe was $\gtrsim 2 \times 10^9$ years old: i.e. not until redshifts $z \simeq 2$, even if it started much earlier.

The above argument depends simply on the existence of an extensive dark halo; but it derives indirect support from models for the formation of discs, and the origin of their angular momentum (Fall and Efstathiou 1980; Gunn 1982). Tidal torques between neighboring bound systems generate rotational speeds only 5–10 % of what is needed for rotational support (Aarseth and Fall 1980; Barnes and Efstathiou 1987). The angular momentum of galactic discs could therefore not have been imparted by tidal torques at anything like their present radius. There must be a much bigger 'lever arm': a tidal couple could only generate the angular momentum in discs if they formed from material 'spun up' at a much larger radii which subsequently fell inward by a factor ~ 10 in r, conserving its angular momentum.

1.2. Emission from Gas in Young Galaxies

If galaxies—or at least their disc components—did indeed form via infall of gas from 100 kpc, we can ask how conspicuous the gas that had not turned into stars would be at $z = 2$. For a total infalling gas mass of $M_g = 10^{11} M_\odot$ the characteristic mean density is $\bar{n} = 10^{-3} \, r_{100}^{-2}$ cm^{-3} where r_{100} is the radius in units of 100 kpc. This density is of course time-dependent; moreover the r-dependence is debatable, because it involves the question of how much gas is free-falling at a given radius. The free-fall time and the virial temperature depend on the total mass of galaxy-plus-halo (which may not all be baryonic). If the material in the outer part of a protogalaxy were at the virial temperature $T \simeq 10^6$ °K its luminosity, primarily in soft X-rays, would be only 10^{42} erg s^{-1}. However, material falling within the radius where $t_{cool} < t_{infall}$ would develop a two-phase structure: gas would condense out as clouds, or in sheets behind radiative shock fronts, cooling to 10^4 °K (or a still lower temperature if heavy-elements were already present), and establish pressure balance with the fraction of gas that remains at $T = T_{virial}$. A fraction $f(r)$ would condense out, the value of f adjusting itself so that the cooling of the remainder $(1-f)$ was balanced by the available heat input. The clouds would have densities of order $(T/T_{virial})^{-1} n(1-f)$. If there were no heating other than from adiabatic compression, $(1-f)$ would drop until the cooling and free-fall timescales were comparable (cf Fall and Rees 1985, 1987). The gas at 100 kpc radius would radiate a luminosity of

$$\sim 10^{45} \, f \, (1-f) \, \left(\frac{M_g}{10^{11} \, M_\odot}\right)^2 \left(\frac{n_e}{n_{HI}+n_e}\right)^2 \, \text{erg s}^{-1} \qquad (1)$$

predominantly in H-recombination lines.

This luminosity is very high if the clouds are fully ionized (implying rapid cooling). Consequently, most of the gas would recombine (i.e. n_e/n_{HI} would become $\ll 1$) unless there were some powerful internal excitation mechanism capable of balancing these radiative losses.

O and B stars could provide enough excitation if early star formation yielded a population with a sufficiently flat IMF, in which case all galaxies at this phase of their evolution would be conspicuous extended emission-line objects. York et al. (1986) make such estimates, on the assumption that a young galaxy is essentially an aggregate of gas-rich dwarfs. However, we do not know enough about the initial IMF and the early rate of star formation to quantify this.

But we do know that some galaxies at $z \gtrsim 2$ possessed active nuclei, and the UV emission from a central QSO would contribute a still more potent input, capable of maintaining gas with the properties above in a fully ionized state, irrespective of the stellar content. Reprocessed or scattered emission of strength quantified by equation (1) would then be detected as narrow lines from a region extended by a few arc seconds. Quantifying this further depends on knowledge of uncertain geometry, covering factors, etc. However, even existing limits on fuzz around high-z QSOs may impose interesting constraints on galaxy formation. Cowie and Hu (1987), in a survey of 11 high-z QSOs, have found no evidence for Ly-α emission even down to a level of around 24.5 mag sec^{-2} around any object except for PKS 1614+051 (Djorgorski et al. 1985; Cowie and Hu 1987) which shows a faint feature displaced by 6 arcsec from the QSO itself. It would be unfortunate if experimental searches for fuzz were to focus only on low-z QSOs, on the alleged grounds that only then was there any significant chance of a positive result.

1.3. Absorption Lines Due to Young Galaxies?

Galaxies at high redshift would still be in the stage where star formation was far from complete, and gas was falling in from distances of 100 kpc. The actual distribution and dynamics of the gas (and the internal cloud structure, covering factor, etc.) depend on uncertain details of the model for a collapsing protogalaxy; but much of the gas would likely be at 10^4 °K, and be predominantly neutral except in special cases when there was a powerful central AGN.

The relevance to absorption lines should now be obvious: if galaxies indeed collapsed from large turnaround radii they would, at $z = 2$, have huge cross sections, and consequently a large 'covering factor' over the sky: this is approximately (cf Hogan 1987)

$$\frac{dN}{dz} = 12\ h^{-1}\ r_{100}^2 \left(\frac{M_g}{10^{11}\ M_\odot}\right)^{-1} \left(\frac{\Omega_{bg}\ h^2}{0.01}\right) \left(\frac{1+z}{4}\right)^{1/2} \qquad (2)$$

where Ω_{bg} is the fraction of the critical density in baryons within galaxies, and h is Hubble's constant in units of 100 km s^{-1}. Several such objects would therefore lie along a typical line-of-sight, intercepting the light from a background QSO. The simple estimate given by equation (2), which could obviously be improved to take account of a range of galactic masses and cross-sections, suffices to show that QSO observations can constrain the amount of clumped HI in typical galaxies at redshifts 2 – 4.

The column density of baryons along a line-of-sight passing a distance r from the centre of a galaxy with $M_g \simeq 10^{11}\ M_\odot$ is 3×10^{20} cm^{-2} even for r as large as 100 kpc. If 10 % of this were in gas at 10^4 °K and density 10^{-1} cm^{-3}, it could not be maintained ionized by the intergalactic UV background (which would have to balance radiative losses of order (1) above). The implications are surprisingly strong—either the covering factor of clouds within each protogalaxy is small, or there is a sufficiently intense heat source to keep the gas at 10^4 °K highly ionized. Most of the gas within protogalaxies must therefore, on cooling from the virial temperature, convert into stars very promptly, or it must be kept ionized by a heat input within the protogalaxy far more intense than the intergalactic background. Otherwise there

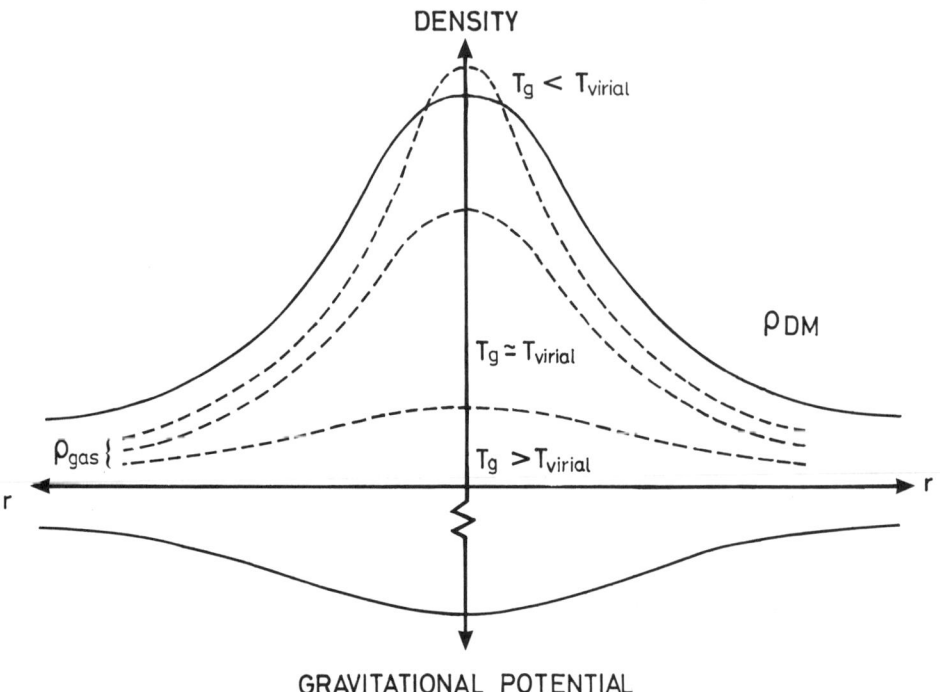

Figure 1. This diagram shows schematically the density profile of the dark matter in a minihalo. The depth of the resultant potential well (whose shape is also indicated) can be characterised by the virial temperature T_{virial}. The baryonic gas, contributing perhaps only 10 % of the total mass, would, at redshifts $z \lesssim 4$, have a temperature T_g controlled by photoionization equilibrium. Under the influence of the minihalo's potential well, the gas settles into a distribution whose profile depends on T_{gas}/T_{virial}. If the potential well were too shallow, the gas density within it would be only slightly enhanced over the mean level. But if the potential well were too deep, the gas would become strongly concentrated near the centre, where its density would dominate that of the dark matter: the gas would then be self-gravitating, and could fragment into stars. Gas could be stably confined in minihalos for which $T_{gas} \lesssim T_{virial} \lesssim 2\,T_{gas}$; such clouds could account for the Ly-α absorption lines if minihalos exist with the properties expected in cold dark matter cosmogony.

would be more systems with neutral hydrogen column density $> 10^{17}$ cm^{-2} in typical QSO spectra than are actually observed.

The high column density systems discussed by Briggs and Wolfe in these proceedings could well be associated with infalling gas in the halos of galaxies, rather than with discs that are rotationally supported. The high column density clouds could resemble the dense filaments found in cooling flows (Crawford et al. 1987). It is quite natural that the distribution of clouds or sheets of gas should be *more extended* in a young galaxy; contrariwise, rotationally-supported discs would grow (via infall) from the centre outwards (cf Gunn 1982) and would therefore tend to be *smaller* at $z > 2$ than at the present epoch.

2. EQUILIBRIUM AND CONFINEMENT OF CLOUDS (ESPECIALLY GRAVITATIONAL CONFINEMENT)

Pressure-confinement of the Ly-α clouds by a hot intergalactic medium was advocated by Sargent et al. (1980), and has been explored by several authors (e.g. Ostriker and Ikeuchi 1983; Atwood, Baldwin and Carswell 1985; Carswell et al. 1987). I am an agnostic on the confinement issue; however, since several other speakers will discuss pressure-confinement, it may serve the interests of overall balance if I focus on *gravitational* confinement. The first quantitative discussion of this possibility was given by Black (1981), whose work highlighted a difficulty; self gravitating isothermal clouds (which respond almost isothermally to changes in density) tend to be unstable—if squeezed, they would collapse; if slightly inflated, they would go into free expansion.

2.1. Gravitationally-Confined Gas in Dark 'Minihalos'

Stable confinement is, however, possible if the gravitational potential well is due not to the gas itself but to something else. One specific possibility suggests itself in the context of the cold dark matter (CDM) cosmogony. According to this model, only 10 % of the matter in the Universe is baryonic, the dynamically-dominant constituent being weakly interacting particles which undergo hierarchical gravitational clustering. The gas participates in this clustering insofar as pressure effects can be neglected (i.e. on scales exceeding its Jeans mass), and thereby accumulates in virialised 'halos' of dark matter. The luminous content of galaxies, embedded within more extensive dark halos, could have formed in this fashion (Blumenthal et al. 1984).

The CDM model is very well-defined: the shape of the initial fluctuation spectrum is specified, and the only free parameter is the amplitude, which must be matched to the present-day clustering scale. Detailed simulations, reported in a series of papers by Davis, Efstathiou, Frenk and White, have achieved gratifying success in accounting for the clustering properties of galaxies. The model is certainly more credible, and has been more fully investigated, than any equally specific alternative; it is therefore interesting that, as a bonus, it offers a natural interpretation of the Ly-α clouds.

The initial fluctuation spectrum, imprinted by processes in the very early universe, has larger amplitudes at smaller scale, and the systems of dark matter that condense and virialise first (i.e. which have the largest turnaround redshift z_{turn}) have low mass. However the characteristic binding energy, proportional to $M^{2/3}(1+z_{turn})$, increases with M. The very small and shallow potential wells that form first would fail to capture any gas at all; eventually, high-M halos form within which gas can collapse and fragment into stars. But there would be some potential wells of intermediate depth, within which gas could be stably confined, being neither able to escape, nor to settle towards the centre and become self-gravitating. As illustrated in Figure 1, this stable equilibrium requires the sound speed v_s in the gas to be somewhat (but not too much) below the virial velocity v_{virial} (Rees 1986; Umemura and Ikeuchi 1986).

The relevant 'minihalos' are those with $v_{virial} \simeq 30$ km s^{-1}; in the CDM cosmogony the corresponding total mass is around 10^9 $M_\odot$, of which 10 % would be in gas, and the turnaround redshifts would be < 5; see Figure 2 and its caption. How these bound clouds could account for the absorption lines is discussed in an earlier paper (Rees 1986), and I will just reiterate the main points here.

The CDM fluctuation spectrum is very flat for subgalactic mass-scales: the amplitude of $\langle(\delta M/M)^2\rangle^{1/2}$ decreases only by a factor 3 between 10^8 and 10^{12} $M_\odot$. This means that the entire hierarchy of non-linear structures built up relatively recently—indeed the characteristic scales of virialised systems would be 100–1000 times smaller at $z = 3$ than at the present. The flat spectrum implies also that there is substantial 'crosstalk' between different scales, in the sense that small-scale fluctuations continue to form from low-amplitude peaks even after much larger systems have formed from regions where the initial overdensity was more extreme. Minihalos would therefore still survive at epochs ($z < 3$) accessible to observations, and even to the present day: not all the bound systems of subgalactic scale would have lost their identity via earlier mergers.

Gas condensing into minihalos would form clouds with roughly the same density and linear dimensions as is postulated in pressure-confined models (Sargent et al. 1980; Ostriker and Ikeuchi 1983). It would therefore be maintained at $\sim 10^4$ °K by a UV background of the same intensity as needs to be invoked in those models. The gravitationally-confined clouds would have a density profile falling off smoothly from a central peak (see §4).

Gravitationally-confined clouds are a necessary consequence of the CDM cosmonogony—had they not been discovered 15 years earlier this could be claimed as a prediction. The dark matter aggregates, as the Universe expands, into progressively larger bound systems, whose binding energy increases with the mass scale. There are therefore inevitably some minihalos with potential wells of the right depth stably to confine photoionized gas. Moreover, the characteristic mass and size are not free parameters, but are determined by the model (see Figure 2).

The evolution with redshift in the cloud density is something that any complete theory should account for. In the CDM model, where clouds are gravitationally confined, this depends on three things:

1. The *ionization level* in the clouds is determined, as in pressure-confined models, by the UV background, whose z-dependence, particularly at $z > 3$, is highly uncertain (see §4 below).

2. *The number of minihalos with v_{virial} in the right range* would depend on z. Such systems would continue to form, as overdense regions in the mass range 10^9–10^{10} $M_\odot$ with progressively smaller initial $\delta M/M$ (i.e. less far out on the Gaussian tail) turn around. But, as a countervailing trend, minihalos would progressively be destroyed as they merged to build up larger scales in the hierarchy for which v_{virial} was larger (Frenk et al. 1985; White et al. 1987; Efstathiou and Rees 1988); gas would concentrate towards the centres of such merged systems, becoming self-gravitating and converting into stars. Isolated minihalos would be less likely to survive unscathed in regions of large-scale overdensity. There would therefore be

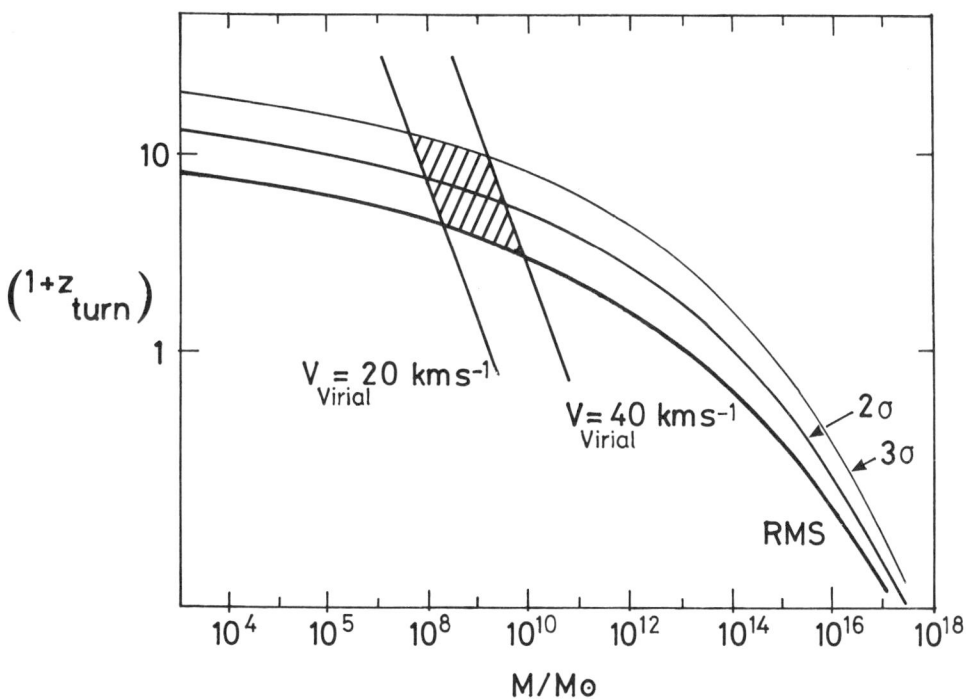

Figure 2. The fluctuation spectrum in the cold dark matter cosmogony has a shape that depends only on Ω and the Hubble constant, for which $\langle(\delta M/M)^2\rangle^{1/2}$ rises very gradually towards smaller scales. The spectrum shown above corresponds to $\Omega = 1$ and $H_o = 50$ km s^{-1} Mpc^{-1}, the amplitude being normalised so as to match the data on galactic halos and how they are now clustered. The spectrum then determines the turnaround redshift z_{turn} for perturbations of a given mass: perturbations corresponding to high-amplitude peaks of an initial gaussian fluctuation field tend to turn around earlier than more 'typical' peaks. The resultant virial velocities scale as $M^{1/3}(1 + z_{turn})^{1/2}$. Minihalos for which v_{virial} lies in the range required to stably confine photoionized clouds (see Figure 1 and its caption) lie in the shaded part of the diagram. The characteristic total masses are $\sim 10^9$ $M_\odot$, of which baryons may contribute only ~ 10 %.

an 'antibias': minihalos would display less clustering than the overall dark matter distribution.
3. Even if a minihalo with v_{virial} in the right range remained undisturbed and unmerged, *its baryonic content could be depleted* by gas-dynamical processes, such have been invoked in connection with biasing mechanisms (Rees 1985; Dekel and Rees 1987).

2.2. Tests of Models for Cloud Confinement, etc.

Obviously any independent evidence on the general intergalactic medium is relevant. Although we have become accustomed to the idea of an all-pervasive intergalactic medium, it is worth remembering that there is no direct evidence that it is hot. Gas with $T = 10^7 - 10^8$ °K definitely exists within clusters of galaxies; and the lack of a Gunn-Peterson effect implies that any diffuse medium between the clusters must be very highly ionized. But if the latter medium were photoionized rather than collisionally ionized, it need be no hotter than 10^4 °K, and could certainly not provide a pressure as high as that within the Ly-α clouds.

The clustering properties of the clouds are another discriminant among possible models. Of special interest is the question of whether there are any gaps in the redshift distribution of the Ly-α lines in any given spectrum. Voids in the cloud distribution as large as the Bootes void (across which the differential Hubble velocity is 6000 km s^{-1}) would cause gaps in the line distribution extending over a range in redshift that would typically contain 10–20 detectable lines. Such gaps have been looked for in a preliminary way (Carswell and Rees 1987), but it is worth thinking more about the optimum statistic for revealing features with such large dz that tests involving the spacing of nearest-neighbor lines are not the most sensitive. Such features would be expected if clouds were nonexistent in void regions; they could also arise if clouds were present, but some effect (e.g. spatial variations in pressure or ionization level) modulated their properties in some large volumes. Lack of such features in high-resolution spectra would imply that there were no large holes in the cloud distribution. Moreover, even if clouds existed everywhere it would have a further implication: it would imply (if they were pressure-confined) large-scale uniformity in the confining pressure, which would tell us something about how the hypothetical intercloud gas was heated. A QSO can enhance the ionizing flux out to distances of ~ 10 Mpc, so apparent voids in the clouds would also be created around powerful QSOs if their individual lifetimes exceeded a few times 10^7 years.

It has been widely recognised that the Ly-α clouds offer clues to galaxy formation—that they may be too small or loosely bound to fragment into stars, or related in some way to dwarf galaxies (e.g. Fransson and Epstein 1982; Ikeuchi and Ostriker 1986). The specific assumptions of CDM cosmogony suggest that there need be no sharp demarcation between the hydrogen-only systems and those displaying heavy-element absorption lines: once systems form with v_{virial} too high for stable confinement (see caption to Figure 1), star formation is inevitable, and could well lead to nucleosynthesis. Dekel and Silk (1986) have discussed in detail how dwarf galaxies might form in the centres of weakly bound CDM halos.

The intergalactic medium was already ionized by $z = 4$. In the CDM cosmogony, a first generation of 'Population III' stars could have formed at redshifts 5–10 from still-neutral primordial gas that condensed in minihalos even smaller than those discussed above (with $v_{virial} = 10$ km s^{-1}). If the 'Population III' IMF were tilted towards massive stars, only $\lesssim 10^{-3}$ of the primordial material, condensing from the 3σ peaks in the density distribution and converted into such stars, would generate enough UV to ionize all the rest of the baryonic material in the Universe (Couchman and Rees 1986). The material would then be put on a higher adiabat, with a consequent rise in the Jeans mass to $10^8 - 10^9 \, M_\odot$. There would consequently be a lull in the cosmogonic process until objects of this mass-scale turned around—it is these that would become the gravitationally-bound Ly-α clouds (though they may have acquired a tiny contamination of heavy-elements from the 'Population III' ejecta).

Bond, Silk and Szalay (1987) have pointed out that any HI clouds that became gravitationally bound but had not yet turned into stars would be disrupted, and go into free expansion, when photoionization occurred. Until they dispersed, such clouds would cause Ly-α absorption; they could therefore constitute a population rapidly evolving with z.

3. PHOTOIONIZATION OF CLOUDS; Z-DEPENDENCE OF THE ULTRAVIOLET BACKGROUND

The arguments that the Ly-α clouds are photoionized by a UV background are strong enough to be uncontroversial; but the physical state of the much larger amount of gas between the clouds is less well pinned down. The radiative cooling and recombination timescales for this intergalactic gas exceed the cosmological expansion timescale; its state at a given epoch therefore depends on its previous history. The same UV flux that ionizes the clouds could also ionize this diffuse medium. But if this were the only heat input, the gas could not be significantly hotter than the clouds (and may indeed be cooler, because of adiabatic expansion). The pressure-confinement hypothesis therefore requires some input of a different kind, which heats the intergalactic gas to $\gtrsim 10^7$ °K. At such high temperatures, collisional processes alone would reduce HI, HeI and HeII to undetectably low levels, and it is then a side-issue whether a UV background (from QSOs or indeed from any other sources) could have photoionized the originally-neutral H and He. Two types of observation have a bearing on the state of the intercloud gas.

1. The lack of Gunn-Peterson (1965) troughs implies that the intergalactic medium has either been heated to $> 10^6$ °K (in which case collisional ionization suffices), or else completely photoionized. Photoionization obviously requires emission of at least one UV Lyman continuum photon for each H-atom in the intergalactic medium; the very stringent limit ($< 5\%$) on the Gunn-Peterson effect claimed by Steidel and Sargent (1987) would place an even stronger demand on the UV background if the intergalactic gas were not hot enough for collisional ionizations to maintain a sufficiently low n_H. The fact that no Gunn-Peterson effects are evident even at $z = 4$, where the QSOs are apparently 'thinning out' if not approaching a cut-off, suggests that this heat input comes from something other than the QSOs themselves, possibilities include young galaxies, pregalactic stars, and the like.

2. Along a typical line-of-sight to $z = 4$ there may be at least one cloud with sufficiently high HI column density to be optically thick in the Ly-α continuum (Bechtold et al. 1987; Tytler 1987). Any UV background present at $z > 4$ would therefore be attenuated by absorption, not merely by the usual cosmological expansion and redshift factors. The UV background at $z = 2$ could therefore come primarily from QSOs even though other types of emitter may have dominated at

earlier times. If the apparent deficit of absorption lines with z_{abs} close to z_{em}—the 'anomaly' or 'inverse effect' (Hunstead et al. 1986)—the results from ionization by the nearest QSO, then its expected z-dependence cannot be calculated without knowing how the UV background depends on z. Indeed, this effect is probably of greater potential interest as a diagnostic of the UV background level.

A third type of potential observation, suggested by Hogan and Weymann (1987), could in principle yield an estimate of I_{UV} at different redshifts. This involves searching for Ly-α *emission* from the clouds, which would contribute a patchy background when the sky was observed through a narrow filter; the intensity would depend on n_e^2, and therefore (for clouds of given HI column density) on I_{UV}^2. A narrow-band search would pick out Ly-α emission from clouds in a thin 'shell' around us in a narrow range of redshifts. Such observations have as yet merely given upper limits to I_{UV}. But if, with improved sensitivity, positive results could be obtained, the characteristic cloud sizes could then in principle be estimated from the angular structure in the sky background.

4. THE DISTRIBUTION OF COLUMN DENSITIES f(N)

Other speakers have discussed the $f(N)$ distribution over a range 10^7 in HI column density, and whether the high-N and low-N systems are different in nature. On the basis of the above discussion, I would conjecture that the high-N systems—those with $N > 10^{17}$ cm^{-2} where self-shielding occurs—involve gas in protogalaxies, either in a disc, in sheets or caustic surfaces behind shock fronts, or in clouds confined by gas at the virial temperature. For $N < 10^{17}$ cm^{-2}, the hypothesis of intergalactic clouds or of subgalactic bound systems is perhaps more attractive. But these systems still span a range of order 10^4 in N (though in an individual case N is hard to determine on the exponential part of the curve of growth).

The $f(N)$ distribution for 10^{13} cm^{-2} < N < 10^{17} cm^{-2} could, to some extent, reflect the distribution of cloud sizes. But this is unlikely to be the whole story: for clouds of given density and ionization level, the column density N would scale only with the cube root of mass, and a range of 10^{12} in cloud masses is too large to be accommodated in most models. But even if the clouds were standardised, there are at least four effects that could yield a spread in N at a given redshift:

1. *Pressure inhomogeneities.* If a cloud exposed to a given UV background is squeezed to a higher pressure P, then the neutral density varies as P^2 (neglecting a small correction due to the dependence of T on n_e/n_{HI} in a photoionized medium). The dependence of cross section on P would be $\propto P^{-2/3}$ for spherical clouds (or for homologous compression of clouds with any shape) but slower if the clouds were filamentary or sheet-like. The value of N for pressure-confined clouds would therefore be correlated with the external pressure. This might be non-uniform if the IGM were heated by large-scale shocks, and the pressure would be higher in the environs of clusters (cf Ikeuchi and Ostriker 1986).

2. *Inhomogeneities in the ionizing background.* One interpretation of the so-called inverse effect (Hunstead et al. 1986; Carswell et al. 1987; Tytler 1987) is that it results from extra photoionization by the nearby QSO (Hunstead et al. 1986; Bajtlik, Duncan and Ostriker 1987). Whether or not QSOs are the prime sources of the UV background, we would expect a deficit of clouds ('clearings' in the Ly-α forest) around intense local UV sources. If individual QSO lifetimes exceeded 10^7 years, there would be cavities (i.e. volumes where the clouds were sparser

than average) around each QSO. These would be spherical if each QSO emitted isotropically and steadily for a time exceeding the light travel time across the cavity; otherwise their shapes would reflect the QSO's 'beam pattern' or its time-variation over the last several times 10^7 years. There might consequently be gaps in the z-distribution of lines in a spectrum, perhaps attributable to other QSOs nearby in the sky and with smaller redshifts, where each cloud yields a reduced N.

3. *Orientation effects.* The HI may be in sheets related to shock fronts, caustic surfaces, etc. For planar slabs of thickness d_s whose column density, for a line-of-sight perpendicular to the plane, is N_o, the distribution of N arising from random orientation is of the form $f(N)dN \propto N^{-3}dN$ (for $N \gg N_o$). This falls off too steeply with N to be a dominant effect on the observed distribution of column densities (cf Barcons and Fabian 1987). Moreover, if the clouds were not exactly flat, but were curved with radius curvature R, then there would be a maximum of order $N_o(R/d_s)^{1/2}$.

4. *Profile effects: gradients across each cloud.* If the clouds were, for instance, gravitationally bound spheres, then the observed N would obviously depend on whether the line-of-sight passed through the centre or grazed the periphery. Randomly placed clouds of identical properties would therefore yield an apparent $f(N)$ depending on the profile across each one. If the total density were to depend on radius as $n_H \propto r^{-p}$, then the neutral density n_{HI} in a photoionized cloud would go as r^{-2p}, and

$$f(N) \propto N^{-(2p+1)/(2p-1)}$$

so $f(N) \propto N^{-1.5} - N^{-2}$ corresponds to p in the range 1.5–2.5. This is just an illustrative example; in a realistic case the power-law decline in n_H would occur only outside of some core radius, and would continue only until the density fell to the intergalactic value. See Umemura and Ikeuchi (1986), Ikeuchi and Norman (1987), and Bond et al. (1987).

If the clouds were actually pressure-confined, these profile effects would not be important because the external pressure would impose a well-defined value of n_H in any near-isothermal photoionized cloud; but density gradients would be inevitable in gravitationally confined or freely-expanding clouds. Such an effect was discussed by Fransson and Epstein (1982) in the context of winds from dwarf galaxies; this particular model predicts a characteristic double-peaked doppler-broadened profile.

Of the above effects (1) and (2) are the prime candidates for accounting for the z-dependence of the line density in models that invoke pressure-confinement.

I am grateful to several colleagues, and especially Bob Carswell and Craig Hogan, for clarifying discussions about topics discussed in this paper.

REFERENCES

Aarseth, S. J. and Fall, S. M. 1980, *Ap. J.*, **236**, 43.
Atwood, B., Baldwin, J. A. and Carswell, R. F. 1985, *Ap. J.*, **292**, 58.
Bahcall, J. N. and Spitzer, L. 1969, *Ap. J.*, **156**, L63.
Bajtlik, S., Duncan, R. C. and Ostriker, J. P. 1987, preprint.
Barcons, X. and Fabian, A. C. 1987, *M.N.R.A.S.*, **224**, 675.
Barnes, J. E. and Efstathiou, G. 1987, *Ap. J.*, **319**, 575.
Bechtold, J., Weymann, R. J., Lin, Z. and Malkan, M. 1987, *Ap. J.*, **315**, 180.

Black, J. H. 1981, *M.N.R.A.S.*, **197**, 553.
Blumenthal, G., Faber, S. M., Primack, J. R. and Rees, M. J. 1984, *Nature*, **311**, 527.
Bond, J. R., Silk, J. I. and Szalay, A. 1987, *Ap. J.*, submitted.
Carswell, R. F., Webb, J. K., Baldwin, J. A. and Atwood, B. 1987, *Ap. J.*, **319**, 709.
Carswell, R. F. and Rees, M. J. 1987, *M.N.R.A.S.*, **224**, 13P.
Couchman, H. M. P. and Rees, M. J. 1986, *M.N.R.A.S.*, **221**, 53.
Cowie, L. L. and Hu, E. M. 1987, *Ap. J. (Letters)*, **317**, L7.
Crawford, C. S., Grehan, D. A., Fabian, A. C., and Johnstone, R. M. 1987, *M.N.R.A.S.*, **224**, 1007.
Dekel, A. and Rees, M. J. 1987, *Nature*, **326**, 455.
Dekel, A. and Silk, J. I. 1986, *Ap. J.*, **303**, 39.
Djorgovski, S., Spinrad, H., McCarthy, P. and Strauss, M. A. 1985, *Ap. J. (Letters)*, **299**, L1.
Efstathiou, G., and Rees, M. J. 1988, *M.N.R.A.S.*, submitted.
Fall, S. M. and Efstathiou, G. 1980, *M.N.R.A.S.*, **193**, 189.
Fall, S. M. and Rees, M. J. 1985, *Ap. J.*, **298**, 18.
Fall, S. M. and Rees, M. J. 1987, in 'Globular Clusters', ed. J. Grindlay, (Reidel), in press.
Fransson, C. and Epstein, R. 1982, *M.N.R.A.S.*, **198**, 1127.
Frenk, C. S., White, S. D. M., Efstathiou, G. and Davis, M. 1985, *Nature*, **317**, 595.
Gunn, J. E. 1982 in 'Astrophysical Cosmology' ed. H. A. Brück *et al.* (Pont. Acad. Sci. Scripta Varia), p. 233.
Gunn, J. E. and Peterson, B. A. 1965, *Ap. J.*, **142**, 1633.
Hogan, C. J. 1987, *Ap. J. (Letters)*, **316**, L59.
Hogan, C. J. and Weymann, R. J. 1987, *M.N.R.A.S.*, **225**, 1P.
Hunstead, R. W., Murdoch, H. S., Peterson, B. A., Blades J. C., Jauncey, D. L., Wright, A. E., Pettini, M. and Savage, A. 1986, *Ap. J.*, **305**, 496.
Ikeuchi, S. and Norman, C. A. 1987, *Ap. J.*, in press.
Ikeuchi, S. and Ostriker, J. P. 1986, *Ap. J.*, **301**, 522.
Ostriker, J. P. and Ikeuchi, S. 1983, *Ap. J. (Letters)*, **268**, L63.
Rees, M. J. 1985, *M.N.R.A.S.*, **213**, 75P.
Rees, M. J. 1986, *M.N.R.A.S.*, **218**, 25P.
Sargent, W. L. W., Young, P. J., Boksenberg, A. and Tytler, D. 1980, *Ap. J. Suppl.*, **42**, 41.
Steidel, C. C. and Sargent, W. L. W. 1987, *Ap. J. (Letters)*, **318**, L11.
Tytler, D. 1987, *Ap. J.*, **321**, 49.
Umemura, M. and Ikeuchi, S. 1986, *Astr. Ap.*, **165**, 1.
White, S. D. M., Davis, M., Efstathiou, G. and Frenk, C. S. 1987, *Nature*, submitted.
York, D. G., Dopita, M., Green, R., and Bechtold, J. 1986, *Ap. J.*, **311**, 610.

DISCUSSION

Bregman: I was wondering how important the lack of metals was in this theory. Suppose you had normal metal abundance or 1/10 metals, would the temperature of the gas drop sufficiently that you would make it difficult to have a lot of absorbing sytems?

Rees: You're talking about the Ly-α cloud model involving the gravitational confinement? Well, the temperature may drop from say 30,000°K to 20,000°K, or a bit less, if metals are present. All that would do is change the range of the masses of the potential wells with the right range of virial temperatures to permit stable confinement. Such a range always exists, because the virial temperature increases more or less monotonically with scale as the clustering hierarchy evolves. Heavy-elements wouldn't change things very much. This relates to the question of dwarf galaxies. Dekel and Silk discussed a model whereby some of these bound systems would evolve into dwarf galaxies, and then of course, you have the possibility of making heavy-elements in situ. The presence of heavy-elements doesn't affect the temperature very much provided the ionization level is high. And that is true whether the clouds are pressure-confined or gravitational-confined.

Norman: Just to comment on that: you want to be careful—you don't want to make too many heavy-elements too fast in these small systems because, you know, the supernova rate will blow them to pieces.

Rees: Yes. That might be another effect that influences the z-dependence of cloud numbers.

Shull: I would like to follow up on Bregman's question. Suppose you did have molecular hydrogen—that could lower the temperature to 100 °K.

Rees: In the cold dark matter model, the first bound systems to form, when the gas is still neutral, would indeed be on still lower mass scales—maybe 10^5 $M_\odot$. My discussion really started after the stage when there had been enough UV background produced to photoionize the entire medium. In a paper with Couchman, we made the point that only about 10^{-4} of the mass of the universe need be converted into OB stars to produce enough UV to ionize everything, including the gas that thereafter gets trapped in minihalos, and maintain the temperature at 10^4 °K. There could be a set of much smaller bound systems formed before photoionization, some of which may get disrupted in the manner suggested by Bond, Silk and Szalay. But after photoionization has occurred, then the temperature of the gas is fairly well-determined. H_2 could also, as you imply, form in the high column density clouds where self-shielding might occur.

J. Bahcall: You explained beautifully why the Lyman lines should arise in postulated 100 kiloparsec halos way back in the dim past, but another thing we postulated for which we had equally little justification at that time, and I wonder if there is better justification now, is that there should be heavy-elements out at 100 kiloparsecs. Is there any reason to think that that's reasonable from present-day galaxy formation?

Rees: I don't know at all. I think that depends on whether large galaxies build up by some hierarchical process where small structures form first. Because of that uncertainty, I was cautious about whether to try and explain the Ly-α system in terms of this model. You can push numbers to do this, and Craig Hogan (*Ap. J. (Letters)* **316**, L59) has done so by postulating radii larger than about 200 kpc. The point I do want to emphasize is that in this class of models one risks getting *too much* neutral hydrogen, because if the gas in a protogalaxy cools down into clouds at 10^4 °K and doesn't immediately turn to stars, then this gas cannot be maintained ionized, and along every line-of-sight through every galaxy you could have a column density above 10^{17} cm^{-2}, and if the protogalaxies really extend out to typical radii exceeding 100 kpc that's more high-N systems than are observed.

Wolfe: How big are those? You said they were pressure-confined.

Rees: The idea here is that if a gas cloud collapses it can't all remain at the virial temperature: it cools down, and one doesn't know what the scale of the blobs in the resultant two-phase medium will be. An upper limit is the Jeans mass, and that's about a million solar masses. In fact Michael Fall and I have a model trying to relate this to globular cluster masses. But if a reasonable fraction of the gas is in this 10^4 °K phase, then if the clouds are well below the Jeans mass the covering factor is larger than unity. So if the clouds are of the order of the Jeans mass, or a bit below, then a typical line of sight will intercept one of them, and the column density will be of order 10^{20} cm^{-2}. Perhaps you are observing clouds like this. The gas would perhaps be on the way to forming a disk, but I don't think there's anything in your data which decides whether you see an already formed disk, rather than clouds or caustic surfaces forming during the collapse.

Wolfe: The VLBI data that Briggs will talk about later says the sizes of these things have to be at least 50 kpc.

Rees: That would suggest it could be something like a shock front. You'll get a high density sheet whose transverse scale would have to be the size you estimate.

Wolfe: But these are on the verge of being self-gravitating.

Rees: Right, yes.

Felten: What about the clustering of these mini-halos (or mini-condensations) on *smaller* scales? Carswell reported that the forest objects show clumping on scales up to 300 km s^{-1} when the background quasar is at $z_{em} \simeq 2$, but that this clumping is absent when $z_{em} \simeq 3$. What is the predicted two-point correlation function for the mini-halos, and can it be compared with these observations?

Rees: The naive first step is to say that in the cold dark matter model you can estimate the *mass* correlation function. In fact the mini-halos could be somewhat *less* clustered than the overall mass distribution, so you would expect a sort of anti-clustering. The simulations by Frenk *et al.* find that the big halos (the ones with high velocity dispersions) tend to be biased in the sense that they form preferentially in the high density regions. But, the places where a small mini-halo is most likely to survive without merging into a larger scale system will tend to be in the middle of a large-scale under-density. The net result is that at the present epoch, all bound systems with velocity dispersions over 100 km s^{-1} show positive biasing whereas the reverse is true for the mini-halos. The small scale halos will be somewhat less clustered than the massive halos. But whether this effect alone is sufficient to explain the data, I don't know. It is very important to look for large-scale clustering in velocity space, which could be attributable to 'void'. Carswell and I have made an attempt but it was a far from optimal analysis of this.

P. Shapiro: When I realized that the observed quasars at $z > 3$ were not sufficient to ionize the IGM in order to explain the absence of the Gunn-Peterson effect at, say, $z \sim 3.5$, I was nevertheless struck by the fact that the observed quasars *do* emit by $z \sim 3$ roughly as many ionizing photons as there are baryons in the universe. Would you care to speculate on why there is this remarkable coincidence? If quasars were *not* responsible for ionzing the IGM at high redshift, for example, and other objects (like primordial galaxies) did this first, why is this quasar UV photon number emitted by $z \sim 3$ even *close* (i.e. within an order of magnitude) to the number of baryons?

<u>Rees</u>: I'm sure you can do much better than me, really! What is important is that despite the presence of the high column density clouds which attenuate the Lyman continuum, this must be enough UV photons to maintain the universe almost completely transparent.

P. Shapiro: It's true, but if the quasars weren't the thing that was after all responsible for making the universe transparent, why did they come within a factor of 10, or 100 of doing so?

<u>Rees</u>: The intergalactic medium is already ionized at $z = 4$, and in your models, you certainly don't have enough quasars. So there's no evidence that the feedback occurs, because something has already ionized the universe at a z of 4. Most of the quasars we observe are at a redshift less than 4. I think you have in mind some sort of negative feedback, but we surely have evidence that this doesn't occur, because the ionization occurred at redshifts larger than the redshifts of most quasars, but that did not stop quasars forming at redshift less than 4.

P. Shapiro: Then it's just a coincidence?

<u>Rees</u>: Yes.

Gondhalekar: A couple of years ago, I calculated the optical depth of the intergalactic medium due to the metal-rich systems, and I assumed that the evolution was similar to the Ly-α clouds. Although the optical depth of the intergalactic medium increases, it doesn't increase sufficiently to destroy all the ionizing photons. If we assume a drop-off in the number of quasars beyond a redshift of 4 they can still ionize the intergalactic medium at a redshift of 4.

<u>Rees</u>: I am not quite sure of the work you are referring to, but I do recall that Bechtold *et al.* considered the high column density systems. They claimed, I think, about 5 along a typical line of sight to $z = 3.5$, but perhaps one of the authors can comment on that?

<u>Ozernoy</u>: Suppose one can improve considerably the lower limit to the Ly-α forest cloud dimensions, D_{min}, which follow from considering a few Ly-α absorption line systems in common between two images of the gravitationally lensed quasar 2345+007 (Foltz et al. 1984, *Ap. J.*, **281**, L1). Recently it has actually been done by Chernomordix and myself (1986, *Astr. Tsirk.*, No. 1436, 1) by accounting for

the fact that the fraction of absorption lines which are *not common* to both spectra is small ($f \approx 0.13$), which increases the lower limit by a factor f^{-1} giving $D_{min} = (80-120)h_{75}^{-1}$ kpc when q_o varies between 1/2 and 0. Could such an improved value of D_{min} be still compatible with your dark minihalo sizes?

Rees: It would be marginally okay. The typical sizes are something like 20–50 kpc. In the detailed calculations by Bond and collaborators where they considered also the possibility that some of these clouds were allowed to expand freely, they do go up to 100 kpc. I think that if it's just a matter of 100 rather than 50 kpc, then that's okay, whereas if you go as big as a Mpc, then of course that would suggest a quite different model and perhaps suggest that you would have to have very extensive sheets, or go back to your kind of explosion model or possibly say that one is seeing Ly-α systems in a single huge galactic halo. But I think for the mini-halo model, as indeed for any pressure-confined clouds other than very big sheets, getting more than 100 kpc or so does become an embarrassment.

Balbus: I'm wondering if you can contrast this theory with the Ostriker and Ikeuchi version of pressure-confined clouds. In their model the clouds form in swept up explosive events rather than swept up cooling shells, and you would expect a quite signficant fraction of the mass of the IGM to actually wind up as Ly-α clouds. Now, in your model, how efficient are the formation of the clouds? Do you get most of the mass forming in these, caught in the wells?

Rees: In this model the fraction is significant. I would guess it's a few percent. It depends on the way the hierarchical clustering proceeds, and on what fraction of the mass is, at redshifts 2 or 3, in potential wells of the right depth. It's of order a few percent. So it's not a particularly small number.

Balbus: Am I right in assuming that in the Ostriker and Ikeuchi model it would be considerably higher?

Rees: It may be a bit higher, yes.

Shull: I wanted to return to this ionization question. One thing I was struck with earlier in the meeting was that the high redshift quasars do not show the Gunn-Peterson dip. When we put this together with the x-ray background (which may or may not come from quasars), don't the x-rays set a fairly low limit to the ionization? The x-rays are more penetrating; shouldn't they keep the intergalactic medium ionized at some low level? Has anyone looked at that?

Rees: The cross-section for photoionization of hydrogen by x-rays is less than by UV photons. So the x-ray background would partially ionize the medium. The best it could do is make the medium transparent to the x-rays. It certainly isn't in itself going to make the medium transparent to Ly-α or Lyman limit photons. One thing it might do of course is to heat up the medium: if you had x-rays with a cut-off below a keV and no UV background then all the photoelectrons would have a keV, and so you could end up with an intergalactic medium with a temperature high enough that it could be kept ionized by collisional ionization. I think models like that have been discussed. But that's a very inefficient model in terms of energy, I believe.

Sargent: Since Martin referred to the Gunn-Peterson effect in his talk, I'd like to show some new data which are now in press (published later in Steidel and Sargent, Ap. J. (Letters), 1987, **318**, L11). In an earlier piece of work (Steidel and Sargent, Ap. J., 1987, **313**, 171), we had obtained absolute spectral energy distributions in the wavelength range 3200–10,000 Å (typically 800 to 2500 Å in the rest frame) for 8 quasars with redshifts in the range $2.6 < z_{em} < 3.38$ with the Double Spectrograph of the Hale telescope. It was found that at the resolution used the true continuum in the region of the Ly-α forest could be represented by an extrapolation of the contiuum defined longward of the Ly-α emission line by $f_\nu \propto \nu^\alpha$, where $0.28 < \alpha < 0.99$. D_A, the fractional flux decrement betweem the Ly-α and Ly-β emission, due to the cumulative effects of absorption lines, is statistically the same in all of the quasars (0.24 ± 0.05), supporting the idea that the absorption is due to material cosmologically distributed along the line of sight.

The spectra obtained in the work described above were combined with high resolution statistical studies of the Ly-α forest in quasars to obtain a new, improved upper limit on the density of generally distributed intergalactic neutral hydrogen. The new limit, $N_{HI}(z=0) < 9.0 \times 10^{-14} h_{100}$ cm^{-3} is approximately 15 times smaller than the limit originally obtained by Gunn and Peterson (Ap. J., 1965, **142**, 1633). If the intergalactic medium at $z = 2.6$ is in ionization equilibrium with the UV background, this new limit implies $n_H(z=0) < 4.6 \times 10^{-7} h_{100}$ cm^{-3} and $\Omega_H < 0.05 h_{100}^{-1}$. Models of Big Bang nucleosynthesis lead to an upper limit on the contribution of baryonic material to be in the range $\Omega_B < 0.034 h_{100}^{-2}$ (Boesgaard and Steigman, Ann. Rev. Astr. Ap., 1985, **23**, 319), while the contribution from the visible parts of galaxies is in the range $0.01 < \Omega_{Gal} < 0.02$ (Yang et al. Ap. J., 1984, **281**, 493). Thus, the new limit on Ω_H is just consistent with Big Bang nucelosynthesis. This result adds to the growing evidence that if $\Omega = 1$ then most of the matter in the Universe cannot be in the form of baryons.

Rees: But this limit on omega could be shifted up if you had a higher UV background than we think comes from quasars, and we have no evidence against that at least until the Hogan-Weymann test is done, because most of the continuum at $z > 3.5$ would be absorbed by clouds along the line-of-sight. Moreover, if the intercloud medium were as hot as it has to be if the clouds are pressure-confined, collisional processes could guarantee high ionization irrespective of the UV background.

Sargent: Yes, but it emphasizes to me the importance of the Gunn-Peterson type of measurement for setting limits on the total ionizing flux from quasars and from young galaxies at these early times.

Rees: Could I mention another indication of this which emerged from discussions I had with Paul Shapiro earlier today. In some pressure-confined models one might envisage clouds which are still smaller than the ones that give a column density of 10^{13} cm^{-2} and which therefore don't produce a line that goes down to zero. The total mass of gas in the form of clouds smaller than the smallest you can observe can't be more than about five times the mass of the larger clouds because those clouds would collectively give too much blanketing: the lines we see in Ly-α give perhaps 20 % blanketing, so five times as much mass in smaller clouds of the same density would give complete blanketing. So one can place a constraint on the number of little tiny clouds, even if one doesn't believe that conductivity and similar effects would rule them out.

N. Bahcall: I have a question or a clarification on the large scale structure that you mentioned, specifically on the question of finding voids with absorption lines. I thought that Wal mentioned this morning that no voids were seen or there were some strong limits on voids from the absorption lines. Did you mention that you can put some limit on that? What are the limits on the voids?

Rees: There's one rather crude limit which was done by Bob Carswell and myself; voids like the Bootes void, if they correspond to a factor of two deficiency in the density of clouds, don't have a filling factor of more than about 5 percent. From the restricted data we had there was no gap of that order. If voids are not manifested in the Ly-α line density, then that implies two things. It implies, firstly, that the clouds must be present everywhere; but, more than that, if the clouds were pressure-confined, it implies that the confining pressure must be the same to within a factor of 2 in the middle of a void as elsewhere, and that's a constraint on how the intercloud gas was heated up.

P. Shapiro: They further imply one other important thing. If the clouds are photoionized by sources which you would not expect to be formed in voids, then that puts constraints on the sources that form outside the void. They would have to have formed early enough to allow their effect to propagate into a void. This pushes their formation back even further, because there's a finite time it takes for the ionization to propagate through a void.

Rees: Yes.

Comment: Does that rule out explosions as a model for ...?

Chairman: (J. Bahcall) —State your name please!

N. Bahcall: Riccardo Giacconi.

Chairman: I always wondered who the boss was! (Laughter)

Rees: It leads to constraints, because if enormous explosions drive out the material, then it does surprise me you don't see evidence for this in the Ly-α forest. An 'escape route' would be to postulate that the intergalactic gas was very hot indeed—so hot that a soundwave can cross a void and the pressure can equilibrate. That requires gas with a temperature which the people who want to explain the x-ray background very much like—more than 40 keV. But, Ostriker, Ikeuchi et al.'s favoured temperature for the hot phase implies a sound speed that isn't high enough for sound waves to equilibrate the pressure in a void. It does seem to me that this search for large Δz-modulation in the Ly-α line density is going to be important.

N. Bahcall: Can I make one more comment on the clustering? Peter Shaver and other workers have shown quite convincingly that there is quite a bit of clustering on the large scale among quasars which is redshift-dependent.

Rees: Sure. Peter Shaver may have found an absolutely colossal cluster about half the size of the universe, and there is certainly growing evidence that quasars are clustered at smaller t.

Shaver: I think the evidence that Neta is talking about comes from two sources. One is catalogs, and this is subject to the usual concerns about dealing in homogeneous catalogs. But the other is from more homogeneous surveys. Work has been done by Boyle and Shanks and others; and we've been able to put together a sample of this kind as well. I think that now we do find clustering at the 5σ level, at least. There's some indication of clustering, but I think also possibly we see evolution in this clustering as well.

J. Bahcall: Peter, are you going to discuss that at this Conference?

Shaver: No; but I will in Hungary. (Laughter)

Rees: It is very important, of course, because that's a discriminant between the pancake-type models where the first galaxies to form are already clustered and the hierarchical type pictures (the cold dark-matter model is one example) where the scale of the clustering in co-moving coordinates was less in the past.

J. Bahcall: Do you have a preprint that those of us that are interested could look at?

Shaver: In a couple of months.

J. Bahcall: And we have to go to Hungary? (Laughter) Well, I think we should thank Martin and all of the people who made remarks and comments.

PROPERTIES OF THE HEAVY-ELEMENT ABSORPTION SYSTEMS

Jacqueline Bergeron
Institut d'Astrophysique
Paris

1. INTRODUCTION

The absorption line systems in the spectra of QSOs have been studied increasingly since it has been realized that they provide direct information on high redshift intervening objects that are not otherwise detectable. The large amount of observational material that is now available reveals the great complexity of the absorbers and the existence of different classes of objects.

We discuss here only the metal-rich systems having absorption lines with small velocity dispersion, typically $b < 100$ km s^{-1}. This class constitutes only a small fraction of the absorbers, about 5 % at $z \sim 2$, the major constituent being the very metal-poor systems. The narrow metal-rich absorption systems represent low density intervening gas, most likely associated with galaxies, as derived from statistical analyses and direct searches for low redshift ($z < 1$) absorbing galaxies. Their study should establish properties of the galaxies and their evolution with redshift.

The strongest metal absorption lines are the CIV and MgII doublets which, for a given observed spectral domain, sample different redshift ranges. This can lead to severe difficulties when comparing properties of low and high redshift absorbers because these lines originate from regions of different ionization level and may be even from different classes of absorbers.

First we describe the CIV and MgII samples that are currently available, and discuss possible cosmological evolution of the density per unit redshift for these systems. Then we outline the properties of the phases with different ionization levels. We describe the constraints on the ionization parameter and density of the absorbers as derived from photoionization models, and also mention problems related to the sphere of influence of individual QSOs to ionizing radiation and to the cosmological evolution of the ionization level of the absorbers. We give some estimate of the heavy element abundances and investigate a possible continuity with the Ly-α forest absorbers. Finally, we briefly present the results of our search for galaxies giving rise to MgII absorption systems.

2. THE MgII AND CIV SAMPLES

We give the bibliography for unbiased and homogenous absorption line surveys because these are the only ones which can be used for statistical analyses. Homogeneous, unbiased samples are obtained from observations of sight-lines that *a priori* are unknown. They are rest equivalent width limited. Multiple systems spanning less than 300 km s^{-1} are counted only as one system. Systems with redshift larger than the QSO emission redshift or smaller by less than 3000 km s^{-1} are excluded since they may be associated with the QSO or its underlying galaxy. The wavelength range shortward of Ly-α emission from the QSO is also excluded to avoid confusion with the Ly-α forest systems.

Large surveys for narrow metal–rich absorption systems were first obtained in the blue for the CIV doublet. The first sample obtained by Weymann et al. (1979) cannot be fully used since it was not obtained with photon counting detectors and reliable equivalent widths could not be measured. To the blue survey of Young, Sargent and Boksenberg (1982), Bergeron and Boissé (1984, hereafter BB84) added observations from about ten sight-lines, mostly from published studies of QSO pairs. For rest equivalent width limits equal to $w_{r,lim}$(CIVλ1548 and λ1550) of 0.3 and 0.2 Å, respectively, this led to 38 systems in 39 sight-lines. Most of the QSOs from the early survey of Weymann et al. (1979) were re-observed by Foltz et al. (1986) and this added 17 systems. Combined with the previous samples this gives 55 systems in 70 sight-lines at an average redshift $\bar{z} = 1.7$. However as shown in Figure 1a the redshift range adequately sampled, $z = 1.40$ to 2.05, is small. Sargent (with Steidel and Boksenberg) announced at this meeting, a not yet published new sample which comprises about one hundred CIV systems down to $w_{r,lim} = 0.15$ Å, and see Sargent (1988) in this volume for more details.

The MgII samples were also first built from blue surveys. Adding to the results of Young, Sargent, and Boksenberg (1982) and Foltz et al. (1986), Tytler et al. (1987) have obtained, for $w_{r,lim} = 0.6$ Å, a sample of 8 MgII systems at $\bar{z} = 0.5$ in 90 sight-lines. For the subsample with $w_{r,lim} = 0.25$ Å there are 14 MgII systems. Preliminary results are also available for an ultraviolet survey in the wavelength range $\lambda\lambda 3150$–3950 by Boulade, Kunth and Tytler (1987).

To study the ionization level of the metal systems Boissé and Bergeron (1985, hereafter BB85) observed 18 QSOs in the red down to $w_{r,lim}$ (MgIIλ2796 or FeIIλ2382) = 0.6 Å. Although the observations spanned a fairly large redshift range $z = 1.2$–1.7, the survey was biased since only those QSOs having a known CIV system were observed for MgII. More recently an unbiased red survey was completed by Lanzetta, Turnshek and Wolfe (1987, hereafter LTW87) who found 16 MgII systems at $\bar{z} = 1.67$ in 31 sight-lines down to $w_{r,lim} = 0.6$ Å. For the subsample with $w_{r,lim} = 0.3$ Å they get 10 MgII systems. This survey gives crucial new information for the determination of the cosmological evolution of MgII systems and in the study of relative properties of systems with different ionization levels. Further, it has led to the discovery of low ionization MgII systems without CIV counterpart. Combining the blue and red surveys gives 24 systems, with $w_{r,lim} = 0.6$ Å, spanning the large redshift range $z = 0.2$–2.05, see Figure 1. Finally, there is an unpublished red survey by Caulet and York which is biased because high spectral resolution observations have been obtained in only small wavelength ranges centered on expected MgII doublets from known CIV systems. Some preliminary results were presented at this Conference (Caulet and York 1987).

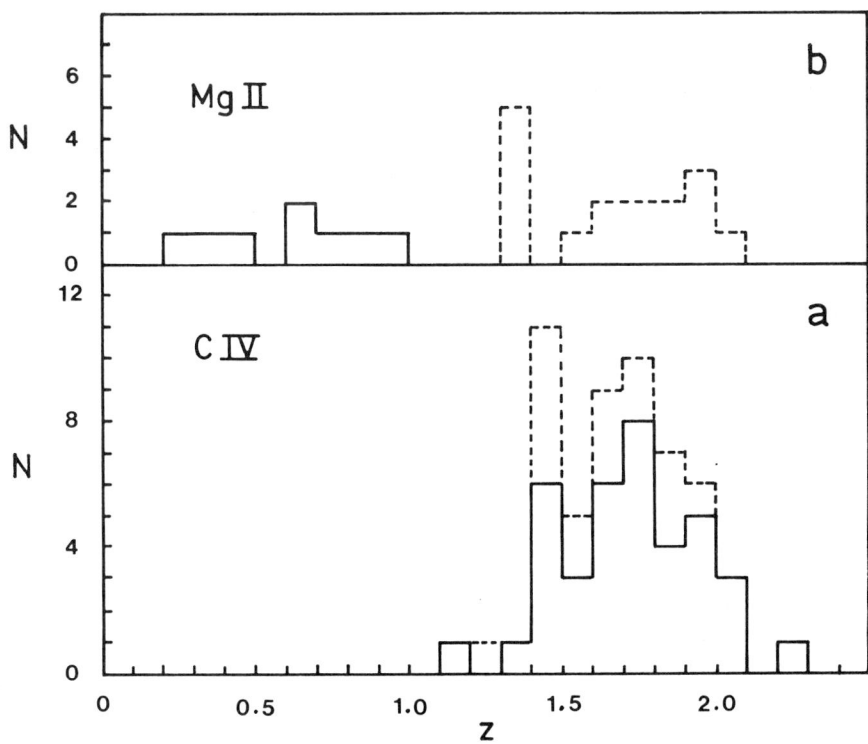

Figure 1. *Histograms of the available CIV and MgII samples. For CIV data are taken from Bergeron and Boissé (1984) (full lines) and Foltz et al. (1986) (dashed lines). For MgII it is those of Tytler et al. (1987) (full lines) and Lanzetta, Turnshek and Wolfe (1987) (dashed lines).*

3. AVERAGE SIZE OF THE ABSORBERS AND COSMOLOGICAL EVOLUTION

For Friedmann models with zero cosmological constant and a deceleration parameter q_o, the number of absorption line systems per unit redshift is given by

$$\frac{dN}{dz} = n_o \sigma_o \frac{c}{H_o}(1+z)^{1+\alpha}(1+2q_o z)^{-1/2}, \qquad (1)$$

where n_o and σ_o are the density and projected cross-section on the sky of the absorbers at $z = 0$, c is the speed of light, H_o the Hubble constant and α parameterizes possible cosmological evolution. This function of z is usually approximated by a single power law

$$\frac{dN}{dz} \propto (1+z)^{\gamma}, \qquad (2)$$

with $\alpha = \gamma - 1$ or $\gamma - 0.5$ for $q_o = 0$ or 0.5, respectively.

The size of the absorbers is obtained, assuming that they are associated with galaxies and averaging over a whole sample, using $\alpha = 0$. Following LTW87 we adopt the Kirshner et al. (1983) galaxy luminosity function and a radius-luminosity scaling law $R \propto L^{0.4}$. With these assumptions, the Holmberg radius of an L_* galaxy is $R_H = 11\ h^{-1}$ kpc, where h is the Hubble constant in units of 100 km s^{-1} Mpc^{-1}. The product $n_o \sigma_o$ has the same value if we assume either absorbing spherical halos around half of all galaxies or absorbing thick discs around every galaxy. The values obtained for R_o for the different samples are given in Table 1. The Lyman limit sample compiled by Bechtold et al. (1984) is also given; it refers to absorbers optically thick at their Lyman limit (LLS).

Table 1. Average size of the absorption line systems

system	$w_{r,lim}(\text{Å})$	$\frac{dN}{dz}$	$\bar{z}$	R_o/R_H	References
MgII	0.6	0.26	0.51	2.8	1
	0.25	0.69	0.5	4.5	1
	0.6	0.85	1.67	3.8	2
	0.3	1.25	1.62	4.6	2
CIV	0.3	1.64	1.70	5.2	3
LLS		1.29	2	4.4	4

References. 1: Tytler et al. (1987). 2: Lanzetta, Turnshek and Wolfe (1987). 3: Bergeron and Boissé (1984). 4: Bechtold et al. (1984).

The size of the absorbing gaseous envelopes is a function of $w_{r,lim}$, as clearly shown in Table 1 for the MgII systems. This is also true for the CIV systems since their equivalent width distribution, is still sharply rising for w_r smaller than 0.3 Å. This excess of small CIV doublet is present in the very high spectral resolution observations of Meyer and York (1987) and Bergeron and D'Odorico (1987). This prevents a straightforward comparison between the sizes given in Table 1. The radius of the ionized envelopes are at least equal to 4 to 5 R_H which is about 3 times larger than the average dimension, $\sim 1.5\ R_H$, of HI discs around spiral galaxies. Further it has been recently discovered that these HI discs have sharp edges at N(HI)$\sim 10^{19}$ cm^{-2}, as discussed by Sancisi (1988) in this volume. Similar information for the ionized absorbing envelopes will be obtained only when a break or turn-over is detected at small w_r in the equivalent width distribution of the CIV or MgII systems.

The cosmological analysis of the metal systems is difficult due to the small size of the samples. Different methods have been used to estimate γ:
1. binning of the data while generating several subsamples to evaluate the uncertainty of the results;

2. building the cumulative distribution and using the Smirnov test statistic; and
3. using the maximum likelihood method.

Smaller uncertainties are obtained when a large redshift interval is sampled. For the MgII systems, LTW87 (see their Figure 8) find

$$\gamma(\text{MgII}) = 2.4 \pm 0.8.$$

Whatever the value of q_o this points towards a positive cosmological evolution. Including the MgII data at very low redshift, $z = 0.13$–0.41, of Boulade, Kunth and Tytler (1987) reduces the uncertainty on γ (Kunth 1987),

$$\gamma(\text{MgII}) = 2.2 \pm 0.5.$$

As will be discussed in a forthcoming section, the MgII systems should constitute a very large fraction of the Lyman limit systems. Thus the smaller value of γ found for the LLS systems by Bechtold et al. (1984),

$$\gamma(\text{LLS}) = 1.3 \pm 1.5,$$

may not be significant, due to the much larger uncertainty, or may reflect some difference arising from the different redshift ranges sampled.

For the CIV systems BB84 found using a cumulative distribution method

$$\gamma(\text{CIV}) = 1.8 \pm 2.0.$$

Adding the sample of Foltz et al. (1986) leads to a very similar value of γ, using either the cumulative distribution or the maximum likelihood method, but reduces a little the uncertainty, $\Delta\gamma = 1.5$. This result, although compatible with no evolution, greatly differs from the unpublished estimate reported by Sargent (1988) in this volume for a new sample, namely $\gamma(\text{CIV}) \sim -1$. Apart from the large uncertainties involved, the origin of this discrepancy will be investigated when this survey becomes available. Further, an unambiguous answer will be obtained for the evolution of both the LLS and CIV systems when a UV absorption line survey is done with the Hubble Space Telescope.

4. IONIZATION LEVEL

The ionization level of the absorbers is well defined only when low and high ions are observable in the optical, i.e. for $z > 1.2$. At these redshifts three phases are present, characterized essentially by the presence and/or absence of the CIV and MgII doublets. We will outline properties related to the opacity of ionizing photons, the multiplicity of the systems and the velocity spread of the subcomponents, the velocity of the absorber relative to that of the QSO, and the redshift range (possible cosmological evolution). We first restrict the discussion to systems at $z > 1.2$.

4.1. Systems of High Ionization Level (H)

In these systems apart from the HI Lyman series, the lines usually detected are the CIV doublet and occasionally the SiIV doublet. In a few cases NV is present even for $z_a < z_e$ systems. Lines from singly ionized and neutral elements are not detected even in high signal to noise spectra. The ionization level can be defined by the value of the line equivalent width ratio

$$\frac{\text{CIV}}{\text{MgII}} = \frac{w_r(\text{CIV}\lambda 1548)}{w_r(\text{MgII}\lambda 2796)}. \qquad (3)$$

For high ionization absorbers the lower limit of CIV/MgII can reach 10 to 40 (see e.g. Bergeron and D'Odorico 1986). These systems are also usually weak with $w_r(\text{CIV})\lambda 1548) < 1$ Å.

The low end of the $w_r(\text{CIV}\lambda 1548)$ distribution has not yet been properly observed. Surveys up to the recent one of Sargent, Steidel and Boksenberg (excluded) had a detection limit of $w_{r,lim}(\text{CIV}\lambda 1548) = 0.3$ Å. Several weaker CIV doublets can be present in one sight-line. Towards PKS2126–15 Meyer and York (1987) found 3 CIV systems with $w_r(\text{CIV }\lambda 1548)$ from 105 to 150 mÅ and Bergeron and D'Odorico (1987), observing in a bluer part of the spectrum, found 3 more systems with $w_r(\text{CIV}\lambda 1548)$ from 58 to 180 mÅ, thus altogether 6 weak CIV systems in a redshift interval $z = 2.45$–2.91 (outside the Ly-α forest). This emphasizes the need for a low w_r CIV survey.

The opacity to ionizing radiation of the absorbers can be derived by comparing statistical properties of CIV and Lyman limit systems, as done by Bergeron, Savage and Green (1987). From the rate of occurrence of optically thick Lyman limits in their LLS sample, Bechtold et al. (1984) infer that systems with $w_r(\text{Ly-}\alpha) > 1.7$ Å have a discontinuity at the Lyman limit. Although the bulk of high ionization level systems have $w_r(\text{Ly-}\alpha) < 1.7$ Å, there are some cases ($\sim$ 10 to 20%) for which $w_r(\text{Ly-}\alpha)$ can reach 2 to 3 Å. It is yet to be investigated whether the latter show weak lines of CII and SiII, although MgII or FeII could remain undetected, as it is the case for the optically thick $z = 1.649$ system in PKS0215+015 (Blades et al. 1985; Bergeron and D'Odorico 1986).

The HI column density distribution is particularly difficult to determine for the high ionization level systems. Most often the HI column density, N(HI), cannot be derived from studies of their optically thin Lyman limit, whereas the Ly-α line is saturated and on the logarithmic part of the curve of growth. Thus it is ususally necessary to estimate the velocity dispersion b from the CIV doublet, assume that the same value of b applies for HI and CIV, and be sure that individual components have been isolated which requires a spectral resolution of at least FWHM = 1 Å (to resolve 2 components at $z = 2$ spanning 80 km s^{-1}). The N(HI) distribution is an important parameter for the investigation of a possible link between the Ly-α forest and the metal systems (Tytler 1987; Bechtold 1987). It should be determined including the weaker CIV systems mentioned above, for which values of $w_r(\text{Ly-}\alpha)$ and N(HI) cover part of the range found for the Ly-α forest sytems.

Some systems with high level of ionization are multiple, but the extreme cases of multiplicity do not belong to this class of absorbers. The velocity spread of the subcomponents is moderate, up to 300 km s^{-1} in some cases, whereas the velocity dispersion of individual components is within the range $b = 5 - 20$ km s^{-1}. The multiple systems are usually those of larger $w_r(\text{CIV}\lambda 1548)$.

4.2. Systems of Mixed Ionization Level (M)

For these absorbers low and high ions are present. The absorption lines are usually strong with $w_r(\text{CIV}\lambda 1548)$ and/or $w_r(\text{Ly-}\alpha) > 2$ Å.

The ionization parameter CIV/MgII has values within the range 0.3–2.5. Therefore, there is a clear discontinuity of CIV/MgII between the high and mixed ionization level systems, and a definite tendency for the systems with increasing $w_r(\text{MgII}\lambda 2796)$ to have a decreasing value of the ionization parameter as found by BB85 and confirmed by LTW87 (see their Figure 11).

As mentioned above from statistical arguments, most if not all of these absorbers should have optically thick Lyman limits. This is verified when observations of the Lyman limit, in the optical or the UV, are available (Bechtold et al. 1984; see also the review of Bergeron, Savage and Green 1987).

These systems often have a complex velocity structure. Extreme cases show 10 to 15 components spanning up to 2000 km s^{-1}. Amongst the subsystems there is often a mixture of high and mixed ionization levels (Pettini et al. 1983; Foltz et al. 1986) and the M systems tend to span a smaller velocity range than the overall complex (LTW87). There are however some multiple M systems with all subcomponents of mixed ionization levels, such as the $z = 1.345$ system in PKS0215+015 (Bergeron and D'Odorico 1986).

The $z_a \sim z_e$ systems, which are typically found in steep-spectrum radio-loud QSOs, often show multiple structure (Foltz et al. 1986; Foltz et al. 1988), including extreme cases such as the two complexes in GC1556+335 (Morris et al. 1986). This tendency to get an excess of $z_a \sim z_e$ systems in steep-spectrum radio QSOs is also present at $z < 1.2$: the 4 $z_a \sim z_e$ systems known at low z are in radio-loud QSOs, 3 being steep-spectrum radio sources (Bergeron, Savage and Green 1987). It is among the $z_a \sim z_e$ systems that the larger spreads of ionization level are found, usually from singly to four times ionized elements, even showing sometimes neutral heavy atoms together with NV as in the $z = 0.401$ system in PKS1912-54 (Bergeron and Boissé 1986). However $z_a \sim z_e$ systems often belong to the H class.

There is no clear difference between the range found for the velocity dispersion of the low and high ions, although for individual components the latter could originate from different regions of the absorbers, as discussed in the photoionization models section, and thus they do not necessarily have identical velocity dispersions.

4.3. Systems of Low Ionization Level (L)

These absorbers have been discovered by LTW87 in their red survey. Absorption lines of CII, MgII and FeII are present but the CIV doublet is undetected. There are only 5 L systems in this red sample. The absorption lines are fairly strong with $w_r(\text{MgII}\lambda 2796)$ in the range 0.4 to 1.7 Å.

The upper limit on the ionization parameter CIV/MgII is 0.5 in 4 cases and 0.2 for the $z = 1.636$ system in PKS0237-233 using the CIV limit given by Boroson et al. (1978). Therefore, there is no clear discontinuity between the M and L systems contrary to what was found for the M and H systems.

About half of the damped Ly-α absorbers (N(HI)$\geq 10^{20}$ cm^{-2}) contain only metals in a low ionization level (Wolfe 1987; Wolfe 1988), and belong to the L class. Do all the L absorbers show damped Ly-α systems? One of the five systems found by LTW87 has a damped Ly-α line. No information on Ly-α is available for the remaining four cases, for which this line is shortward of 3300 Å. UV observations are needed to investigate the link between L and damped Ly-α systems.

Too few L absorbers are known and they have not been observed at high enough spectral resolution to conduct a proper study on their multiplicity. However, for a few multiple systems of mixed ionization level there are components of low ionization state.

In summary, the stronger systems at $z > 1.2$ are:
1. of mixed or low ionization level;
2. optically thick to ionizing radiation; and
3. have large column densities and/or multiple components spanning a large velocity range.

In our Galaxy only the M and L classes have been detected.

We give in Table 2 the number and percentage of the different classes present at $\bar{z} \sim 1.7$ in the survey of BB85, LTW87 and Caulet and York (1987, CY87 in the Table). Contrary to our knowledge prior to LTW87 survey, only about half of the metal-rich absorbers are of high ionization level.

Table 2. Fraction of the different phases at $\bar{z} \sim 1.7$

survey	H	M	L	$w_{r,lim}(\text{Å})$
		Number		
BB85	8	7	0^a	0.6
LWT87	17	13	5	0.3
CY87	23	9	excluded	0.3
		Percentage		
BB85[b]	53	47	0^a	0.6
LWT87	49	37	14	0.3
CY87[b]	72	28	—	0.3

[a] although the survey was biased, since we observed only sight-lines with expected MgII or FeII lines from known CIV systems, the redshift interval sampled was large enough to detect L systems. In fact one of the L system found by LTW87 is marginally present in the BB85 data. This does not apply to the CY87 survey which sampled only small wavelength interval around expected MgII doublets.

[b] the fraction of large w_r(CIV) systems is different in the two samples.

4.4. Systems at Low Redshift

At $z < 1.2$, systems of only mixed or low ionization level are known since no UV absorption line survey is yet possible. Four of these systems, observed with IUE, show a strong discontinuity at the Lyman limit and strong but not damped Ly-α lines. However, there are two fairly weak MgII systems which do not show any break at the Lyman limit. In one case the system is certain, with saturated MgII lines of moderate strength ($z = 0.852$ in AO 0235+164), but the UV data are very noisy (Snidjers et al. 1982). Although UV observations of better S/N ratio are needed for confirmation, this suggests that at low redshift some of the weaker systems of low ionizing level are optically thick to ionizing radiation which should imply a cosmological evolution of the ionization state of the absorbers.

There is an excess of weak MgII systems in the samples of Tytler et al. (1987) and Boulade, Kunth and Tytler (1987) at $0.25 < w_r(\text{MgII}\lambda 2796) < 0.6$ Å, which is not present in the survey of LTW87. Although the latter does not go as deep as the lower z surveys, this could also point towards a cosmological evolution of the ionization state of the weaker systems (see next section).

Some of the low z MgII systems have an associated CIV absorption as shown by IUE observations of bright QSOs (Bergeron, Savage and Green 1987) but the low sensitivity of IUE prevents any conclusion when the CIV doublet is undetected (Bergeron and Kunth 1983). Further low redshift absorbers of high ionization level remain to be discovered.

Complex velocity structure, difficult to reconcile with motions within a single isolated galaxy, are also found in low z MgII absorbers as for the $z = 0.8526$ system in PKS1327-206 with 6 components spanning 560 km s^{-1} (Bergeron, D'Odorico and Kunth 1987: see their Figures 1 and 2).

Multiple systems spanning at most 300 km s^{-1} are also often present in low z MgII absorbers (Bergeron and D'Odorico 1987).

5. PHOTOIONIZATION MODELS

The ionization state of the absorbing gas can be estimated assuming that photoionization is the dominant process. The metal absorption line widths imply a radiation ionizing source at least for ions of higher ionization level, such as CIV, for which the observed velocity dispersions $b = 5$ to 8 km s^{-1} (see Blades 1988, this volume) are incompatible with collisional ionization by thermal electrons. However, for MgII the smallest values found for b are 8 to 10 km s^{-1} which only gives a fairly high upper limit for the gas temperature, T(MgII) $< 1 \times 10^5$ °K, and does not exclude thermal collisional ionization.

The basic assumptions introduced in photoionization models are:
1. ionization by the diffuse UV background;
2. transfer of the ionizing radiation in the regions of the absorber optically thick to the UV flux; and
3. given metal abundances.

For these models, the ionization level of the elements is mainly a function of the ionization parameter alone,

$$U = N(\nu > \nu_{LL})/(nc), \tag{4}$$

where $N(\nu > \nu_{LL})$ is the total flux of ionizing photons incident on the cloud and n the gas density. The values for n and for the size ℓ of the absorber are also a function of the absolute value of the ionizing radiation flux.

A grid of photoionization models were built by Bergeron and Stasinska (1986) for U in the range 10^{-5} to 10 and abundances relative to the solar ones from 10^{-4} to 1. The spectrum of the radiation flux was assumed to be a power law $J(\nu) \propto \nu^{-\alpha}$ with $1 \leq \alpha \leq 2$. Most models involve clouds with constant gas density but constant gas pressure was also investigated. The main results are as follows:
1. There is a strong discontinuity of singly ionized element to CIV ionic ratios at $N(HI) \sim 2 \times 10^{17}$ cm^{-2} for $U > 1 \times 10^{-3}$. This result is independent of the heavy element abundances and the power law index α. For a given value of U this discontinuity is more pronounced for MgII and FeII than for CII and SiII. For the likely value of the diffuse UV background at $z \simeq 2 - 2.5$ due to QSO emission (Sargent et al. 1980; Gondhalekar 1983)

$$J(\nu_{LL}) = 10^{-21} \; erg \; cm^{-2} s^{-1} Hz^{-1} \; steradians^{-1} \tag{5}$$

the inferred value of n for an homogeneous cloud is small: $n = 1 \times 10^{-2}$ cm^{-3} for $U = 2 \times 10^{-3}$.

2. There is a possibility of deriving either U or N(HI) directly from some ionic ratios. A few of these ratios are independent of N(HI), as N(SiIII)/N(SiIV) for $U \leq 10^{-1}$ and N(NV)/N(CIV) for $U \leq 3 \times 10^{-3}$ (see Figures 4 and 5 in Bergeron and Stasinska 1986). Others remain roughly contant for N(HI) $> 3 \times 10^{17}$ cm^{-2}. This occurs when the ions originate essentially from the optically thin part of the absorbing cloud, which is the case for ions at least 3 times ionized whenever $U < 10$. The N(SiIV)/N(CIV) ratio can thus also be used to fix U.

For absorbers optically thin to ionizing radiation, U can be derived from any ionic ratio since they are all independent of N(HI). Knowing several ionic ratios would then help to determine α. Unfortunately these absorbers have typically small column densities and their metal absorption lines are weak and difficult to detect. We will come back to this point while discussing the heavy element abundances for the absorbers of high ionization level.

The N(OI)/N(CII) is a good indicator of large N(HI) and is of particular interest for smaller z absorbers if both Ly-α and the Lyman limit cannot be observed. It is close to unity for N(HI) $> 10^{19}$ cm^{-2} for $U > 10^{-4}$ (see Figure 7 in Bergeron and Stasinska 1986).

3. The range of U found for both the high and mixed ionization level absorbers is two decades around U$\sim(1-3) \times 10^{-3}$. This is true for $1 \leq \alpha \leq 2$, but this range is slowly shifted towards higher values of U when α increases. The newly discovered low ionization level systems do not yet significantly increase the allowed domain for U since the lowest upper limit of 1/5 found for the N(CIV)/N(MgII) ratio is close to values observed for the M systems. For $\alpha = 1.5$ this limit implies $U < 1 \times 10^{-4}$ for N(HI) $= 1 \times 10^{20}$ cm^{-2}. Using the value of $J(\nu_{LL})$ given in equation 5 still leads to low gas densities even in the optically thin case, $n > 0.2$ cm^{-3}.

The H systems have on average an ionization parameter about one order of magnitude higher than the M systems. The domain of U common to both types of absorbers is a function of the abundances. It is small for clouds of constant density, roughly a factor of 2 for solar abundances, $Z_\odot$, and no common range exists for $Z < Z_\odot/20$. For clouds with constant pressure the common range is increased by a factor of about 3.

4. The hydrogen ionization level, averaged over a whole cloud, is high even for large N(HI). Thus the size of the absorbing cloud is a function of the assumed ionization source and cannot be derived directly from N(HI) and n.

For high ionization parameters, the HII/HI ratio sharply drops at N(HI)$\sim 2 \times 10^{17}$ cm^{-2} and the size of the opaque core is small compared to the optically thin outer parts. This is no longer true for $U < 10^{-3}$. Using $J(\nu_{LL})$ given by equation 5, $\alpha = 1.5$ and $U = 2 \times 10^{-3}$ leads to a size for absorbers of mixed ionization level of 10 kpc, wherever $10^{18} <$ N(HI) $< 10^{20}$ cm^{-2}, and the HII/HI ratio, averaged over the whole cloud, is then in the range 250 to 4.

Absorbers of low ionization level, without damped Ly-α line, should have smaller sizes. For $U < 10^{-4}$ and N(HI) $< 10^{20}$ cm^{-2} clouds have dimensions $\ell < 0.2$ kpc.

As mentioned above, N(HI) is not well determined for high ionization absorbers, but some clouds should have N(HI) $\simeq 3 \times 10^{16}$ cm^{-2}. Using $U = 1 \times 10^{-2}$ leads to HII/HI $= 4 \times 10^3$ and $\ell = 20$ kpc.

The sizes inferred for the M and L absorbers are somewhat smaller than derived from statistical analyses of MgII and CIV samples, as discussed above and from direct observations of the absorbing galaxies (see last section). For $H_o = 100$ km s^{-1} Mpc^{-1}, this discrepancy would disappear if $J(\nu_{Ll})$ was roughly 3 times larger than given in equation 5. A more complex spectral dependence of the ionizing flux should also be investigated.

5.1. Sphere of Influence of Individual QSOs

The high rate of dectection of NV absorption in $z_a \sim z_e$ systems can be accounted for by the proximity of an additional ionizing source: the QSO itself. The sphere of influence of a given QSO can be defined as the volume surrounding the QSO in which the ionizing flux from the QSO is larger than the diffuse UV flux. For $J(\nu_{LL})$ given in equation 5 and the "outward only" approximation for the diffuse UV background (Bergeron and Stasinska 1986), the radius of the sphere of influence equals

$$r = 13.3(L_i/10^{48} \ erg \ s^{-1})^{1/2} \ Mpc, \qquad (6)$$

where L_i is the luminosity of the QSO at $\nu_{LL} \leq \nu \leq 10 \ \nu_{LL}$, with a spectral dependence of the UV source taken equal to $J(\nu) \propto \nu^{-1}$. If for $z_a \sim z_e$ absorbers the velocity difference Δv between the absorbing cloud and the QSO is only a consequence of the expansion of the universe, Δv is a simple function of r. For $q_o = 0.5$ and $z_e = 2$, one obtains

$$\Delta v = 3\sqrt{3} \times 10^2 hr = 6900(L_i/10^{48})^{1/2} \ h \ km \ s^{-1}. \qquad (7)$$

Typical cases of $z_a \sim z_e$ systems with strong NV absorption doublets which are most probably influenced by the QSO UV source can be found in the sample of Young, Sargent and Boksenberg (1982). For two such systems, $z = 1.917$ in Q0848+163 and $z = 1.974$ in Q1157+014, Δv is smaller than 1000 km s^{-1} but more extreme cases exist, for example the two systems in the bright ($L_i \sim 10^{48}$ erg s^{-1}) QSO Q1309–056 with Δv of 4600 and 13,300 km s^{-1}.

5.2. Cosmological Evolution of the Ionization Level

If, for $z_a < z_e$ systems, the dominant ionization source is the UV background due to QSO emission, this should result in some cosmological evolution of the ionization level of the absorbers, assuming that the gas density of the absorbing clouds is independent of z. The ionizing diffuse UV flux should decrease by about a factor of 20 ($\alpha = 1$ to 1.5) between z of 2 and 0.5 if the contribution to the UV background is mostly due to QSOs at $z > 2$. For values of n at $z \sim 0.5$ similar to those found at $z \sim 2$ the ionization parameter should then be at most 10^{-3}, and a large majority of the absorbers at $z \sim 0.5$ should have low or mixed ionization levels. This result assumes that strong UV sources within the absorbing galaxy do not dominate over the diffuse UV background, an assumption which should not hold at least at $z = 0$.

Is there any evidence that either the gas density and/or the ionization level of the absorbers are a function of z? The assumption of a lack of correlation between the density of interstellar matter and z is probably verified for the inner parts of galactic discs but this is not so obvious for the outer parts of the ionized envelope. The existence of an intrinsic evolution of $dN(\text{MgII})/dz$ can be accounted for by a decrease in the size of the absorber for decreasing z which would involve an increase of n in MgII envelopes. This would also contribute to lower the absorber ionization level for decreasing z. A possible evidence of cosmological evolution of the ionization level is given by the evolution with z of the fraction of weak MgII systems. The ratio of MgII absorbers with $0.3 < w_r(\text{MgII}\lambda 2796) < 0.6$ Å to the total number of systems with $w_r(\text{MgII}\lambda 2796) > 0.3$ Å is equal to 0.30 and 0.65 in the sample of LTW87 at $\bar{z} = 1.7$, and the combined samples of Tytler et al. (1987) and Boulade, Kunth and Tytler (1987) at $\bar{z} = 0.50$ (Kunth 1987). This evolution will be confirmed when other MgII samples at large z with $w_{r,lim} = 0.3$ Å become available. Since the weaker systems at $z \sim 2$ essentially belong to the high ionization class, the large fraction of weak MgII systems at low z, alone, suggests an evolution of the ionization level. Of course a UV survey of low z CIV systems would directly clarify this point.

6. HEAVY ELEMENT ABUNDANCES

There are a number of uncertainties involved in the determination of the heavy element abundances:
1. the HI content of the absorbing cloud;
2. the ionization state of the heavy elements; and
3. the velocity dispersion of individual components.

For mixed and low ionization level systems the dominant ions are observed for some elements (in particular Si) thus no ionization correction factor is needed to determine the metal content. Further, if N(HI) is very large ($\geq 10^{20}$ cm^{-2}, the HII/HI ratio averaged over the whole cloud should be small: for the photoionization models discussed in the previous section $\overline{\text{HII/HI}} < 1$ for $N(\text{HI}) > 1.5 \times 10^{20}$ cm^{-2} and $U < 0.3$ (constraint always satisfied for M systems at $z \sim 2$).

An example of such cases is the $z = 1.345$ system in PKS0215+015 for which data over a large wavelength range (Blades et al. 1985) and at high spectral resolution (Bergeron and D'Odorico 1986) are available. In this system the averaged HII/HI ionic ratio is roughly equal to 2 and the metal abundances are equal to 0.1–0.2 $Z_\odot$. Similar values have been derived for other large N(HI) systems.

The absorbing galaxy associated with mixed ionization level absorbers has been identified in several cases at $z < 1$ (see next section), and most probably all M and L systems are gaseous envelopes around galaxies whatever their redshift. The abundances inferred for these systems are within the range found for the outer part of nearby galaxies from the analysis of HII regions at radial distances close to R_H.

The situation is not as clear for the high ionization level absorbers. Although the intervening assumption is strongly supported by the statistical analysis of CIV samples, it has not yet been unambiguously shown that, as for the M and L systems, they are associated with galaxies (this is a harder task since only H systems at $z > 1.1$ are known). If they do, there should be some continuity in the metal abundances between those different classes of absorbers.

At present, the heavy element abundances of the H systems are poorly known. High spectral resolution data give the velocity dispersion, but the ionization degree of the metals and both the HI and HII content are difficult to ascertain. What are the ions most easily detectable aside from CIV for absorbers with $N(\text{HI}) < 2 \times 10^{17}$ cm^{-2}? Assuming a relative abundance between the heavy elements equal to the solar value, Bergeron and Stasinska (1986) found that wherever $U < 10^{-1}$:
1. the N(CII)/N(SiIV) ratio is always much larger than unity; and

2. the $N(CII)/N(NV)$ ratio is smaller than unity for $U > 7 \times 10^{-3}$, with $N(CII)/N(CIV) \simeq N(NV)/N(CIV) \simeq 0.1$ at $U = 7 \times 10^{-3}$.

Thus, to get the ionization level of the H systems ($U > 10^{-3}$) optically thin to ionizing radiation, it appears necessary to detect either CII or NV. This requires high signal to noise ratio data with $w_{r,lim}$ roughly one order of magnitude smaller than $w_r(CIV\lambda 1548)$. Such surveys are just beginning to become available (Meyer and York 1987; Bergeron and D'Odorico 1987).

For the weak $z_a < z_e$ systems of high ionization level, there is a possible confusion between the NV doublet and the Ly-α forest which may account for the rareness of reported NV detections. If this bias was indeed preventing the discovery of NV doublets associated with CIV systems, this could explain why the range found for the ionization parameter, or density of the absorbers, is only two decades for the H and M systems at $z \simeq 2$. The existence of some observational bias due in part to confusion with the Ly-α forest for the CII absorption has already been demonstrated by the recent discovery of the L sytems.

Even if only CIV and HI are observed, it is however possible to place some rough limit on $C/C_\odot$ from the mere detection of CIV. In the optically thin case the $N(CIV)/N(HI)$ ratio goes through a maximum equal to 0.6 ($C/C_\odot$) at $U = 2 \times 10^{-2}$ (Bergeron and Stasinska 1986). If the velocity dispersion of both HI and CIV is given by the observations, the estimate of $N(HI)$ and $N(CIV)$ leads to a lower limit on $C/C_\odot$. Using the column densities derived by Atwood, Baldwin and Carswell (1985) for the $z = 2.24626$ and 2.26187 weak CIV systems in Q0420–388, one gets $C/C_\odot > 1.6 \times 10^{-4}$ and 6.1×10^{-2}. Although the uncertainties on $N(HI)$ are large, this shows that meaningful limits can be obtained.

Although very metal deficient, the Ly-α forest clouds may not be primordial gas. If so an important question would be the continuity of $Z/Z_\odot$ between the H metal-rich systems and the Ly-α forest. The heavy element abundances can be more easily estimated for optically thick Ly-α forest absorbers, since, whatever the ionization state in their optically thin outer regions, CIV and/or CII absorption should be present in the optically thick core and lines from these ions can be outside the Ly-α forest range. Ly-α forest absorbers of very large $N(HI)$ are rare but do exist and a good case is the $z = 2.0769$ system in Q2206–199N in which Robertson, Shaver and Carswell (1983) have detected a weak CIV line. New high S/N ratio data of this QSO confirm the presence of a weak CIV doublet and could give some clue to the value of $Z/Z_\odot$ in Ly-α forest clouds (Carswell, Robertson and Shaver 1988).

7. PROPERTIES OF GALAXIES GIVING RISE TO MGII ABSORPTION SYSTEMS

We briefly summarize the results of our search for galaxies giving rise to MgII absorption systems, a more detailed description of the results being given in another review (Bergeron 1987).

The first attempt to detect an absorbing galaxy associated with low z MgII absorbers was made by Weymann et al. (1978), but the sensitivity of the detectors used then was too low to unambiguously find these galaxies. Later another unsuccessful attempt was made by Carswell et al. (1984) for the $z = 0.356$ and 0.359 MgII absorbers in Q1101–264, although more sensitive detectors were used. Recently a MgII system, $z = 0.430$ in PKS2128–12, was successfully identified (Bergeron 1986).

To understand whether that detection was fortuitous or not, we have estimated the expected average magnitude of an absorbing galaxy and its average angular separation from the sight-line to the QSO. We have used a Schechter type luminosity function with $M_* = -20.6$ for $h = 0.5$, a scaling law $R \propto L^{5/12}$ and assumed that half of the galaxies have spherical halos. We find that 50 % of the absorbing galaxies must then be brighter than $m = 22.1, 22.8$ and 24.5 for $z = 0.2, 0.4$ and 0.8, respectively. Further, in roughly 50% of the cases the expected angular separation between the galaxy center and the QSO must be at least 8.5, 5.4 and 3.8 arcsec for $z = 0.2, 0.4$ and 0.8, respectively. Thus, the absorbing galaxies should be easily detectable with present day techniques and the properties of the galaxy G2128-12 at $z = 0.430$, which is 8.6 arcsec away from the sight-line to the QSO and has a magnitude $m(r) = 21.0$, are close to the predicted ones.

Table 3. Properties of MgII Absorbing Galaxies

QSO	z_a	type	θ (arcsec)	R/R_H	$M(r)$	reference
0151+04	0.160	em	6.4	1.1	−20.7	1
		em	10.9	1.9	−19.6	1
1127−14	0.313	em	9.5	2.7	−21.2	2
2128−12	0.430	em	8.6	2.9	−21.3	3
1511+10	0.437	em	6.7	2.4	−20.7[a]	2
1038+06	0.441	em	9.6	3.3	−21.3[a]	2
0109+20	0.535	abs	7.0	2.7	−21.0	2
2145+06	0.790	em	5.9	2.7	−21.7	2
0952+17	0.238	no galaxy found				2
1229−02	0.395[b]	no galaxy found				2
1332+55	0.373	abs	5.0	1.6	−21.3	4
1209+07	0.393	HII	7.1	2.3	−20.2	5
0235+16	0.524[b]	HII	2.3	0.9		6
1101−26	0.356	no galaxy found				7
	0.359					

[a] V magnitude
[b] sytems with 21 cm absorption detected
references − 1: Bergeron et al. (1987c). 2: this work, 3: Bergeron (1986), 4: Miller, Goodrich and Stephens (1987), 5: Cristiani (1987), 6: Smith, Burbidge and Junkkarinen (1977), 7: Carswell et al. (1984).

We summarize in Table 3 the results of our survey and include other searches made with similar sensitivity. The linear separation between the galaxy and the sight-line to the QSO is given in units of the Holmberg radius at $z = 0$ and the type refers to the properties of the galaxy spectrum. One of the cases previously published (Smith, Burbidge and Junkkarinen 1977) concerns the $z = 0.524$ system in the BL Lac object 0235+164 ($z_e = 0.94$); we consider that it belongs to the intervening galaxy sample although the authors favor the alternative assumption of a nebulosity associated with the BL Lac object.

There are altogether 13 $z_a < z_e$ MgII systems for which a deep search for the galaxy has been made, out of which 9 belong to our sample. In 10 cases, there is a successful identification and the redshift range of the absorbing galaxies is large, from 0.16 to 0.79. The angular separation is always smaller than 10 arcsec and the galaxies are bright, M(r) < −19.8. In our survey the limiting m(r) magnitude was 2 to 3 magnitudes lower than observed for the absorbing galaxies. Since our detection rate was higher than expected if faint galaxies, as the Magellanic Clouds, had gaseous envelopes, we infer that only bright galaxies, L≥ L_*/3, can give rise to MgII absorption systems (a detailed analysis is given in Bergeron 1987). Another striking property of the galaxies with gaseous envelopes is their spectral type: 8 out of 10 show emission lines, among which 2 resemble giant HII regions with very strong [OII] and [OIII] emission lines and a weak continuum. Further, all the spectra, even when only absorption lines are present, show a very blue continuum with $J(\lambda)$ roughly constant shortwards of [OII]λ3727. This is true even for the highest z galaxy G2145+06 for which a constant $J(\lambda)$ continuum is detected down to $\lambda_r = 2140$ Å. This suggests recent stellar formation activity for all the absorbing galaxies identified so far.

The smallest angular separation found concerns a system ($z = 0.524$ in BL 0235+ 164) with detected 21 cm absorption. The linear separation of 0.9 R_H is in agreement with the average size found for HI discs R(21 cm)~1.5 R_H (see Sancisi 1988, this volume). The detection of this galaxy was done under good seeing conditions and when the BL Lac object was not in a very bright state. We suggest that for the other system with 21 cm absorption, $z = 0.395$ in PKS1229–02, the non detection of the absorbing galaxy is a consequence of its small angular separation with the QSO. Indeed 21 cm detections are rare for MgII systems (Briggs and Wolfe 1983), thus the 21 cm absorption regions are smaller than MgII envelopes and observations either under excellent seeing or with the Hubble Space Telescope are then needed to identify 21 cm absorbing galaxies.

The high luminosity and specific spectral type of the MgII absorbing galaxies suggest that the fraction of galaxies with large MgII gaseous envelopes is less than 50%. To evaluate this number with some accuracy requires finding high-z galaxy-QSO pairs with an angular separation less than 10 arcsec, making sure that the galaxy and the QSO do not belong to the same cluster, i.e. $z_g \ll z_e$, and then searching for MgII absorption in the QSO spectrum. One of the absorbing galaxy presented in Table 3 was the first case of such a search (Bergeron et al. 1987c).

There is also the open question of gaseous envelopes of high ionization level. At present systems of high ionization level are only known at $z > 1.1$ and identifying L_* galaxies at $z > 1.1$ is a very difficult task. Thus, one needs an ultraviolet CIV survey with the Hubble Space Telescope to find low z systems with only high ions.

REFERENCES

Atwood, B., Baldwin, J. A., and Carswell, R. F. 1985, *Ap. J.*, **292**, 58.
Bechtold, J., Green, R. F., Weymann, R. J., Schmidt, M., Estabrook, F. B., Sherman, R. D., Wahlquist, H. D., and Heckman, T. M. 1984, *Ap. J.*, **281**, 76.
Bechtold, J. 1987, In *High Redshift and Primeval Galaxies*. Eds. J. Bergeron, D. Kunth and B. Rocca-Volmerange. Editions Frontières, in press.
Bergeron, J. 1986, *Astr. Ap. (Letters)*, **155**, L8.
──────── . 1987, In *Evolution of Large Scale Structure in the Universe*, IAU Symp. 130. Eds. J. Audouze and A. Szalay, in press.
Bergeron, J., and Kunth, D. 1983, *M.N.R.A.S.*, **205**, 1053.
Bergeron, J., and Boissé, P. 1984, *Astr. Ap.*, **133**, 374.
──────── . 1986, *Astr. Ap.*, **168**, 6.

Bergeron, J., and D'Odorico, S. 1986, *M.N.R.A.S.*, **220**, 833.
Bergeron, J., and Stasinska, G. 1986, *Astr. Ap.*, **169**, 1.
Bergeron, J., Savage, B., and Green, R. F. 1987, In *The Scientific Accomplishments of the IUE*, ed. Y. Kondon. Reidel, Dordrecht, p. 703.
Bergeron, J., D'Odorico, S., and Kunth, D. 1987, *Astr. Ap.*, **180**, 1.
Bergeron, J., Boulade, O., Kunth, D., Tytler, D., Boksenberg, A., and Vigroux, L. 1987, *Astr. Ap.*, in press.
Bergeron, J., and D'Odorico, S. 1988, in preparation.
Blades, J. C., Hunstead, R. W., Murdoch, H. S., and Pettini, M. 1985, *Ap. J.*, **288**, 580.
Boissé, P., and Bergeron, J. 1985, *Astr. Ap.*, **145**, 59.
Boroson, T., Sargent, W. L. W., Boksenberg, A., and Carswell, R. F. 1978, *Ap. J.*, **220**, 772.
Boulade, O., Kunth, D., and Tytler, D. 1988, in preparation.
Briggs, F. H., and Wolfe, A. M. 1983, *Ap. J.*, **268**, 76.
Carswell, R. F., Morton, D. C., Smith, M. G., Stockton, A. N., Turnshek, D. A., and Weymann, R. J. 1984, *Ap. J.*, **278**, 486.
Carswell, R. F., Robertson, J. G., and Shaver, P. A. 1987, in preparation.
Caulet, A., and York, D. G. 1987, in *QSO Absorption Lines: Probing the Universe, A Collection of Poster Papers*, eds. J. C. Blades, C. A. Norman and D. A. Turnshek, (ST ScI Publication), p. 76.
Cristiani, S. 1987, *Astr. Ap. (Letters)*, **175**, L1.
Foltz, C. B., Weymann, R. J., Peterson, B. M., Sun, L., Malkan, M. A., and Chaffee, Jr., F. H. 1986, *Ap. J.*, **307**, 504.
Foltz, C. B., Chaffee, Jr., F. H., Weymann, R. J., and Anderson, S. A. 1988, *This volume*.
Gondhalekar, P. M. 1983, *M.N.R.A.S.*, **204**, 997.
Kirshner, R. P., Oemler, A., Schechter, P. L., and Schectman, S. A. 1983, *Astron. J.*, **88**, 1285.
Kunth, D. 1987, in *High Redshift and Primeval Galaxies*. Eds. J. Bergeron, D. Kunth and B. Rocca-Volmerange. Editions Frontières. in press.
Lanzetta, K. M., Turnshek, D. A., and Wolfe A. M. 1987, *Ap. J.*, **322**, 739.
Meyer, D. M., and York, D. G. 1987, *Ap. J. (Letters)*, **315**, L5.
Miller, J. S., Goodrich, R. W., and Stephens, S. A. 1987, *A. J.*, **94**, 633.
Morris, S. L., Foltz, C. B., Weymann, R. J., Schectman, S., Price, C., Boroson, T. A., and Turnshek, D. A. 1986, *Ap. J.*, **310**, 40.
Pettini, M., Hunstead, R. W., Murdoch, H. S., and Blades, J. C. 1983, *Ap. J.*, **273**, 436.
Robertson, J. G., Shaver, P. A., and Carswell, R. F. 1980, *XXIV Colloque International de Liège*. ed. J. P. Swings, p. 602.
Sancisi, R. 1988, *This volume*.
Sargent, W. L. W., Young, P. J., Boksenberg, A., and Tytler, D. 1980, *Ap. J. Suppl.*, **42**, 41.
Sargent, W. L. W. 1988, *This volume*.
Smith, H. E., Burbidge, E. M., and Junkkarinen, V. T. 1977, *Ap. J.*, **218**, 611.
Snijders, M. A. J., Boksenberg, A., Penston, M. V., and Sargent, W. L. W. 1982, *M.N.R.A.S.*, **201**, 801.
Tytler, D. 1987, *Ap. J.*, **321**, 49.

Tytler, D., Boksenberg, A., Sargent, W. L. W., Young, P., and Kunth, D. 1987, *Ap. J. Suppl.*, **64**, 667.
Weymann, R. J., Boroson, T. A., Peterson, B. M., and Butcher, H. R. 1978, *Ap. J.*, **226**, 603.
Weymann, R. J., Williams, R. E., Peterson, B. M., and Turnshek, D. A. 1979, *Ap. J.*, **234**, 33.
Wolfe, A. M. 1987, *Proc. Phil. Trans. R. Soc.*, in press.
Wolfe, A. M. 1988, *This volume*.
Young, P., Sargent, W. L. W., and Boksenberg, A. 1982, *Ap. J. Suppl.*, **48**, 455.

DISCUSSION

E. M. Burbidge: Don York raised the question about the presence of MgI in the low-ionization systems, and you commented that it's a weak line. It is seen in some systems.

York: The only point I was making is that it is almost always seen at some level. It's very common; MgI is present, but it's just not strong.

E. M. Burbidge: Yes, now, do you know how many of these systems in your samples are radio sources in which 21 cm is detected?

Bergeron: Only two. Further I know that Art Wolfe is trying very hard to find 21 cm absorption as well as Patrick Boissé, John Dickey and collaborators. The latter have a poster at this conference. We began to have the feeling that for those systems with an HI column density of only 10^{19} cm^{-2} there is no hope to detect 21 cm absorption. Higher column densities are needed, N(HI) $\sim 10^{20}$–10^{21} cm^{-2}.

Concerning MgI I don't think, unfortunately, that it is a good indicator of high N(HI), because charge-transfer dominates the ionization equilibrium and the MgII/MgI ionic ratio should always be close to 100 in high N(HI) absorbers. Indeed, for all the absorption systems with a MgI detection, you get a very small span in the MgII/MgI column density ratio from 50 to 100 or 200. I think that a good indicator of 21 cm absorption is OI, but unfortunately it is only available for high redshift absorbers and not for those which we are investigating.

Blades: In your fields, how many galaxies do you commonly find, and how many redshifts have you obtained of those?

Bergeron: I would say that when I started, I got the spectrum of only one object; now usually I obtain spectra for at least two objects per field. In those where no galaxy was found, I tried very hard, and I obtained spectra for up to five objects in the field of the lowest redshift absorber in Q 0952+17. The thing which is striking is that there is no object within 10 arcsec, as also found by Carswell *et al.* in 1984 for Q 1101−264. The closest object is 15 to 20 arcsec away. The 5 galaxies observed in the field of Q 0952+17 and the 3 by Carswell *et al.* in the field of Q 1101−264 have redshifts different from that of the absorber, with several redshifts per field. For systems with 21 cm absorption there may be some chance of not finding the galaxy. In fact, for 0235+16 where there is a 21 cm absorption, it implies that the galaxy is right on the line-of-sight to the QSO and cannot be detected. It could be resolved only with HST. I am now including all previous negative cases in my survey, because when I started this program I had always positive identification and other researchers in the past had always negative ones, which I didn't understand. Some of them turned out to be positive as Q1038+06, other confirmed negative as 0952+17. Finally for some there is a candidate but the spectroscopy has to be completed.

G. Burbidge: How do you define a galaxy?

Bergeron: I define a galaxy as an object, being resolved, and having absorption and/or emission lines.

G. Burbidge: Well, for the case of 0235+164, there is no evidence that that's a galaxy!

Bergeron: OK, typically the spectrum would have a Balmer discontinuity, CaII H and K in absorption and possibly O[II] emission. It is true that the one published by S. Cristiani and 0235+164 look more like HII regions ionized by bright stars: like a starburst galaxy spatially resolved. I want in fact to re-observe 0235+164.

G. Burbidge: So, going back to your table where you list them all, in how many of them can you actually see evidence for stars?

Bergeron: All of them except those marked HII region. Where I put spiral it is just a definition which means both stellar CaII H and K in absorption and O[II] in emission is observed; elliptical means only CaII in absorption is observed; and HII region means there are strong O[II] and O[III] emission lines and a weak continuum with no absorption lines.

G. Burbidge: Well, that is your definition. That is your belief.

Kronberg: I just want to comment that my collaborators and I have just done a rotation measure of 1229–02 which does have faint emission around the QSO, extended enough so that we can actually get a Faraday rotation map across it, and the intent of this was to try to detect the intervener in rotation measure. And, in fact, we did see some interesting variations in rotation measure across the source. I can show the map if anybody is interested. But at the moment it is not entirely clear that we can assign the measure entirely to the intervener. So it's a little bit of an open question. But I just wanted to mention that we have tentatively discovered something possibly interesting that is extended between 1229 and us.

Bergeron: A short comment. That's very interesting, because as I said we expect the 21 cm absorber to be fairly close on the sky to the quasar, and not easily detectable at La Silla where usually the seeing is moderate. These cases should be done in very good seeing and they would then most probably be detected.

Ostriker: Perhaps a naive question. In your metal line systems, you find low states of ionization. When we look at the interstellar medium towards stars, we also see absorption by dust. Now, have you checked as to the amount of dust you would expect if the ratio of metal line absorption to dust in these systems were the same as it is in our Galaxy? And at that level would it have been detected?

Bergeron: In fact, Boissé and myself (IAP preprint no. 195) have searched for dust in the BL Lac object 0215+015 which had a spectrum void of any emission lines. At the time of the observation it was in a low state and weak emission lines were present. These lines have also recently been reported by Foltz and Chaffee (1987, *A. J.*, 93, 529). We obtained a spectrum of very high signal to noise in the continuum with the aim of finding either a bending of the spectrum or the 2200 Å diffuse absorption band. We find no deviation of the continuum from a power law to a very high degree. We just do not see any dust.

Blades: Let me just make a comment on the $z = 1.345$ system in 0215+015. We find for this system the strongest CI absorption seen in any QSO. It has a very large equivalent width. We also found MgI, and there is very strong MgII; all the low ions are present. This system looks very much like gas in our own Galaxy.

Ostriker: My question is, if you calibrated it, given the absorption line strength, how much dust would you have expected to see? And is that amount precluded?

Bergeron: It is excluded that the dust to gas ratio is like that in our Galaxy. It is at least 7 times smaller. LMC or SMC type dust are possible. The best cases now to search for dust are those of Wolfe, where $N(HI) \sim 10^{21}$ cm^{-2}.

Wolfe: But do you know the MgII column density? A problem here is that it is very difficult to get the column density of metals.

Bergeron: No, for this system in 0215+015 it is very accurately known. In this case there is no doubt about the column density.

Wolfe: Oh, OK. But there are problems for high column density systems in general.

Ostriker: Changing the calibration for the SMC won't help. Because we are comparing with the amount of metals. So it's not a question of metal depletion.

Morton: That only samples the 2200 Å band type of dust, but there is the rest of the dust that knocks off the ultraviolet.

Boksenberg: But fitting a power law to the continuum would tell you that.

Bergeron: It's a problem. Towards 0215+015 the HI column density is not large enough to produce a significant bending of the spectrum in the UV and Galactic type dust cannot be excluded from the overall shape of the continuum alone.

Savage: I would like to change the topic to the hot gas. Features you don't see can sometimes be as important as features you do see. And you didn't say anything about OVI. I fully realise it's in the Ly-α forest, but it is *the* best hot gas tracer we have. And if there are hot halos out there, you might expect occasionally to see OVI absorption. So I would, as a start, encourage everyone who is looking at these lines, to make a comment on whether they can say anything about OVI, even an upper limit is very valuable. What is the situation here?

Bergeron: I don't know. I do not have any observations in the OVI region. You had better ask those who look a the Ly-α forest region.

Savage: Occasionally, it is relatively clear in the Ly-α forest. It is very important to set limits in those situations where you can.

Bergeron: I agree with you. But my high spectral resolution sample concerns only wavelength ranges to the red of the Ly-α forest.

E. M. Burbidge: About the object that you labelled HII in 0235+164, all you see there is a little companion object that is indistinguishable from the star about 2.5 arcsecs away from the BL Lac object. The interesting thing is that its emission lines do occur in the spectrum of the BL Lac objects itself. So there certainly is gas around but no evidence for stars.

Bergeron: I think this object must be re-observed in good seeing condition to try to resolve it. It is difficult this object being so close to the BL Lac.

E. M. Burbidge: We have a paper in press about this object.

Bergeron: I would like to see it.

Wolfe: I wanted to get back to the MgI again. I was very interested in the one object where you saw MgII absorption, but which was optically thin in the Lyman limit. And what interested me there is the way you explain it by a lower ionization parameter, right? My question then is, you lower the ionization parameter for a reasonable spectrum, then you would expect more MgI than usual. It seems to me you would expect MgI to MgII to be harder than in the normal case. My question is have you looked into that?

Bergeron: Not really, because I have not yet pushed my grid extensively for an ionization parameter smaller than 2×10^{-4}, but even for $U = 2 \times 10^{-4}$ the MgI to MgII ratio is still only 1%.

HIGH RESOLUTION OBSERVATIONS OF HEAVY-ELEMENT SYSTEMS

J. Chris Blades
Space Telescope Science Institute
Baltimore

Abstract. Observations at high resolution ($R > 10^4$) can provide a detailed understanding of the various absorption phenomena in QSOs. At the lowest redshifts, there is direct evidence that normal galaxies produce heavy-element absorption; in these cases, QSOs can be used as effective probes of the outer regions of galaxies. New work on redshifted MgII suggests that active galaxies, such as starburst and emission-line galaxies, also produce the heavy-element absorption. Extension of these studies to $z \sim 1$ is urgently needed. At higher redshifts, the strong CIV systems break up into many narrow components spread over large velocity ranges. Line width measurements show that at least some clouds consist of warm (5×10^4 °K) photoionized gas, and large variations in the strengths of species occur from component to component. At $z > 3$, heavy-elements continue to be found in high column density HI system, although with low abundance. New studies of low column density HI clouds which make up the Ly-α forest suggest that some contain heavy-elements and, thus, are not primordial.

1. INTRODUCTION

Our current understanding of redshifted absorption is based largely on spectroscopic observations taken at low resolution, typically between 1.5 – 4Å FWHM. Such observations identify the absorbers, give a general idea of ionization level, and yield information that is statistical in nature. However, redshifted lines—because they are interstellar-like—have small intrinsic widths, with b-values ($b = \sqrt{2}\sigma$) that are often less than 10 km s^{-1}. Frequently, the lines show complex structure, consisting of many components spread over large velocity ranges, typically from 100 to 1000 km s^{-1}. Morever, large variations in the strengths of components occur from species to species. High resolution is needed for precise measurement of narrow lines to determine the physical conditions of the absorbing gas, which otherwise can not be established. Figure 1 compares two observations taken at widely different resolutions of the same redshifted CIV doublet. The gain in information provided by the high resolution observation is dramatic: the apparently single CIV doublet at low resolution (typical of that used in the better CIV surveys) is shown to consist of at least three separate components for which accurate column densities, b-values and relative

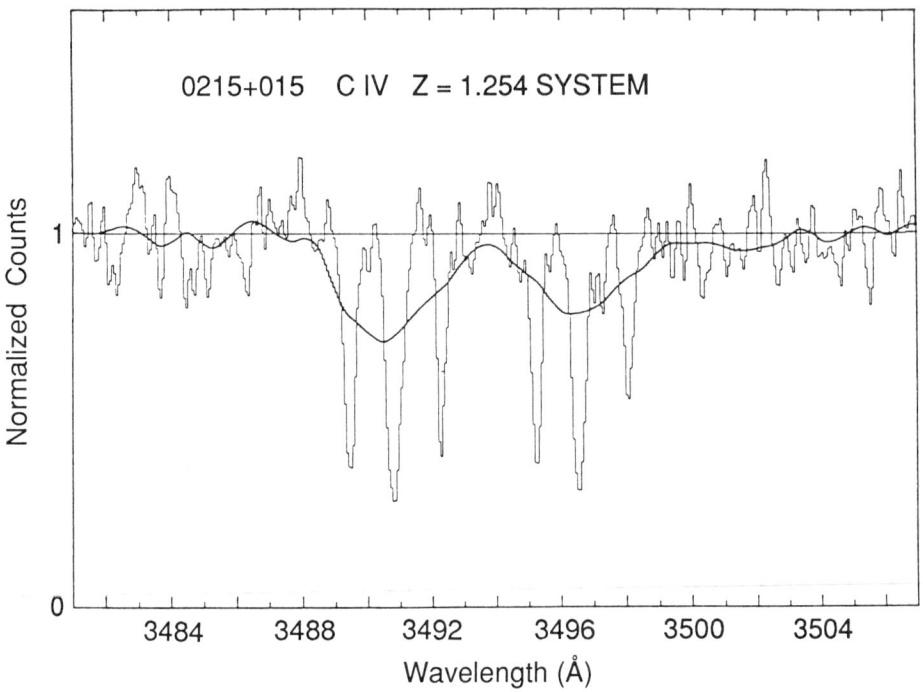

Figure 1. *A comparison of the same CIV doublet in 0215+015 obtained at two different resolutions, first shown by Pettini (1985). The lower resolution spectrum (from Blades et al. 1982) is at 120 km s^{-1} FWHM and only reveals the presence of a single CIV doublet. Our more recent spectrum has about six times the resolution and reveals more precisely the nature of the absorption.*

velocities can be determined.

Since the discovery of redshifted absorption over twenty years ago by Burbidge, Lynds and Burbidge (1966) and Stockton and Lynds (1966) in 3C 191, a major goal of any study has been to establish the origin of the features. Although historically divided into two distinct camps—the so-called intervening versus intrinsic hypothesis—redshifted absorption may originate through several, perhaps many, unrelated phenomena, and any particular QSO sight-line may penetrate more than one example. At very low redshift, there is unambiguous evidence that normal galaxies produce intervening absorption, and there are examples, equally convincing, of associated absorption (Foltz et al. 1988). Other systems have been proposed to account for the lines, such as dwarf irregular galaxies (Silk 1985), starburst galaxies (York et al. 1986; York 1988), normal galaxies in clusters (e.g. Pettini et al. 1983), cooling flows in clusters (Crawford et al. 1987), proto-galaxies (Rees 1988; Wolfe 1988) and even ejecta from BAL QSOs (Turnshek 1988). Many of these proposals were discussed during the Workshop. The discovery of complex absorption that may be common to widely separated QSOs (Jakobsen et al. 1986, 1987; Sargent 1988) could cause all the ideas to be revised. The challenge is to establish which phenomenon is occurring for any particular redshift system and then use the data as a probe of that system.

Apart from a few instances, it is difficult to point to a specific example and say, 'here is a clear case of absorption by the disk of spiral galaxy—now, what does this say about the galaxy in question?' or, 'here is an unambiguous example of absorption caused by gas in a proto- galaxy, what can we deduce about abundances at that epoch?' Important details are not established, in part because of a lack of high resolution observations. The recent supernova in the Large Magellanic Cloud provides a topical and dramatic illustration. Figure 2 shows several interstellar profiles found towards SN1987A. The sight-line penetrates low density, high latitude gas in the outer regions of our Galaxy and gas near 30 Doradus before reaching the supernova. A rather unexceptional line-of-sight, by comparison with many of the exotic systems mentioned previously, and yet high resolution observations (~ 3 km s^{-1}) by Vidal-Madjar et al. (1987) reveal at least 24 separate velocity components spread over 300 km s^{-1}! At the lower resolution of IUE ($\sim 20 - 30$ km s^{-1}) the number reduces to about 9 or 10 components, integrating over all detected species (Blades et al. 1988). However, at the much lower resolution typically used for QSO absorption line studies (100 km s^{-1} or lower) the velocity structure would reduce to 2 or 3 components.

The nature of the absorbing gas can not be established on the basis of low resolution work, as emphasized time and again during the Workshop. Here is the crux of the problem faced by observers of redshifted absorption. For progress to be made, more emphasis needs to be placed on high resolution work. This review highlights studies which are based on high resolution work and which point the way to future and more comprehensive observations. To provide order to a rather disparate set of observations, the review is structured into low, intermediate, and high redshift sections.

2. LOW REDSHIFT ABSORPTION: QSO-GALAXY SYSTEMS

At the lowest redshifts, i.e. $z < 0.3$, an important aspect of high resolution work is the study of intervening CaII $\lambda\lambda 3933.7$, 3968.5 and NaI $\lambda\lambda 5890.0$, 5895.9 in QSOs that are directionally close to nearby galaxies on the plane of the sky. There have been a number of notable successes in detecting galaxian absorption which are reviewed in the section that follows.

The first detections were important because they provided direct evidence that the outer regions of galaxies can produce the narrow absorption lines found in QSOs, which is crucial evidence in favour of an intervening origin. High resolution observations are necessary because the CaII doublet is weak and because precise velocity measurements are needed to establish where the gas occurs in the intervening galaxy.

However, the subject has not been without controversy. One reason is that non-detections have been slow in getting into print, leading Morton, York and Jenkins (1986) to opine, 'well, you always show us the successes but never say how many other cases there are which do not show galaxian absorption'. Non-detections are as significant as detections: without a large and unbiased sample the types of systems and the typical extent of intervening gas can not be established, which are major goals of the work.

QSOs can be effective probes of galaxian absorption. As the emphasis moves from detection to detailed study, results on the intervening absorption should be combined with direct observations of the galaxy in question, including derivation of a rotation curve, not only to establish the location of the gas but also to derive its chemical

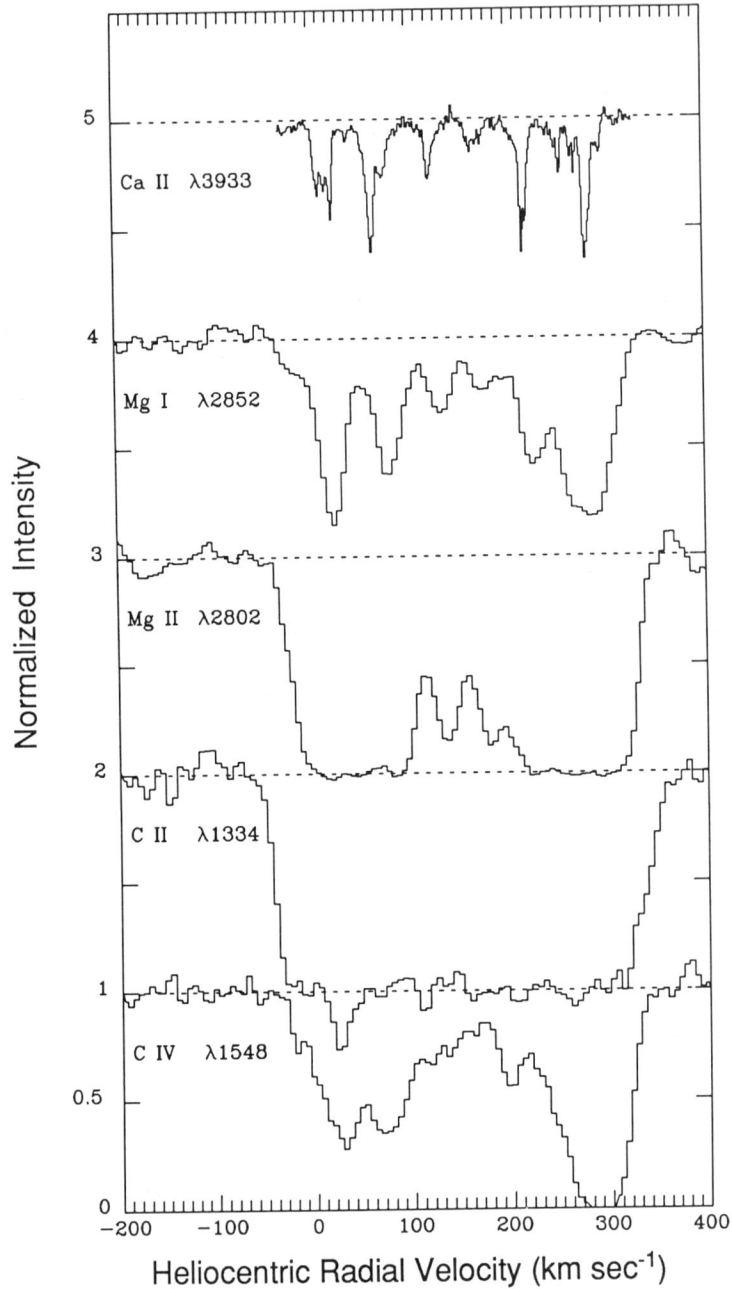

Figure 2. *A comparison of selected interstellar profiles found towards SN1987A in the LMC. The optical CaII K profile is taken from Vidal-Madjar et al. (1987) and the four ultraviolet profiles are from the spectral atlas of Blades et al. (1988). The resolution of the optical spectrum is about 10 times higher than the ultraviolet spectra.*

composition and physical condition.

2.1. Detections

There are 5 QSO–galaxy pairs which show intervening CaII and 4 of these show NaI.

2.1.1. 3C 232 and NGC 3067

was the first published case of a foreground galaxy (NGC 3067) producing absorption in the spectrum of a QSO (3C 232). The precursor observation, by Haschick and Burke (1975), had found HI absorption at 1418 km s^{-1} in the direction of this radio-loud QSO. They attributed the absorption to gas associated with NGC 3067, an inclined spiral galaxy, with a velocity of 1456 km s^{-1} (Rubin, Thonnard and Ford, 1982), which is situated 1.9 arcmin from 3C 232 on the plane of the sky. Prompted by this discovery, Boksenberg and Sargent (1978) searched for galaxian absorption in the optical spectrum of 3C 232, and found the CaII doublet at 1406 km s^{-1}. They searched for but did not find NaI. In their spectrum the CaII lines are broadened with a FWHM = 90 km s^{-1}, after allowing for the instrumental width, and they estimated that $b \sim 50$ km s^{-1}. These values strongly suggest the presence of unresolved velocity structure.

Of course, the significance of the first observations went far beyond actual detection. The most dramatic aspect is that the line-of-sight to 3C 232 passes well outside the optical extent of NGC 3067 (see Plate I) which has a diameter $D_{25} = 2.46$ arcmin (corresponding to an isophote of $B = 25$ mag arcsec^{-2}). CaII is ubiquitous in interstellar gas in our Galaxy; and, as Boksenberg and Sargent (1978) pointed out, its strength in NGC 3067 was consistent with that seen for galactic halo gas, based on studies of high-latitude stars. They emphasized that the discovery added support to the idea that QSO absorption lines arise in the extended halos of normal galaxies as proposed by Bahcall and Spitzer (1969).

Recently, John Stocke (personal communication) using high resolution facilities at the MMT has re-observed the optical lines in 3C 232. Significantly, he resolves CaII into three velocity components; and he also detects NaI in at least two components. The strongest feature of both species has a velocity similar to the HI, while the other Ca II components are displaced by −46 and +128 km s^{-1}. HI counterparts have not been reported for the new components. Although Stocke's results are preliminary and must await more data, his work should resolve the uncertainty of whether the absorption occurs in the disk or halo (or both) of NGC 3067.

In galactic gas, the ratio N(CaII)/N(NaI) can be diagnostic of halo or disk gas, having a value ~ 0.2 in low-velocity disk gas and factors of 10 or so larger in high-velocity and halo gas (presumably linked to the level of depletion of CaII which is high in the disk). Stocke's detection of NaI in two clouds, including the strongest feature that also shows HI, may well be an indication that they are disk clouds. A stronger statement will have to wait for measurement of the column densities of CaII, NaI and HI for each component.

If located in the disk of NGC 3067 then the material is situated at a distance of ~ 20 kpc ($H_o = 100$ km s^{-1} Mpc^{-1}, $q_o = 0$) from the centre of the galaxy, according to a detailed and careful analysis by Rubin, Thonnard and Ford (1982). A distance of 20 kpc is not exceptionally large. However, as pointed out by Carilli and van Gorkom (1987), a difficulty is that the rotational velocity of the disk of NGC 3067 has to increase by a factor of two from its visible edge to the position of the absorbing cloud to account for the observed velocity (and a larger factor for the other clouds).

Such an increase is not seen in other galaxies, see for example Kent (1987). However, 3C 232 is situated very close to the minor axis of NGC 3067 which makes the velocity calculation uncertain. The discrepancy could also be removed by requiring the clouds to have substantial peculiar motions (not an uncommon situation for clouds in our Galaxy) or by a warp in the disk.

Much more can be learned about the system, and progress can be made because the QSO is reasonably bright. Ultraviolet observations of 3C 232 with the Faint Object Spectrograph on Space Telescope will be of great interest.

2.1.2. PKS 2020–370 and Klemola 31.

The second published detection was in 2020–370, a faint QSO (17.8 mag; $z_e = 1.050$), which as Peterson and Bolton (1972) first noted is situated close to a group of six galaxies, Klemola 31 (Klemola 1969), on the plane of the sky. Boksenberg et al. (1980) reported the detection of CaII at $z_a = 0.02865$ in 2020–370, very similar in redshift to the two nearest galaxies in the group. A spiral galaxy with $z = 0.0288$ is 20 arcsec from the QSO; the next nearest is an elliptical, $z = 0.0285$, which is 45 arcsec away. The K line may be marginally resolved, and Boksenberg et al. argue that, almost certainly, the lines have several unresolved components. They do not detect galaxian NaI; and neither do Boissé et al. (1988) who obtain a better equivalent width limit, see Table 1. The line of sight to the QSO intercepts the outer projections of both the disk and halo of the spiral.

Recently, Carilli and van Gorkom (1987) reported detection of 21 cm absorption towards 2020–370, coincident in redshift with galaxian CaII. The HI line is unresolved, and so they can only set an upper limit of 10.6 km s^{-1} to the width of the line. Boissé et al. (1988) also detect 21 cm in this direction, as well as HI emission from the nearby spiral. Without more detailed information, including higher resolution CaII observations and mapping of the 21 cm emission, as well as a rotation curve for the spiral galaxy, an origin for the absorbing gas can not be determined. Although an association with the spiral seems most plausible, especially in view of the HI emission, association with the elliptical galaxy or even with intercluster gas can not be ruled out.

2.1.3. 0446–208 and a Galaxy in A514.

Here a high redshift QSO, 0446–208, $z_e = 1.896$, lies 13 arcsec from a probable S0 galaxy. The system was initially discovered by Richard Hunstead in a program dealing with radio source identifications in rich clusters of galaxies (Mills, Hunstead and Skellern 1978). In a remarkable field the QSO lies behind a rich Abell cluster, A514 (Abell 1958) about 26 arcmin southeast of its centre. Blades, Hunstead and Murdoch (1981) first discovered strong CaII K and H in the spectrum of 0446–208 at a very similar redshift ($z_a = 0.0667$) to that of the galaxy, and Baldwin, Phillips and Carswell (1985) subsequently detected galaxian NaI.

The system has the distinction of showing very strong absorption, see Table 1. S0 galaxies are generally considered to have a low gas content, and the presence of such strong absorption is unexpected. The lines of both CaII and NaI are broad, suggesting unresolved structure. The NaI lines in particular look very wide, although the spectrum is noisy. The N(CaII)/N(NaI) ratio is near unity which led Baldwin et al. to attribute the system to disk gas which would be at a distance of ~ 12 kpc from the centre of the intervening galaxy. Improved spatial imaging is required in order to establish the galaxy type which is still in doubt. However, higher resolution spectroscopic work will prove to be difficult because of the faintness of 0446–208.

2.1.4. 1327–206 and ESO 1327–2041.

In this system the QSO 1327–206 ($z_e = 1.1696$) and an almost face-on spiral galaxy ($z = 0.0180$) are separated by 38 arcsec on

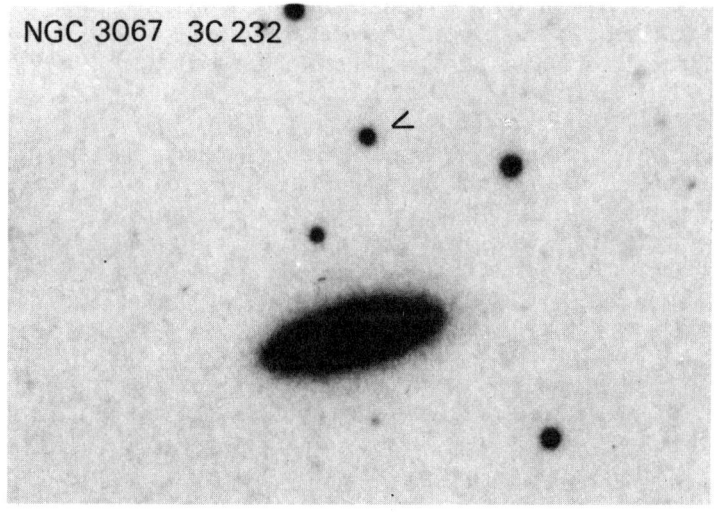

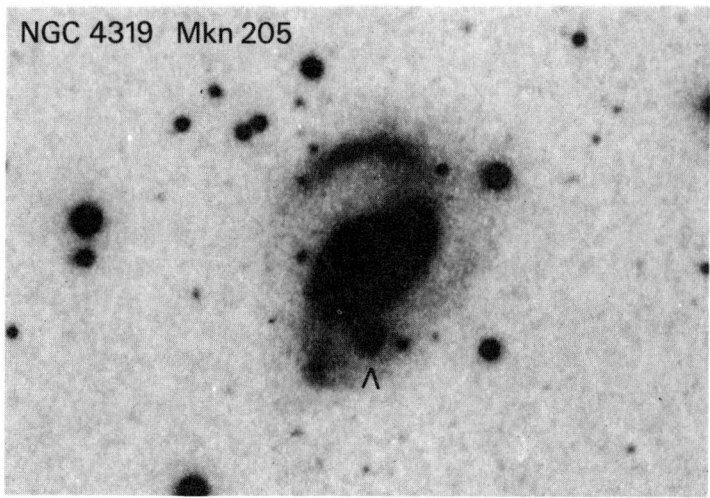

Plate I. *The fields of two QSO-galaxy systems taken from enlargements of POSS prints, with NE at the top left corner. The QSOs are indicated by arrows. The scale is the same in each field (1 cm ~ 35 arcsec). The NGC 3067–3C232 system shows intervening galaxian absorption while NGC 4319–Mkn205 does not.*

the plane of the sky. Preliminary observations by Kunth and Bergeron (1984) showed the presence of strong galaxian NaI but an (apparent) absence of CaII—an unexpected reversal of the more usual pattern of line strengths. Subsequent observations by Bergeron, D'Odorico and Kunth (1987b) have since shown the presence of galaxian CaII (in fact, quite a strong line with an equivalent width of 0.5 Å) and their higher resolution study of NaI shows complex velocity structure with at least two clouds, separated by 236 km s^{-1}. Although the two species are difficult to compare directly because of different resolution, CaII appears to be associated with both NaI velocity components. The CaII/NaI equivalent width ratio is small, ~ 0.34, typical of the value for this ratio in galactic disk clouds.

Additional NaI clouds may occur but detector artifacts make it difficult to be certain about weaker components (see Figure 4 in Bergeron, D'Odorico and Kunth 1987b). Their verification would be very important. The low level of ionization as demonstrated by the very strong NaI absorption does suggest a disk origin, with the absorbing clouds being part of the east arm of the spiral galaxy at a distance of ~ 21 kpc from the centre. Recently, Boissé et al. (1988) have successfully detected a broad (FWHM = 70 km s^{-1}) 21 cm feature toward 1327–206, implying a substantial HI column (3×10^{20} cm^{-2} for an HI spin temperature of 100 °K) and with a redshift similar to both CaII and NaI.

Bergeron, D'Odorico and Kunth (1987b) suggest that the large velocity spread seen for the NaI doublet implies a very chaotic velocity field—possibly due to a tidal interaction with another member of the group. Long-slit spectroscopy of the galaxy to determine its velocity field is required to establish whether this is a reasonable explanation. The galaxy is known to have a composite spectrum, exhibiting both absorption and emission, with a very broad and unusual 21 cm emission profile. Detailed spectroscopy of its outer regions should be straightforward. The velocity range shown by the NaI lines is large and may be much larger, if the weaker components are verified. Within our Galaxy, one would be hard pressed to find such a range within a single sight-line, except in rare cases, such as towards the Carina nebula (Walborn and Hesser 1975) for example.

2.1.5. PHL 1226 and IC 1746. In the fifth and most recent case, Bergeron et al. (1987a) have detected CaII K (but not H) in the low redshift QSO PHL 1226 ($z_e = 0.404$), which they attribute to the edge-on spiral galaxy IC 1746 situated 55 arcsec from the QSO. [As a historical curiosity I note that Kunth and Bergeron (1984) did not find CaII in a previous high resolution spectrum of PHL 1226 obtained with Boksenberg. In a brief review, Cristiani, Danziger and Shaver (1987) list this system as a null result, quoting from Kunth and Bergeron, 1984.]

The line-of-sight to PHL 1226 intercepts IC 1746 close to its major axis at a radial distance of ~ 14 kpc. Bergeron et al. 1987a detect a line with a strength of 0.6 Å and a velocity relative to the centre of IC 1746 of 309 km s^{-1}, which is compatible with the velocity gradient along its major axis.

2.2. Summary of Results

The five detections are listed in Table 1 which gives a summary of rest equivalent widths of CaII K and NaI D2 and column densities estimates. Sargent mentions a sixth system on page xxxx of this volume, as yet unpublished. Three separations are given: the first is the angle subtended by the QSO line-of-sight with the centre of the intervening galaxy on the plane of the sky, and the second, ρ, is the distance the angle corresponds to at the galaxy, i.e. the minimum distance for the cloud from the center of the galaxy ($H_o = 100$ km s^{-1}, $q_o = 0$). The third value, d, is the estimated distance if the cloud were located in a putative galactic disk. Of course, these distances are estimates only and should not be taken too literally. Nevertheless, the consistency of the separations and their range, from 8 – 20 kpc, is very striking. There is nothing here to support claims of extensive halos, 50 – 100 kpc in extent; nevertheless, distances of 20 kpc or so are significant.

Table 1
Detections of galaxian CaII and NaI in QSO-galaxy systems

QSO	System galaxy	CaII W_r(K) (Å)	CaII N (cm^{-2})	NaI W_r(D2) (Å)	NaI N (cm^{-2})	Separation ρ (arcsec)	Separation ρ (kpc)	Separation d (kpc)	Ref.
PHL 1226	IC 1746	0.58	$\geq 6 \times 10^{12}$	...	...	55.4	13.8	~14	1
0446–208	anon	0.90	11×10^{12}	1.41	8×10^{12}	13	11.5	14	2, 3
3C 232	NGC 3067	0.43	6×10^{12}	~0.16	...	114	7.7	20	4, 5
1327–206	ESO 1327–2041	0.49	5×10^{12}	1.14‡	21×10^{13}	38	9.6	21	6
2020–370	Klem 31A	0.34	6×10^{12}	≤ 0.16	$\leq 8 \times 10^{11}$	20	8.0	17	7, 8

References: 1. Bergeron et al. (1987a), 2. Blades, Hunstead and Murdoch (1981), 3. Baldwin, Phillips and Carswell (1985), 4. Boksenberg and Sargent (1978), 5. Stocke (1987; personal communication), 6. Bergeron, D'Odorico and Kunth (1987b), 7. Boksenberg et al. (1980), and 8. Boissé et al. (1988).

‡ includes only the two well-established lines at $z = 0.0175$ and 0.0183 from Table 3 in reference 6.

Apart from 0446–208, the CaII lines are weak, and are similar to those expected for halo gas from studies of high latitude absorption in our Galaxy (e.g. Morton and Blades 1986). Alternatively, the weak lines could arise in disk gas where the sight-line makes only a brief encounter with the disk. Much stronger lines, 1 – 2 Å or larger, with extensive velocity structure, would be anticipated for an intercept penetrating substantial disk material. NaI is either strong (0446–208, 1327–206) or weak (3C 232, 2020–370). Again by analogy with our Galaxy, strong NaI is indicative of low ionization, disk material. On the basis of line strength, the most promising candidates for disk absorption are towards 0446–208, 1327–206 and (probably) 3C 232; a further candidate for disk absorption is PHL 1226 because of the orientation of the line of sight with the galaxy, IC 1746. There is not a clear-cut example of halo absorption. I find the evidence is mildly persuasive that all detections to date are examples of disk absorption with an absorption cross-section for CaII and NaI in normal spirals of $\lesssim$ 20 kpc.

Without exception, the lines reveal considerable velocity structure either directly or indirectly, because of large b-values. When resolved, the velocity range is between 150 – 250 km s^{-1}. Extensive velocity structure of this scale is clearly not uncommon in galaxies, as a growing body of absorption-line data demonstrates, including galaxies IC 4329A (Wilson and Penston 1979), M83 (Jenkins et al. 1984; D'Odorico et al. 1985) and NGC 5128 (Rich 1987). The CaII K profile towards SN1987A in Figure 2 is also an indication of the range of velocity that can be expected.

The velocity ranges are substantial but not on the same scale as found in some of the higher-redshift MgII and CIV systems which are described later.

2.3. The Curious Case of Mkn205–NGC 4319

The ambiguity over where the absorption arises for the five QSO-galaxy systems is because, in each case, the QSO lies well outside the optical image of the intervening system. However, an unambiguous case is provided by Mkn 205, a low redshift QSO, and NGC 4319 an almost face-on spiral where the sight-line passes within ~ 4 kpc of the nuclear region of the galaxy, and is close to the long, trailing southern arm. The two objects are at different redshifts. The field is shown in Plate I where it makes an interesting juxtaposition with that of 3C 232—NGC 3067. The Mkn 205–NGC 4319 field has a certain notoriety because of claims of a physical connection between the galaxy and QSO (Cecil and Stockton 1985; Sulentic and Arp 1987a; and references therein), and because of unpublished reports that galaxian CaII is not present in the spectrum of Mkn 205. Of all QSO-galaxy systems, NGC 4319 *should* be a textbook case of intervening CaII absorption.

A high resolution spectrum obtained at La Palma Observatory (Bowen et al. 1987) is shown in Figure 3. A prominent galactic absorption is seen in CaII with $W(K) = 0.38$ Å. The sight-line to Mkn 205 is at a galactic latitude of $42°$ and so a contribution to the observed line strength may come from high-latitude galactic gas. However, there is no evidence of CaII in NGC 4319 to an equivalent width limit of 0.055 Å! As already stressed, CaII is ubiquitous in our Galaxy, and sight lines from the solar neighbourhood very quickly show strong CaII whatever the galactic direction (Morton and Blades 1986), and for NGC 4319 the sight-line passes through both halo and disk components. The non-detection of CaII is anomalous in NGC 4319. Subsequently, I was astonished to see in a low-resolution but high signal-to-noise spectrum of Mkn 205 published by Sulentic and Arp (1987b, their Figure 2) a weak absorption line just shorward of 5900 Å which may be interstellar NaI gas in NGC 4319, slightly blueshifted with respect to the strong stellar NaI absorption (their Figure 1). Could there be strong NaI absorption, within 4 kpc of the center of NGC 4319, but no detectable CaII, as seems to be implied by these observations? Mkn 205 is an important object waiting for Space Telescope.

2.4. Null Detections

In marked contrast to the five detections, a larger number of QSO-galaxy systems do not show galaxian CaII or NaI. The QSO IE 0104.2+3153, which on the plane of the sky is 10 arcsec (16 kpc) from a giant elliptical galaxy, was initially thought to show galaxian CaII (Stocke et al. 1984) on the basis of a low resolution spectrum. Subsequently, their higher resolution observations failed to detect CaII to the equivalent width limit, $W(CaII\ K) \leq 0.2$ Å. [As an illustration of how hard it is to remove an erroneous result from the literature, Cristiani, Danziger and Shaver (1987) refer to this system as showing probable intervening absorption as did Morton, Jenkins and York (1986) in their discussion but not in their table of results.] Bothun, Margon and Balick (1984) searched unsuccessfully for CaII ($W(K) < 0.1$ Å) in two spiral galaxies, NGC 1300 and M61, using background QSOs having projected separations of 45 and 22 kpc, respectively. In the most detailed study published so far, Morton, Jenkins and York (1986) studied seven QSOs shining through the outer regions of 12 galaxies, the majority of which were spirals. They failed to find CaII with equivalent width limits ranging from $W(K) \leq 0.06$ to 0.24 Å. The conclusions of this important study are discussed later.

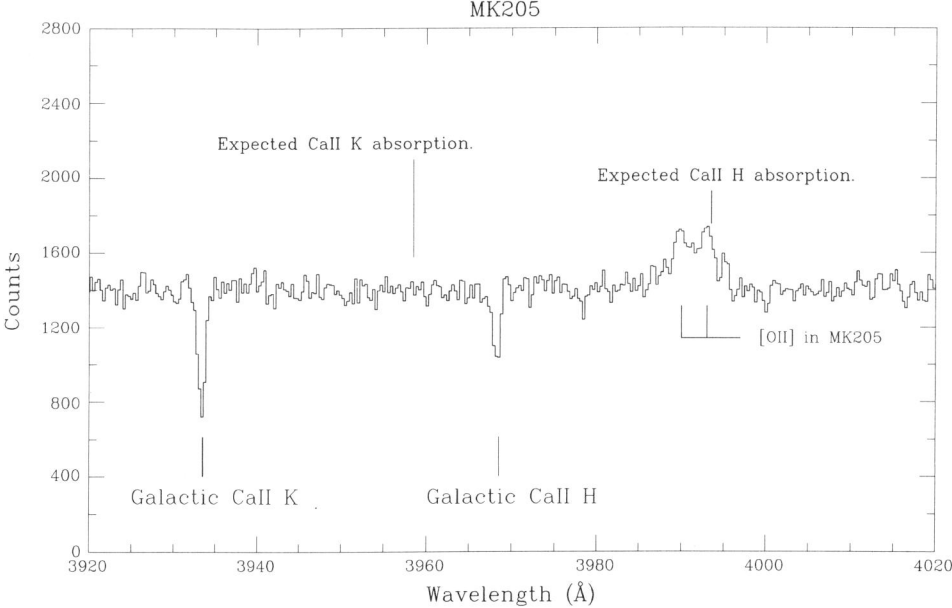

Figure 3. *A high resolution spectrum of Mkn 205 obtained at La Palma Observatory, Bowen et al. 1987. Although prominent galactic CaII is found, there is no evidence of absorption from NGC 4319.*

The systems 3C 268.4/NGC 4318 (Haschick, Crane and Baan 1983) and 3C 178/NGC 2377 (Haschick et al. 1980) are also noteworthy in that both systems show 21 cm absorption due to the foreground galaxies. I am not aware of any optical studies of NaI or CaII in these directions. Unfortunately, 3C 268.4 is faint at 18.4 mag and the other system is extended.

There are many other examples of QSO-galaxy systems and Monk et al. (1986) gives a detailed list. To find further cases, Ann Monk has been examining the fields around selected galaxies and clusters, mainly located in the southern hemisphere, and has had considerable success in locating QSOs. So far there are 87 new QSOs and other background objects (Monk et al. 1988), including 7 in the NGC 253 and NGC 1291 fields and 5 around NGC 300. We expect that many of these will turn out to be useful probes to explore the distribution of gas in these and other galaxies. Meanwhile, Bowen, Penston, Pettini and I have completed high-sensitivity observations for 5 new QSO-galaxy systems, including the case of Mkn205–NGC4319 mentioned above. We do not detect CaII K to a 3σ limit of 0.06Å in any galaxy (Bowen et al., in preparation). The systems have minimum separations of between 4 (Mkn205 – NGC 4319) and 65 kpc.

Another ten or so QSO-galaxy systems that cover a wide range of projected separations have been studied by Turnshek, Bohlin, and me using the MMT at 1 Å resolution. Our target selection was based on objects to be observed with the Space Telescope immediately after launch, as part of the guaranteed observations. Our data are not yet fully analysed, but the first plots show that we do not detect intervening CaII to a limit $W(K) < 0.2$ Å.

The Morton et al. (1986) study is worth returning to because of the important conclusions they were able to reach. Although they did not detect CaII in any of their sight-lines, they derived interesting limits for the sizes of the outer absorbing regions of galaxies by the expedient of using results from other studies having similar sensitivities, including detections mentioned previously. They found that all the observations of CaII were consistent with this species being detected out to a projected distance from the centre of the galaxy of between 8 – 18 kpc, (also see the data in Table 1). The new surveys that I have mentioned should define this distance more precisely, by virtue of better detection limits and smaller QSO-galaxy separations.

The important question that Morton et al. (1986) addressed was: how does the CaII projected distance compare with that for MgII absorption, the latter being derived indirectly on the basis of its frequency of occurrence in QSO spectra (Young, Sargent and Boksenberg 1982; Tytler et al. 1987). By normalizing the projected distance for the luminosity of the galaxy, they concluded that CaII is observable out to about the same distance that MgII is detectable. For several reasons, MgII should be much easier to detect than CaII, hence MgII should occur much further out than CaII. (The CaII, MgII comparison in Figure 2 is an illustration of the relative strengths of these two species.) They concluded that either galaxies have sharp cut-offs in their outer envelopes at ~ 20 kpc or that there has been a change in size from redshifts ~ 1, which apply to the MgII data, to the recent epochs applicable for the CaII data. The issue has not yet been resolved.

3. INTERMEDIATE REDSHIFT SYSTEMS: MGII ABSORPTION

At $z \gtrsim 0.3$ the MgII $\lambda\lambda 2795.5, 2802.7$ doublet moves into the optical domain and becomes accessible for higher resolution studies. It is a strong doublet, easy to detect and study. Like CaII and NaI, this species has been studied extensively in interstellar gas in our Galaxy and in the Magellanic Clouds, which provides an important set of data for comparison with redshifted MgII especially when CaII also is detected. Other lines become accessible at intermediate redshifts, including MgI $\lambda 2852.1$ which is generally quite weak, the MnII triplet, $\lambda\lambda 2576.1, 2593.7$ and 2605.7, and a large number of FeII lines including the strong $\lambda 2382.0$ and $\lambda 2599.4$ lines. Unfortunately these other species have received little attention, and systematic work is needed on their detection and analysis for comparison with galactic gas.

There have been many low-resolution surveys for MgII, including several that were mentioned during the Workshop. Surprisingly, low redshift surveys are not very successful in detecting the doublet. As Tytler et al. (1987) point out, although over 50 MgII systems are known, very few have been found as a result of homogeneous surveys. Tytler et al., found only three MgII systems in 24 QSOs, all of which were known previously. In a larger survey at higher redshift, Lanzetta, Turnshek and Wolfe (1987) detected 22 MgII doublets in 32 QSOs and were able to compare the properties of the MgII- and CII-selected systems. One important result was that they found most of the MgII-selected systems are dominated by low-ion absorption [i.e. $W(\text{MgII}) > W(\text{CIV})$], confirming earlier work by Boissé and Bergeron (1985), a property which the lines share with disk and halo galactic gas. The MgII surveys must be followed by high resolution programs to obtain complete results for the distribution of ions.

A case in point are the three MgII doublets studied by Tytler et al. which, at resolutions of 0.6–1.2 Å, do not show velocity structure. The widths are too large to be thermal and the authors attribute them to large turbulent motions. Large scale turbulence is generally not seen in galactic absorption, where typically the absorption strength arises from blending of clouds at slightly different velocities, as well as curve of growth effects, see the discussion in Savage (1988). The large widths may be due to insufficient resolution. If not, the broad MgII absorption may be an example of $z \sim 0.5$ material in a different physical condition compared with galactic gas.

One final point: like CaII, systems containing MgII are accessible to imaging studies. For many, the nature and size of the absorber can be determined by deep imaging. Progress is summarized briefly in the next section and described in more detail by others in the book (Bergeron 1988; York 1988).

3.1. MgII Absorption—Detection of Intervening Systems

The advent of CCD cameras has facilitated detection of galaxies out to redshifts of 1 (and even greater) and work has started to detect and measure the systems responsible for redshifted MgII absorption. Such data provide a critically important link between present-day galaxies and systems causing the higher redshifts. An ideal program would be to obtain deep images of all QSO fields with known MgII absorption, drawn from a compilation such as that by Hewitt and Burbidge (1987), and covering a wide range in z, from 0.2 to 1. A quick perusal of the Hewitt and Burbidge catalogue shows that there are 59 QSOs with at least one low-redshift MgII absorption system in this redshift range, and 17 have two or more. Follow-up spectroscopy of *all* candidate galaxies in each field would identify the likely intervener on the basis of its redshift, the type of galaxy and the minimum projected distance. The sample should be large to catch the different types of intervening sytems that might occur. Bergeron (1988) calculates the feasibility of detecting MgII galaxies. The survey would probably settle, once and for all, the nature of intervening systems out to $z = 1$.

The research area is developing rapidly. Before the Workshop, five detections had been published: AO 0235+164 situated near an HII emission galaxy (Smith, Burbidge and Junkarinen 1977); PKS 2128–123 near a probable spiral galaxy (Bergeron 1986); 1209+107 near an HII emission galaxy (Cristiani 1987); IC 55.27 near a luminous elliptical galaxy (Miller, Goodrich and Stephens 1987); and PHL 1226 near two galaxies (Bergeron et al. 1987). A further five detections were announced during the Workshop (Bergeron and Boissé 1987; Bergeron 1988); and Don York and his collaborators are also reporting detections, using a technique of narrow-band CCD imaging centered on redshifted emission lines of the putative galaxy (York 1988; Yanny et al. 1987).

A majority of the intervening systems identified with MgII absorption show emission lines and have very blue continua, although, by contrast, one is an elliptical galaxy. Whether this result is confirmed as more systems are identified and studied remains to be seen. Observation of CaII is limited, so far, to studies of normal galaxies, and clearly needs to be expanded to other, more active systems.

I cannot overemphasize the need for high resolution spectroscopic observations of MgII and other lines to complement the CCD imaging work to elucidate the physical and chemical conditions of the gas and to allow comparative studies between intermediate redshift galaxies and current epoch systems, especially our Galaxy. Existing absorption line data are uninformative about the absorbing material. A recently acquired high resolution spectrum of AO 0235+164 (Crotts 1987) points the way, it shows that the MgII lines are strong and broad but better signal-to-noise ratio is required.

3.2. CaII in MgII Systems

There have been several searches for CaII in systems that show MgII, but they have been largely unsuccessful. Redshifted CaII is detected in 0454–220 (Robertson et al. 1988) and is reported in 2128–123 (Bergeron, D'Odorico and Kunth 1987). Unsuccessful searches for CaII in prominent low-ionization systems, to good detection limits, include the $z_a = 0.395$ system in 1229–021 by Briggs et al. (1985), the $z_a = 0.359$ system in 1101–264, and in 0735+178 at $z_a = 0.424$ (Robertson et al. 1987). Table 2 gives a complete list.

Table 2.
Summary of CaII and MgII in intermediate redshift systems

QSO	z_a	CaII $W_r(k)$ (Å)	CaII $N(cm^{-2})$	MgII $N(cm^{-2})$	Ref
0002–422	0.84	0.3(?)	...	1.6×10^{16}	1
0215–015	1.34	<0.1	$< 1 \times 10^{12}$	$> 2.4 \times 10^{14}$	2
0454–220	0.47	0.13	1.6×10^{12}	3×10^{15}	3
0735+178	0.42	<0.07	$< 7.5 \times 10^{11}$	1×10^{14}	3
1101–264	0.36	<0.4	$< 5 \times 10^{11}$	4×10^{13}	3
1229–021	0.39	<0.2	$< 1 \times 10^{13}$	$< 1 \times 10^{16}$	4
2128–123	0.43	0.11	$< 1 \times 10^{12}$	1.7×10^{15}	5

References: 1. Boisse and Bergeron, 1985 (detection uncertain, only K line found), 2. Blades et al. in preparation, 3. Robertson et al. 1987, 4. Briggs et al. 1985, and 5. Bergeron, D'Odorico and Kunth, 1987.

In 0454–220, weak CaII H & K is detected in a redshift system at $z_a = 0.4745$ that also contains MnII, MgI and strong MgII and FeII lines. On detailed analysis, the five FeII lines define a single-component curve-of-growth, with $b = 30$ km s^{-1}, and lines of MgI, MgII and CaII were fitted with the same b-value. The column densities that were obtained showed that CaII is depleted by a factor of 35 relative to FeII and by a factor of 120 relative to MgII. Both CaII and MgI lie on the linear part of the curve-of-growth and are well determined, the model may have underestimated FeII and MgII, in which case the depletions would be larger.

If at all typical of intermediate redshifts, the ratios show why CaII has not been readily detected. The pronounced depletion of Ca, coupled with the relatively small natural abundance of the element, leads to very small equivalent widths in most systems. We were successful in 0454–220 because of high column density in the system and a good equivalent width sensitivity, provided by the high resolution observation.

The depletion levels are comparable with those found in interstellar clouds in the Galaxy. For example, towards ζ Oph, Morton (1975) found depletions of Ca relative to Fe and Mg by factors of 42 and 130 respectively, if the first ions are the dominant stages (virtually identical depletions to those in the $z_a = 0.4745$ system). Shull et al. (1977) found $N(\text{FeII})/N(\text{CaII}) \sim 50$ to 100 for diffuse high-latitude clouds; and for clouds near the sun, the ratio is nearer 200 (York and Kinahan 1979; York 1983;

Hobbs 1978; Ferlet et al. 1985). The similarity of the ratios in the $z_a = 0.4745$ system and in interstellar clouds in the Galaxy suggests that a possible explanation of the depletion is condensation of the refractory elements onto grains. Additional support for the interpretation is provided by an IUE spectrum of 0454−220. Although noisy, the spectrum shows a strong HI Lyman-limit at $z_a = 0.4745$. The large HI column implied ($> 10^{17}$ cm^{-2}) indicates that the sight-line is traversing a substantial column of neutral gas, perhaps typical of HI clouds in a galactic disk. Other possible explanations of the low ratios of CaII/FeII and CaII/MgII include differences in the ionization parameter or even differences in the chemical abundances between galactic gas and material responsible for the 0.4745 absorption.

In 0454−220 a second but much weaker MgII system occurs, separated by 1816 km s^{-1} from the MgII−CaII system. Tytler et al. (1987) argue that there are good statistical reasons for believing the two systems are physically related, possibly as a cluster of galaxies. Their deep imaging reveals the presence of several field galaxies but not a cluster.

Further searches for CaII in QSO absorption systems are desirable to establish whether or not strong depletion of CaII is typical. Certainly, the observational evidence so far points to a deficit of CaII relative to MgII and FeII. Any dependence of the degree of depletion on the ionization state of the absorption system would be most important in pointing to a choice between condensation onto grains or effects of ionization equilibrium as the cause. No high-ionization lines are within the available spectral range, with the exception that CIV $\lambda\lambda1548$, 1550 is just becoming accessible at $\lambda3500$ as CaII K approaches the useful limit at $\lambda9000$, for $z = 1.3$. If the strong depletion found for 0454−220 is typical, then sensitivity limitations will confine detections of CaII K to those systems with the highest MgII and FeII column densities.

In Table 2, the CaII strengths and upper limits are somewhat smaller than for high-latitude galactic gas (Morton and Blades 1986) and are considerably smaller than K line strengths in the QSO-galaxy systems in Table 1. Indeed, a K line as weak as that towards 0454−220 would only have been discovered in the best quality data. But MgII is strong in all systems listed in Table 2, and the lack of CaII may not be due solely to depletion but may indicate that the redshifts are caused by objects other than normal (spiral) galaxies. Recall that Morton, York and Jenkins (1986) found CaII, as measured from present day galaxies, and MgII, as determined from redshifted systems, appear to reside in gaseous envelopes of comparable sizes. In any particular system the two species should occur together. Hence, another explanation for the absence of CaII is a possible evolutionary change, with CaII envelopes being larger at current epochs. Such hand-waving arguments highlight the urgent need for imaging programs of intermediate redshift systems.

3.3. MgII Velocity Structure in Redshifted and Galactic Gas

In the few redshift systems that have been observed at high resolution, the strong MgII lines break up into many narrow components. An early example of this phenomenon was the $z_a = 0.424$ MgII redshift system in 0735+178 observed at 30 km s^{-1} by Boksenberg, Carswell and Sargent (1979). At this resolution MgII is spread over 165 km s^{-1} and shows four velocity features. Their study is revealing because it shows that a large number of different velocity models can fit the data successfully, hence, calling out the need for even higher resolution to resolve the ambiguity. The total column densities that they obtain from quite different model fits—which appear to match the data equally well—range from 6.9×10^{13} to 5.0×10^{16} cm^{-2}! Their chosen model has seven components. Other examples include the $z_a = 0.8526$ system towards 1327-206 (Bergeron, D'Odorico and Kunth 1987) where the data reveal the presence of at least 6 components spread over 560 km s^{-1} and the $z_a = 0.524$ absorber towards AO 0235+164 (Crotts 1987), which shows several components over a velocity range of ~ 350 km s^{-1}. In 0215+015, observations at 10 km s^{-1} resolution of FeII in the $z_a = 1.345$ system shows ~ 12 components spread over 250 km s^{-1} and we expect the MgII absorption is expected to show similar structure.

What is the velocity structure saying about the type of system causing the absorption? The velocity structure and line strengths present in the redshifted MgII systems are different from that seen for galactic gas. Within our Galaxy, the MgII tends to be extremely strong, with individual velocity components quickly blending with one another to produce a single, heavily saturated profile spread over a small (50–150 km s^{-1}) velocity range. Again, the sight-line to the LMC supernova is a good illustration, see Figure 2. Here, the outer regions of two galaxies produce an almost black absorption profile, with little discernible velocity structure.

The broader and more structured, redshifted MgII complexes are qualitatively different. York et al. (1986) have suggested that starburst galaxies can provide the right conditions leading to complex MgII profiles (see York's review in this volume); and preliminary evidence from both the imaging studies and the CaII/MgII systems points away from normal galaxies.

There is need for high resolution spectroscopy of MgII—especially for those systems where the intervening system has been detected. Unfortunately, there is no way of knowing from the low resolution surveys whether the complex MgII systems are typical at these redshifts, because weaker lines, especially those with less velocity spread, will go unnoticed.

4. ABSORPTION SYSTEMS AT HIGH REDSHIFT

At redshifts above 1 the CIV $\lambda\lambda 1548.2, 1550.8$ doublet starts to dominate the picture; the species has been surveyed extensively—see the reviews by Sargent (1988) and Bergeron (1988). Many other important heavy-elements can be detected at the higher redshifts, including highly ionized species, SiIV $\lambda\lambda 1393.8, 1402.8$, AlIII $\lambda\lambda 1854.7, 1862.8$ and many lower ionization species of which CII $\lambda 1334.5$, SiII $\lambda\lambda 1304.4, 1526.7$, FeII $\lambda 1608.5$ and AlII $\lambda 1607.8$ are among the stronger. At still higher redshift, lines of CIII, NV, OVI, and SiIII become accessible.

For many systems, especially those that were discovered because of prominent CIV, the relative strengths of the redshifted lines, especially the high- to low-ion strengths, are quite different to those found for galactic gas, as Wolfe (1983) and Danly, Blades and Norman (1987) have shown. For example, CII is always much stronger than CIV, in both low and high latitude gas, whereas the reverse is true for the majority of high redshift systems. Additionally, there may be a significant population of systems dominated by low-ionization absorption (Lanzetta, Turnshek and Wolfe 1987), more closely resembling galactic material. Examples of complex CIV absorption are discussed in the next section.

Abundance studies of high redshift systems show that heavy elements are underabundant, often by a factor of 10 or so, when compared with solar values. There are huge uncertainties in such work, due to, among other things, hidden velocity components containing significant columns and ionization states that are unaccounted; and few QSOs are bright enough to allow detailed study. A system in 0528−250 has been investigated in considerable detail and is described later. Some results on the presence (or otherwise) of heavy elements in $z > 3$ systems, including low column density HI clouds, are presented in the final sections.

4.1. Complex CIV Absorption

One of the most noticeable aspects of high redshift systems is the strength and velocity complexity of the absorption, especially for that seen in CIV. Two examples, which have proved amenable to high resolution work are systems in B2 1225+317 and 0215+015.

4.1.1. The $z_a = 1.79$ Absorption System in B2 1225+317

Although there have been a number of intermediate resolution studies of this redshift system, by Wilkerson et al. (1978), Grandi (1979), and Sargent et al. (1980), since discovery by Ulrich (1976), a detailed understanding has been acquired only through high resolution work by York et al. (1984) and Bechtold et al. (1987). The high resolution data was obtained with the MMT and echelle spectrograph coupled to an intensified reticon at a resolution of ~ 10 km s^{-1}. A remarkable set of velocity profiles covering a wide range of ions were obtained.

The complex CIV profile is shown in York et al. (1984). Its profile, previously known to contain three components, consists of at least eleven distinct velocity components over a total velocity range of ~ 500 km s^{-1}. Profile fits to individual $\lambda\lambda 1548$, 1550 pairs yield b-values in the range 6.4 to 13.2 km s^{-1}. Since the lowest b-values are comparable to the instrumental resolution, individual components could still be multiple, in which case the b values are upper limits. The widths imply temperatures between 3.4 to 14.5×10^4 °K, if they are thermally broadened.

In a more complete study, Bechtold et al. (1987) compare profiles for a range of species including MgI, II, FeII and SiII, as well as the high-ions CIV and SiIV. Although the comparison is hindered somewhat by the relatively poor signal-to-noise ratios of the spectra, the velocity profiles differ in strength from species to species, with some velocity features showing both low and high ions.

4.1.2. CIV Systems in the BL Lac 0215+015

At last count (Blades et al. 1985) seven redshift systems from $z_a = 1.254$ to 1.719 occur in 0215+015, and most are complex. The object faded recently, and turned into a weak-lined QSO with an emission redshift of $z_e = 1.715$ (Foltz and Chaffee 1987). Over the last few years, Hunstead, Murdoch, Pettini and I have been acquiring high resolution spectra at the AAT with the high dispersion Cassegrain spectrograph and IPCS: a combination which gives small spectral ranges but has many virtues, including a low background level, minimal scattered light, and photon-counting.

Initial work on CIV in the $z_a = 1.549$ and 1.649 systems (Pettini et al. 1983) revealed seven and nine velocity components spread over 300 and 900 km s^{-1}. Our new (and unpublished) higher resolution observations of these and other species show additional velocity components. Examples of our new data and model fits are shown in Figures 4 and 5. For the $z_a = 1.549$ system we identify 9 CIV components, labelled A

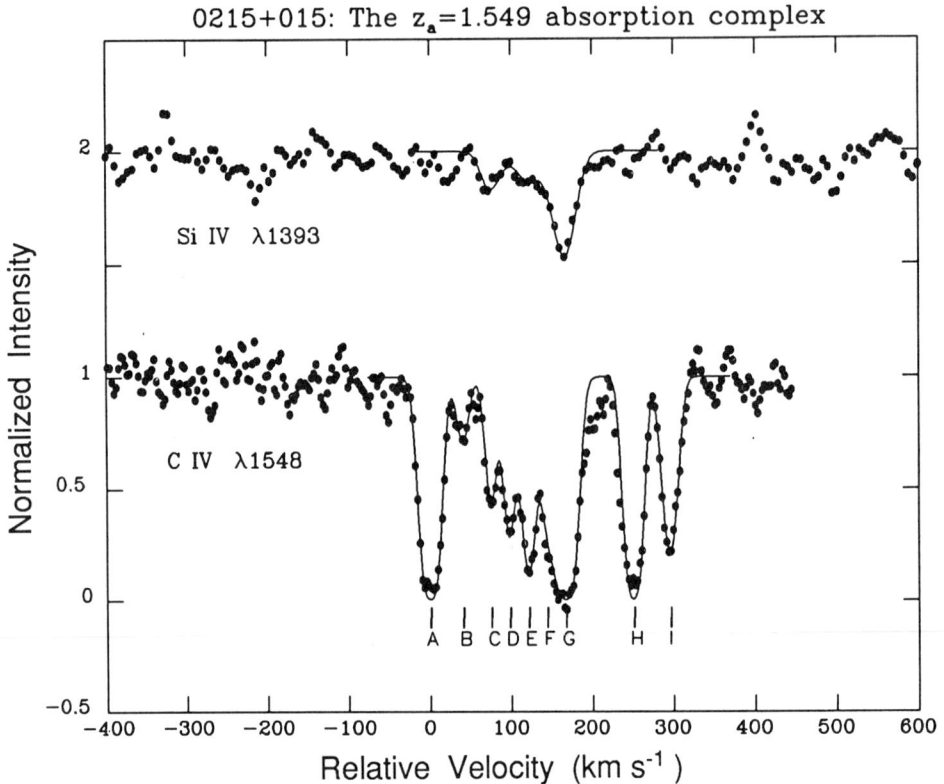

Figure 4. *High resolution observations of CIV λ1548 and SiIV λ1393 in the z = 1.549 system in 0215+015 at b_{instr} = 6 and 14 km s^{-1} respectively. At least nine clouds occur for CIV over ~300 km s^{-1}; note the absence of SiIV λ1393 in most components. The multi-component model fits are shown.*

through I in the figure. Only the strong component, G, appears to have a definite SiIV counterpart, where we obtain N(CIV)/N(SiIV) = 16. Other strong CIV components, such as A, D, E and H, show no SiIV and have lower limits for the column density ratio $\gtrsim$ 50. For most components, b-values are in the range 7 – 10 km s^{-1}, which is the same range as reported by York et al. (1984) for CIV in B2 1225+317, and again implying the presence of warm (5 × 10^4 °K) gas that is photoionized rather than hot coronal-type gas ($T > 10^5$ °K). In the z_a = 1.649 system, for which there is a more complete set of data, we identify 14 CIV components, labelled A through N in Figure 5. Again, large variations in the ion ratios occur from species to species, as can be seen from the comparison in the figure.

As a result of high resolution work on 0215+015 we discovered, by chance, two very weak metal-containing systems, at z_a = 1.686 with N(HI) $\lesssim$ 10^{18} cm^{-2}, and N(CIV) = 1.6 × 10^{13} cm^{-2}, and at z_a = 1.719 with N(HI) $\leq$ 2.5 × 10^{17} cm^2, and N(SiIII) > 8 × 10^{12} cm^{-2}. Unlike the other strong and complex systems occurring in 0215+015, which are typical of CIV systems detected in 1.5 – 4Å survey programs, the very weak systems would have gone unnoticed (apart from their Ly-α lines). How often do systems like z_a = 1.686 and 1.719 occur: are they, perhaps, more common

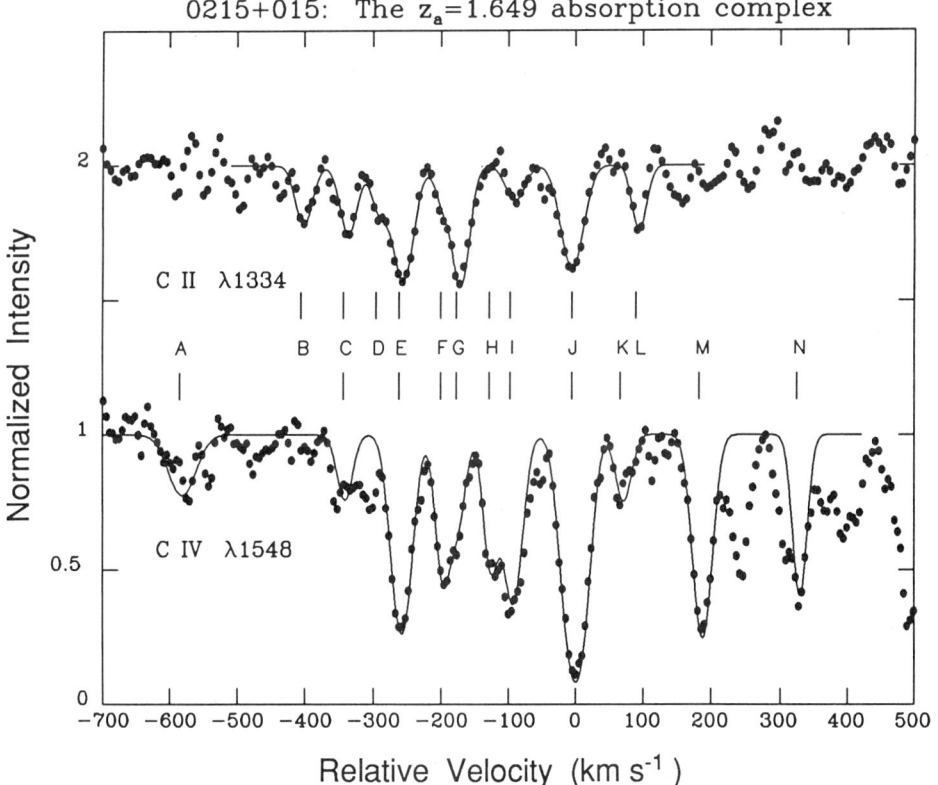

Figure 5. *High resolution observations of CIV $\lambda1548$ and CII $\lambda1334$ in the $z = 1.649$ system in 0215+015, at $b_{instr} = 12$ and 14 km s^{-1}, respectively. We identify 14 different velocity components over ~ 950 km s^{-1} with widely different N(CIV)/N(CII) ratios. The multi-component fits to the data are shown; CIV $\lambda1550$ components appear beyond $+200$ km s^{-1}; but for illustration purposes only fits to $\lambda1548$ are shown.*

than the complex CIV systems in which so much attention is placed? That they represent a significant population is gaining support from other studies capable of detecting weak lines, such as Meyer and York (1987a), who find very weak heavy-element absorption associated with high redshift HI clouds and Bergeron (1988), who states that the CIV equivalent width distribution rises sharply for $W_r(\lambda1548) < 0.3$ Å.

4.1.3. Origins

Bechtold, Green and York (1987) propose that a single galaxy at $z = 1.79$ is responsible for the strong absorption complex towards B2 1225+317, arguing that the low ionization species show a similar velocity structure as the same species in sight-lines through the Milky Way. The high-ions could originate in a photo-ionized halo—with a very large velocity dispersion—that is corotating with the disk. Nevertheless there are differences: the high-ion strengths are very much larger than found for our Galaxy, suggestive of a considerably higher level of ionization, and the velocity extent is about 5 times larger than found for the same species in the galactic halo.

To account for the complex CIV systems in 0215+015, Pettini et al. (1983) examined the possibility that they both originate in overlapping clusters of galaxies at $z = 1.549$ and 1.649. Under the assumption that two CIV components occur in each intervening galaxy, the observed multiplicity would require that the line-of-sight passes near the center of a rich cluster, twice! Morris et al. (1986) further pursue the possibility that intervening clusters cause the complex absorption. They studied two strong CIV absorption complexes in GC 1556+335, which span velocity ranges of 988 and 1677 km s−1. The existence of significant numbers of strong CIV systems makes the cluster hypothesis unlikely, unless the standard cluster parameters are radically different at redshifts between 1–2. Morris et al. demonstrate that for these types of multi-component systems, the mean time between cloud collisions—assuming each cloud represents a galaxy in a cluster of size 1 Mpc—is much less than the Hubble time, and less than the age of the system, if the epoch of cluster formation is $z \gtrsim 2$. Of course, the relative motions may not be random but regular, if the objects giving rise to the absorption were in a state of collapse. Morris et al. are unable to settle on a satisfactory model for the absorption they find in GC 1556+335. One idea is that the material is associated with the QSO, possibly entrained in a radio jet, and a similar model could also be put forward for the $z_a = 1.649$ system in 0215+015 which is blueshifted by ~ 7400 km s^{-1} with respect to the central source.

Crawford et al. (1987) have proposed that cooling gas in intervening cooling flows, associated with the cores of clusters and groups of galaxies could produce complex CIV absorption. In their scheme, component multiplicity is caused by numerous filaments of cooling gas suspended in a highly subsonic gaseous flow. Providing all clusters and sub-clusters contain flows $> 100\ M_\odot$ yr^{-1}, their calculations show that the covering factor can account for the common occurrence of CIV systems. There are many uncertainties, however. Whether substantial and numerous flows existed in the past is not yet established. The ionization level of the cooling gas may be too low (Crawford et al.) because most of the carbon is CII not CIV in the line-emitting regions. The large differences in the observed strengths of the various absorbing species, especially CIV:SiIV:CII, are probably caused by changes in ionization levels—perhaps by sources local to the absorbing cloud (such as would be found in a starburst galaxy); it remains to be seen whether cooling flows provide the right ionizing environment to produce the observed ion ratios. Dust is not expected to be present in cooling flows, so they could be useful probes of elemental abundance, without the worry of unknown depletion factors. Detection and high resolution spectroscopy of a QSO behind a known cooling flow would be of great interest.

Other proposals include star-burst galaxies (York et al. 1986), gas-rich dwarfs (York et al. 1988) and proto-galaxies, and these alternatives are discussed elsewhere in the book. The paucity of good quality, high resolution data is a major obstacle in determining the origin of the complex CIV systems. For example, to establish the ionization level both low and high ions need to be measured at the same resolution in a wide range of systems. The detailed imaging and spectroscopy of lower redshift systems, especially MgII, will provide important insights to the origin of CIV systems.

4.2. The $z_a = 2.812$ System in PKS 0528−250

This highly-variable, high-redshift QSO ($z_e = 2.77$) has been the subject of numerous absorption studies. Three definite absorption systems are recognized along with a number of unidentified lines (one of the very few QSO spectra showing unidentifed lines). The $z_a = 2.812$ system has the distinction of being redshifted by 3500 km s^{-1} relative to the QSO *and* to show damped Ly-α absorption, which Wolfe (1988) and his collaborators argue are indicative of disks in proto-galaxies. Over the years, there have been a number of abundance studies in this system. Because the Lyman continuum is optically thick the system will be predominantly neutral, thus facilitating abundance determinations in the HI gas.

At the low resolution of 3 Å, Smith, Jura and Margon (1979) found that both sulphur and silicon were significantly underabundant. They interpreted their results to imply an absence of dust and found that the actual abundances were a factor of 10 less than solar.

At somewhat higher resolution (2 Å) Morton et al. (1980) found the system to be double with a splitting of 175 km s^{-1}. They derived an HI column density of 1.9×10^{21} cm^{-2}. Their analysis showed that the low ionization species could be fitted with b-values in the range 80–150 km s^{-1}—a sure indication of unresolved velocity structure, and a clear warning not to trust the column densities for abundance studies! Nevertheless their results showed depletions similar to Smith, Jura and Margon, factors of 8 to 160 relative to solar in SII, OI and NI as well as the typical factor of 10 for SiII and FeII. Morton et al. (1980) also reported the presence of 17 unidentified lines longward of Ly-α emission. In a follow-up program, Chen and Morton (1984) re-examined these unidentified lines; on the basis of new data they rejected 10 lines, but added 11 new ones, most of which they attributed to weak CIV redshifts.

In a high signal-to-noise but low resolution spectrum (2.5 Å) Meyer and York (1987b) detected NiII in the same system and attributed two of the unidentified lines to further transitions from NiII. Their interpretation shows that the abundance of NiII relative to HI indicates that nickel is much less depleted than in galactic clouds. They also remark upon the apparent lack of dust in this system and find that the N(FeII)/N(NiII) ratio is almost certainly greater than the solar Fe/Ni abundance ratio.

Making this redshift absorption system even more remarkable is the recent report by Foltz, Chaffee and Black (1987) of the detection of H$_2$—after an original claim for this molecule in the redshift system by Levshakov and Varshalovich (1985). Using new data at 1 Å resolution, the model fit is N(H$_2$) = 10^{18} cm^{-2}, $b = 5$ km s^{-1}, and $T_{ex} = 100$ K. The presence of molecular hydrogen is exceedingly difficult to explain in the absence of dust. Indeed, for a typical galactic ratio for the dust-to-gas, the H$_2$ column density would require a visual extinction A_v of 0.8 mag! One possible explanation is the non-equilibrium formation of H$_2$ in cooling zones behind shocks.

I have drawn attention to the $z_a = 2.812$ absorption system in 0528–250 because of its many peculiarities: the unidentified lines, the abundance anomalies, probable detection of H$_2$, its apparent proximity to the QSO, the issue of whether or not there is dust extinction—in an effort to stimulate further work on the system. It is a unique system and one worthy of special attention. Good as the current observations are, higher resolution work is required before further progress can be made. Such observations will help in resolving the velocity structure and yield much improved column densities, as well as sorting out the extraordinarily large number of unidentified lines in the spectrum longward of Ly-α emission.

4.3. Heavy-Elements at High Redshift

Although only a few cases have been studied so far, that strong HI systems con-

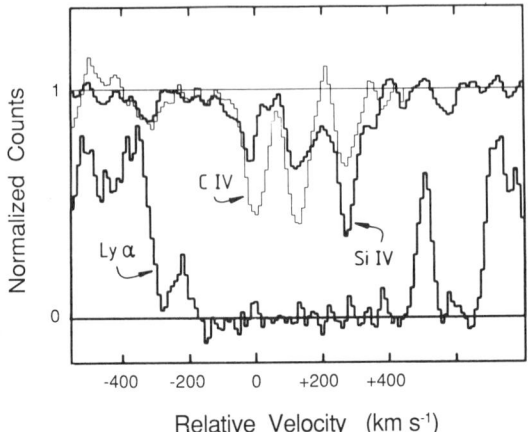

Figure 6. *Superimposed spectrum plots of Ly-α, CIV $\lambda1548$, and SiIV $\lambda1393$ in the $z_{abs} = 3.55$ system in 2000–330 from Hunstead et al. (1986a). The velocity origin has been arbitrarily set at $z = 3.5481$. The ratio of CIV/SiIV is seen to vary significantly among the components.*

tain heavy-elements, even for $z > 3$. Such high redshift systems are awkward to investigate because lines longward of Ly-α are shifted into the red and infrared while those shortward are inevitably confused with the Ly-α forest. Three very strong HI complexes occur in the QSO 2000–330 at $z > 3$, corresponding to five heavy-element redshifts (Hunstead et al. 1986b; Hunstead et al. 1987). The $z_a = 3.55$ Lyman limit system has a damped Ly-α line and shows three velocity components in CIV and SiIV and strong SiIII $\lambda1206$—see Figure 6. The $z_a = 3.332$ system is based mainly on heavy-element identifications in the Ly-α forest, but Meyer and York (1987a) report the presence of SiII $\lambda1526$. Finally, the $z_a = 3.18$ system shows two blended damped Ly-α profiles and three metal line systems contained within the overall complex. One of these, at $z_a = 3.1723$, appears to be a single low-ionization system, which facilitates the determination of abundances shown in Table 3 (Hunstead et al. 1987). We note the weakness or absence of high ionization species (e.g. SiIII weaker than SiII; SiIV not detected). Measurements imply a more-or-less uniform underabundance in C, N, O and Si of ~ 2.2 dex relative to solar values. If similar systems are common at $z \sim 3$, this would provide important constraints for models of galactic evolution.

4.4. Are There Metals in the Ly-α Only Clouds?

Evidence is starting to accumulate which challenges the widely-held belief that the Ly-α forest lines in high redshift QSOs are devoid of heavy elements. The original proposal that the numerous unidentified lines shortward of the Ly-α emission line in QSOs are due to HI clouds was put forward by Lynds (1971) and developed by Sargent et al. (1980) who proposed that the Ly-α clouds were not associated with galaxies but represent primordial material. The origin and evolution of these clouds are issues that are central to our understanding of the early Universe.

The task of identifying heavy-elements associated with low column density HI clouds is difficult for a number of reasons. High redshift QSOs are faint objects not easily amenable to high resolution and high signal to noise observations. The

TABLE 3.
Element abundances in the $z_a = 3.1723$ system in 2000−330

Ion	Ab$_\odot$	log N(X)	log depletion
HI	12.00	19.78	...
CII	8.57	≤13.95*	≤−2.40
NI	8.06	≤13.65*	≤−2.18
OI	8.83	14.40	−2.21
AlII	6.40	<12.1**	<−2.1
SiII	7.55	13.15	−2.18
SiIII	7.55	13.08	...

* Upper limit due to line blending;
** Upper limit based on nearby spectral features.

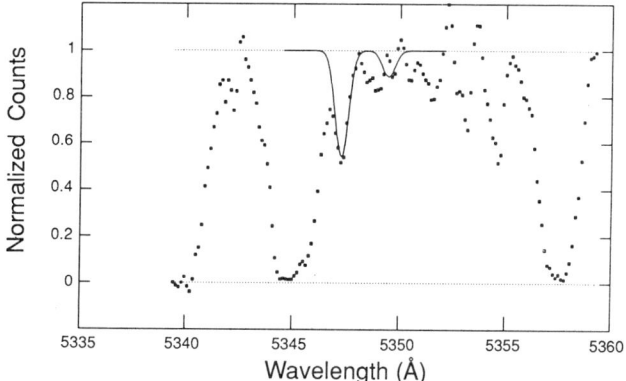

Figure 7. *SiIII λ1206 model fits to Ly-α forest data in 2000−330 at wavelengths corresponding to two well-established HI clouds with $z > 3$. The weaker component gives an upper limit to the SiIII column of $<10^{12}$ cm^{-2}, the stronger component at $z_a = 3.434$ has a SiIII column of 7×10^{12} cm^{-2} and an HI column of only 3×10^{15} cm^{-2}.*

HI columns are generally low ($\leq 10^{17}$ cm^{-2}) and the heavy-element lines will be correspondingly weak, requiring detection limits of 50–100 mÅ. Chaffee et al. (1986) have shown that either OVI (low density regimes) or CIII will set the lowest possible metal-to-hydrogen ratio—but both occur in the Ly-α forest region where line blending is extremely severe.

The composite spectra produced by Norris, Hartwick and Peterson (1983) by averaging a large number of individual Ly-α systems in two high-z QSOs showed some evidence for the presence of the O VI λλ1031.9, 1037.6 doublet. The abundance they determined was indicative of Population II material rather than with primordial (H, He only) un-enriched clouds. Others have tried the technique of summing many Ly-α redshift systems to improve the signal-to-noise but have failed to detect the OVI doublet or other heavy-element lines.

In a more direct approach, Chaffee et al. (1986) studied a pair of Ly-α clouds having N(HI) $\sim 5 \times 10^{16}$ cm^{-2}. They failed to find the corresponding CIII $\lambda 977$ line; their observation set an upper limit for the metal-to-hydrogen ratio of $10^{-3.5}$, an extremely low value. On the other hand, very high signal-to-noise spectra of 2000–330 and 2126–158, and subsequent detection of very weak CIV and SiIV lines corresponding to Ly-α lines, led Meyer and York (1987a) to propose the existence of an appreciable population of weak, heavy-element systems that could blur the distinction between the Ly-α only and heavy-element systems, a proposal supported by the work of Bergeron (1988) in this volume.

Tytler (1988) has pointed out that, to first order, the N(HI) distribution, which covers an enormous column density range, from $\leq 10^{13}$ cm^{-2} for the weakest Ly-α line, up to $\sim 10^{22}$ cm^{-2} for damped systems, could be represented by a single power law, suggesting some continuity in properties. It may not be unreasonable to expect heavy-element absorption for HI clouds with column densities $\gtrsim 5 \times 10^{15}$ cm^{-2}. Observations of heavy-element lines with smaller HI columns are beyond the reach of present day equipment, unless a drastic over abundance inflates the expected line strengths.

For 2000–330 we have been acquiring high resolution data covering the forest region with the aim of detecting heavy-element systems. Although data collection and analysis are far from complete, there is evidence of heavy-element absorption corresponding to some Ly-α lines. There are 21 HI clouds with column densities in the range of $10^{15} - 10^{16}$ cm^{-2}, covering the redshift range 3.34 to 3.78. For these clouds, the high-resolution data covers at least Ly-α, -β, -γ for each individual HI cloud; and consistent profiles can be fitted to the series lines. The reliable columns and precise redshifts that result from the model fitting are essential when searching for possible associated heavy-elements. For the 21 redshifts, excluding regions that were obscured by strong absorption, we found a significant number of SiIII $\lambda 1206$ detections. Figure 7 shows an example of a fit. Newer observations confirm the existence of SiIII in the $z = 3.3910$ and 3.4340 systems. The detection of heavy-elements in these low column HI clouds would imply significant enrichment of the material, and suggest that rather than representing intergalactic material, at least some of the Ly-α forest lines are galaxies.

I thank Richard Hunstead, Hugh Murdoch, Mike Penston, Max Pettini, Don Morton and Don York for productive collaborations and discussions over the years. The manuscript was typed beautifully by the indefatigable Dorothy Whitman; I am very grateful to both Ralph Bohlin and Dave Turnshek for comments on an early draft of the manuscript.

REFERENCES

Abell, G. O. 1958, *Ap. J. Suppl.*, **3**, 211.
Bahcall, J. N., and Spitzer, L. 1969, *Ap. J.*, **156**, L63.
Baldwin, J. A., Phillips, M. M., and Carswell, R. F. 1985, *M.N.R.A.S.*, **216**, 41P.
Bechtold, J., Green, R. F., and York, D. G. 1987, *Ap. J.*, **312**, 50.
Bergeron, J. 1986, *Astr. Ap.*, **155**, L8.
Bergeron, J. 1988, *This volume*.
Bergeron, J. and Boissé, P. 1987, in preparation.
Bergeron, J., Boulade, O., Kunth, D., Tytler, D., Boksenberg, A., and Vigroux, L. 1987a, *Astr. Ap.*, accepted.

Bergeron, J., D'Odorico, S., and Kunth, D. 1987b, *Astr. Ap.*, **180**, 1.
Blades, J. C., Grewing, M., Panagia, N., Pettini, M., Wamstecker, W., and Wheatley, J. 1988, in preparation.
Blades, J. C., Hunstead, R. W., and Murdoch, H. S. 1981, *M.N.R.A.S.*, **194**, 669.
Blades, J. C., Hunstead, R. W., Murdoch, H. S., and Pettini, M. 1982, *M.N.R.A.S.*, **200**, 291.
Blades, J. C., Hunstead, R. W., Murdoch, H. S., and Pettini, M. 1985, *Ap. J.*, **288**, 580.
Boissé, P., and Bergeron, J. 1985, *Astr. Ap.*, **145**, 59.
Boissé, P., Dickey, J. M., Kazès, I. and Bergeron, J. 1988, *Astr. Ap.*, in the press.
Boksenberg, A., Carswell, R. F., and Sargent, W. L. W. 1979, *Ap. J.*, **227**, 370.
Boksenberg, A., Danziger, I. J., Fosbury, R. A. E., and Goss, W. M. 1980, *Ap. J. (Letters)*, **242**, L145.
Boksenberg, A., and Sargent, W. L. W. 1978, *Ap. J.*, **220**, 42.
Bothun, G. D., Margon, B., and Balick, B. 1984, *Pub. A.S.P.*, **96**, 583.
Bowen, D., Penston, M. V., Pettini, M. and Blades, J. C. 1987, *QSO Absorption Lines: Probing The Universe, A Collection of Poster Papers*. eds. J. C. Blades, C. A. Norman and D. A. Turnshek, (ST ScI Publication), p. 106.
Briggs, F. H., Turnshek, D. A., Schaeffer, J., and Wolfe, A. M. 1985, *Ap. J.*, **293**, 387.
Burbidge, E. M., Lynds, C. R., and Burbidge, G. R. 1966, *Ap. J.*, **144**, 447.
Carilli, C. L., and van Gorkom, J. H. 1987, *Ap. J.*, **319**, 683.
Cecil, G., and Stockton, A. N. 1985, *Ap. J.*, **288**, 201.
Chaffee, F. H., Foltz, C. B., Bechtold, J., and Weymann, R. J. 1986, *Ap. J.*, **301**, 116.
Chen, J. and Morton, D. C. 1984, *M.N.R.A.S.*, **208**, 167.
Crawford, C. S., Crehan, D. A., Fabian, A. C., Johnstone, R. M. 1987, *M.N.R.A.S.*, **224**, 1007.
Cristiani, S. 1987, *Astr. Ap.*, **175**, L1.
Cristiani, S., Danziger, I. J., and Shaver, P. A. 1987, *M.N.R.A.S.*, **227**, 639.
Crotts, A. P. S. 1987, *QSO Absorption Lines: Probing the Universe, A Collection of Poster Papers* eds. J. C. Blades, C. A. Norman and D. A. Turnshek, (ST ScI Publication), p. 119.
Danly, L., Blades, J. C., and Norman, C. A. 1987, *QSO Absorption Lines: Probing the Universe, A Collection of Poster Papers* eds. J. C. Blades, C. A. Norman and D. A. Turnshek, (ST ScI Publication), p. 88.
D'Odorico, S., Pettini, M., and Ponz, D. 1985, *Ap. J.*, **299**, 852.
Ferlet, R., Vidal-Madjar, A., and Gry, C. 1985, *Ap. J.*, **298**, 838.
Foltz, C. B., and Chaffee, F. H. 1987, *A. J.*, **93**, 529.
Foltz, C. B., Chaffee, F. H., and Black, J. H. 1988, *Ap. J.*, **324**, 267.
Foltz, C. B., Chaffee, F. H., Weymann, R. J., and Anderson, S. F. 1988, *This volume*.
Grandi, S. A. 1979, *Ap. J.*, **233**, 5.
Haschick, A. D., and Burke, B. F. 1975, *Ap. J. (Letters)*, **200**, L137.
Haschick, A. D., Crane, P. C., and Baan, W. A. 1983, *Ap. J. (Letters)*, **269**, L43.
Haschick, A. D., Crane, P. C., Greenfield, P. E., Burke, B. F., and Baan, W. A. 1980, *Ap. J.*, **239**, 774.
Hewitt, A., and Burbidge, G. 1987, *Ap. J. Suppl.*, **63**, 1.
Hobbs, L. M. 1978, *Ap. J.*, **222**, 491.

Hunstead, R. W., Murdoch, H. S., Blades, J. C., and Pettini, M. 1986a Proceedings of IAU Symposium 119, Quasars, ed. G. Swarup and V. K. Kapahi (Dordrecht: Reidel), p. 569.
Hunstead, R. W., Murdoch, H. S., Peterson, B. A., Blades, J. C., Jauncey, D. L., Wright, A. E., Pettini, M., and Savage, A. 1986b, *Ap. J.*, **305**, 496.
Hunstead, R. W., Pettini, M., Blades, J. C., and Murdoch, H. S. 1987, in IAU Symposium 124, Observational Cosmology, eds., A. Hewitt, G. Burbidge and L. Z. Fang (Dordrecht: Reidel), p. 799.
Jakobsen, P., Perryman, M. A. C., and Cristiani, S. 1987, *QSO Absorption Lines: Probing the Universe, A Collection of Poster Papers*. Eds. J. C. Blades, C. A. Norman, D. A. Turnshek (ST ScI Publication), p. 110.
Jakobsen, P., Perryman, M. A. C., Ulrich, M. H., Macchetto, F. and di Serego Alighieri, S. 1986, *Ap. J. (Letters)*, **303**, L27.
Jenkins, E. B., Rodgers, A. W., Harding, P., Morton, D. C., and York, D. G. 1984, *Ap. J.*, **281**, 585.
Kent, S. M. 1987, *A. J.*, **93**, 816.
Klemola, A. R. 1969, *A. J.*, **74**, 804.
Kunth, D., and Bergeron, J. 1984, *M.N.R.A.S.*, **210**, 873.
Lanzetta, K. M., Turnshek, D. A., and Wolfe, A. M. 1987, *Ap. J.*, **322**, 739.
Levshakov, S. A. and Varshalovich, D. A. 1985, *M.N.R.A.S.*, **212**, 517.
Lynds, C. R. 1971, *Ap. J. (Letters)*, **164**, L73.
Meyer, D. M., and York, D. G. 1987a *Ap. J. (Letters)*, **315**, L5.
_____. 1987b, *Ap. J. (Letters)*, **319**, L45.
Miller, J. S., Goodrich, R. W., and Stephens, S. A. 1987, *A. J.*, **94**, 633.
Mills, B. Y., Hunstead, R. W., and Skellern, D. J. 1978, *M.N.R.A.S.*, **185**, 51P.
Monk, A. S., Penston, M. V., Pettini, M. and Blades, J. C. 1986, *M.N.R.A.S.*, **222**, 787.
_____. 1988, preprint.
Morris, S. L., Weymann, R. J., Foltz, C. B., Turnshek, D. A., Shectman, S., Price, C., and Boroson, T. A. 1986, *Ap. J.*, **310**, 40.
Morton, D. C. 1975, *Ap. J.*, **197**, 85.
Morton, D. C., and Blades, J. C. 1986, *M.N.R.A.S.*, **220**, 927.
Morton, D. C., Chen, J., Wright, A. E., Peterson, B. A., and Jauncey, D. L. 1980, *M.N.R.A.S.*, **193**, 399.
Morton, D. C., York, D. G., and Jenkins, E. B. 1986, *Ap. J.*, **302**, 272.
Norris, J., Hartwick, F. D. A., and Peterson, B. A. 1983, *Ap. J.*, **273**, 450.
Peterson, B. A., and Bolton, J. G. 1972, *Ap. Letters*, **10**, 105.
Pettini, M. 1985, ESO Workshop on Production and Distribution of C, N, O Elements, Garching 1985, eds. J. J. Danziger, F. Matteucci and K. Kjär, p. 355.
Pettini, M., Hunstead, R. W., Murdoch, H. S., Blades, J. C. 1983, *Ap. J.*, **273**, 463.
Rees, M. J. 1988, *This volume*.
Rich, R. M. 1987, *Astron. J.*, **94**, 651.
Robertson, J. G., Morton, D. C., Blades, J. C., York, D. G., and Meyer, D. 1987, *Ap. J.*, in the press.
Rubin, V. C., Thonnard, N., and Ford, W. K. 1982, *Astron. J.*, **87**, 477.
Sargent, W. L. W. 1988, *This volume*.
Sargent, W. L. W., Young, P. J., Boksenberg, A., and Tytler, D. 1980, *Ap. J. Suppl.*, **42**, 41.

Savage, B. D. 1988, *This volume*.
Shull, J. M., York, D. G., and Hobbs, L. M. 1977, *Ap. J. (Letters)*, **211**, L139.
Silk, J. 1985, *Ap. J.*, **297**, 9.
Smith, H. E., Burbidge, E. M. and Junkkarinen, V. T. 1977, *Ap. J.*, **218**, 611.
Smith, H. E., Jura, M., and Margon, B. 1979, *Ap. J.*, **228**, 369.
Stocke, J. T., Liebert, J., Schild, R., Gioia, I. M., and Maccacaro, T. 1984, *Ap. J.*, **277**, 43 (Erratum, 1985, *Ap. J.*, **295**, 685).
Stockton, A. N., and Lynds, C. R. 1966, *Ap. J.*, **144**, 451.
Sulentic, J. W., and Arp, H. C. 1987a, *Ap. J.*, **319**, 687.
_____. 1987b, *Ap. J.*, **319**, 693.
Turnshek, D. A. 1988, *This volume*.
Tytler, D. 1988, *This volume*.
Tytler, D., Boksenberg, A., Sargent, W. L. W., Young, P., and Kunth, D. 1987, *Ap. J. Suppl.*, **64**, 667.
Ulrich, M.-H. 1976, *Ap. J. (Letters)*, **207**, L73.
Vidal-Madjar, A., Andreani, P., Cristiani, S., Ferlet, R., Lanz, T., and Vladillo, G. 1987, *Astr. Ap.*, **177**, L17.
Walborn, N. R., and Hesser, J. E. 1975, *Ap. J.*, **199**, 535.
Wilkerson, M. S., Coleman, G., Gilbert, G., Strittmatter, P. A., Williams, R. E., Baldwin, J. A., Carswell, R. F., and Grandi, S. A. 1978, *Ap. J.*, **223**, 364.
Wilson, A. S., and Penston, M. V. 1979, *Ap. J.*, **232**, 389.
Wolfe, A. M. 1983, *Ap. J. (Letters)*, **268**, L1.
Wolfe, A. M. 1988, *This volume*.
Yanny, B., Hamilton, D., Schommer, R. A., Williams, T. B. and York, D. G., 1987, *Ap. J. (Letters)*, **323**, L19.
York, D. G. 1983, *Ap. J.*, **264**, 172.
York, D. G. 1988, *This volume*.
York, D. G., Caulet, A., Rybskii, P., Gallagher, J., Blades, J. C., Morton, D. C., and Wamsteker, W. 1988, preprint.
York, D. G., Dopita, M., Green, R. F., and Bechtold, J. 1986, *Ap. J.*, **311**, 610.
York, D. G. Green, R. F., Bechtold, J., and Chaffee, F. H. 1984, *Ap. J. (Letters)*, **280**, L1.
York, D. G., and Kinahan, B. F. 1979, *Ap. J.*, **288**, 127.
Young, P., Sargent, W. L. W., and Boksenberg, A. 1982, *Ap. J. Suppl.*, **48**, 455.

DISCUSSION

Boksenberg: A difficulty with CaII in QSO-galaxy systems is that the lines are weak, and that they are all similar in value. They probably lie on the linear part of the curve-of-growth. The numbers for the equivalent widths and column densities that you showed are all within a tight range. The point I want to make is that you only have to be a few times, not many times, lower in column density and you won't detect the lines at all. So it doesn't surprise me that you detect CaII sometimes, but not in other cases where you might have expected to find it.

Blades: Yes, it is frustrating that even for significant column densities, say 10^{12} to 10^{13} cm^{-2}, the lines are relatively weak. The detections indicate that the lines are broad or resolve into velocity structure, so that generally high resolution work is needed to yield good column densities, and I suspect in some cases the CaII may move off the linear part of the curve of growth. For the non-detections there are good limits in many cases. I still think that the non-detection of CaII in Mkn 205, which shines through NGC 4319, is very unexpected. The Mkn 205 sight-line penetrates halo gas, front and back, and the disk, without intercepting a single CaII cloud.

Wolfe: Chris, I have a number of questions. First of all, on the 0528–250 underabundant system, this was studied with about 2 Å resolution by Morton et al. (1980, *M.N.R.A.S.*, **193**, 399). How confident are you that at that resolution they really have the right b-values and columns? There could be a lot of unresolved components.

Blades: Yes; that's absolutely correct. That is why I stressed the need for higher resolution observations. The b-values are probably way off, perhaps by as much as a factor of ten too large. The column densities will not be as much in error except in the unlucky situation where you have an extremely narrow component having lots of column that is masked by a broader but lower density column. My guess is that the columns may be off by factors of 2 to 3 say, but not 10 times.

Wolfe: The other thing I wanted to say is that Frank Briggs and I looked for 21 cm absorption in the 0454–220 system which you now have damped Ly-α and a CaII detection. We didn't find anything to a good upper limit as I recall.

Blades: That's interesting. There is velocity structure in the absorption system, with the CaII seeming to reside in only one of the components.

Foltz: In 0215+015 where you have detected strong CI lines in your high-resolution spectrum, did you look for fine structure lines?

Blades: Yes, we looked hard but didn't find convincing evidence for fine structure. We are hoping to get out a paper soon on this and the other redshift systems.

Ostriker: What was the column density of the N(HI) clouds where you were looking for SiIII and CIII?

Blades: It's in the range of $N(HI) = 10^{15}$ to 10^{16} cm^{-2}.

Ostriker: I am not sure how significant your detections of metals are in these clouds.

Blades: Well, we are certainly not 100 percent certain either. But we do find an excess of SiIII detections compared with our control line, which is marginally significant. It is sufficiently encouraging to us to attempt further observations to confirm these weak lines, and this work is in progress.

Barthel: I have a general question. Has similar work been done in other QSOs, on OQ172 for example?

Blades: I don't know of any other observations, they would certainly be worthwhile.

From existing spectra of Ly-α forest regions in high-z QSOs it may be possible to identify spectral regions where there are few HI lines and which would therefore be accessible to detecting CIII and SiIII.

E. M. Burbidge: I wish to emphasize that OQ172 would certainly be interesting to examine, because it's got no Lyman limit break at all. Its continuum runs right through, so you would get a good HI column for that system.

Bregman: This is just a comment regarding dust. The column densities that are seen are typically less that 10^{20} cm^{-2}, but looking through the disk in our Galaxy the total column is about 3×10^{20}. But, the extinction is so small perpendicular to the disk that you see no evidence of UV absorption by dust against extra-galactic objects due to our galaxy, so it's very unlikely that you'd see such evidence for dust in intervening absorbing systems.

Blades: I don't agree entirely. Reddening of the order of $E(B-V) = 0.1$ mag by galactic dust is seen toward Magellanic Cloud stars. There is evidence for dust at higher latitude. I think that some of the limits toward QSO absorption systems are comparable (see, for example, Boissé and Bergeron (1987 preprint) study on the 1.345 absorption in 0215+015).

Wampler: I don't believe the galactic reddening is as high as you say.

Shull: I can comment more precisely because I was looking at this. For $N(HI) \sim 10^{20}$ cm^{-2}, the extinction would be 0.3 times the abundance of the metallicity. That's certainly detectable. Towards the supernova or in the LMC or stars in its vicinity the magnitudes of extinction are on the order of 0.1 to 0.2. That is detectable, and Ed Fitzpatrick has done a lot of work on this.

Ostriker: You do have variations in metallicity, and there is no reason to assume these QSO absorption systems have the same metallicity. Therefore, the more reasonable thing always is to normalize on the observed heavy-element systems rather than on the neutral hydrogen.

Norman: Chris, you brought up the question about molecular Hydrogen. Could you comment on the reality of that. Why don't we always see H$_2$ when you have damped Ly-α systems? It seems obvious that anywhere the sight line intercepts a disk you should have H$_2$. And what about other molecules, like CO—have these been seen?

Blades: The problem with H$_2$ is that it occurs in the Ly-α forest region, and so there is always going to be a confusion problem. And the H$_2$ lines are likely to be weak in comparison with the HI lines. Its a very difficult observation to make. There have been a few claims of CO detection, but none of these have been confirmed. Again, higher resolution and good signal-to-noise is needed.

Spitzer: Remember, the total HI column in an absorption system may be for several clouds—so you can't always expect it to be a good indicator for H$_2$ in a single cloud. The situation is not always all that clear.

Black: For H_2 in 0528–250 (Foltz et al. 1987, poster paper) we were struck by the fact that there were no strong anti-coincidences (between the observations and the expected H_2 lines) at the expected redshift and many anti-coincidences at other redshifts.

G. Burbidge: Is your study similar to that of Levshakov and Varshalovich?

Blades: The advantage John Black and his colleagues had was that they were using new higher resolution data.

Norman: Look, CO observations are very important! If you have CO you must have dust.

Shull: That's not strictly true—you can make CO in the gas phase.

Jenkins: I think another very important test is to make sure that the relative line strengths of the H_2 agree with the Frank Condon factors for the different vibrational systems. And that's probably a much better test than just looking for coincidences in frequency space.

Boksenberg: Wouldn't that come out of the cross-correlation analysis?

Jenkins: Well, no, because you can have random strengths and get a good cross correlation peak, but if you make sure that, for example, the 5–0 line is much stronger than the 0–0 line, and do a ranking of the line strengths with the theoretically expected intensities acknowledging that there's still some saturation that's probably occurring, then I think that's an important test to apply to get a secure identification.

Black: The simulations were worked out for a range of b values, excitation temperatures and column densities.

York: If I understand the simulations, you've taken the weakest lines that could fit and then the rest of the model shows where all the lines should be. But clearly a large part of the spectrum is useless because of very strong Ly-α absorption lines. How much of your spectrum really gives you an independent look at H_2? It looks like it might be only 10 %?

Black: I see what you mean. And that's difficult to quantify.

Blades: I hate to keep banging away on the same point, but there are clear spectral regions where if you had two or three times the resolution, which would be feasible if you worked very hard, you could really nail it down.

York: I want to emphasize that there is an excellent upper limit for CI in this absorption system in 0528–250. Carbon is ionized at 1100 Å, similar to H_2. And I find it very peculiar to see one but not the other.

Shull: If CI is really that low, there is no doubt, CI would be destroyed in the absence of dust, but that H_2 could survive.

York: But we don't know any places in our galaxy like this. Its worth calculating.

Black: One other comment about molecular H_2. Another QSO, PHL 957, has a well-known low ionization system with an even stronger HI column, and one of the other things that we've done recently is to look very hard for H_2. We don't find it to a very good limit. So there are cases where you have strong HI but not molecular hydrogen.

Sargent: To change the subject, do you or anybody else know anything about deuterium, from high resolution observations?

Carswell: Not that I'm willing to talk about. (Laughter) In this viewgraph I shall lay the evidence before you! In 0420–388 there is a damped HI system. We have fitted a velocity model to the Lyman and metal lines. A component in the system has a very low b-value of 10 km s^{-1} so if this were reflected in HI we thought we might have some chance of seeing deuterium in the wing of the higher Lyman series lines. And there is something that corresponds to D in a few of the profiles. We place an upper limit on the D/H ratio of 10^{-4}.

IN SEARCH OF A UNIFIED DESCRIPTION OF THE NARROW ABSORPTION LINE SYSTEMS

David Tytler
Department of Astronomy
Columbia University

Abstract The properties of the Ly-α and the narrow metal line systems in QSO spectra are compared. While they clearly differ in their clustering behaviour, they might nevertheless be similar in many ways, including their origins, metal abundances and rate of evolution. Much information can be gained from comparative studies.

1. INTRODUCTION

Stars can be sorted into about half a dozen categories using their individual values for two basic parameters to determine their location on the Hertzsprung-Russell diagram. The importance of this classification derives in part from its observational simplicity, but especially from the underlying theoretical framework which allows one to deduce many of the likely physical characteristics, including the evolutionary status of a star directly from the observable characteristics. It is with such an elegant and powerful scheme in mind that we should attempt to classify the absorption line systems of QSOs.

The absorption systems would obviously be expected to include a wide variety of species, simply because absorption will arise in any and all gaseous structures which are numerous, spatially extended, or located preferentially close to QSOs. Great variety is indeed being found, but unfortunately current observations are only sufficient to provide an extremely restricted range of parameters which can be used in a classification scheme. The scheme which has been devised has become the starting point for all discussions of the systems, and consequently it is vital that its strengths and weaknesses be thoroughly understood.

The velocity distribution of the gas causing the absorption is regarded as the primary characteristic. Systems which have a velocity spread in excess of about 2000 km s^{-1} are known as the trough or broad absorption lines (BAL) systems. They display a high level of ionization, are frequently very complex, and are most naturally interpreted as arising in gas ejected from QSOs. Nearly all of the remaining systems have velocity widths of under 200 km s^{-1}. Statistical analyses of the distribution of these narrow line systems indicates that they must predominantly arise in intervening material which is at cosmological distances from the QSOs. This distinction into intervening and ejected systems is seriously incomplete because systems with intermediate properties are found, and we do not know the minimum velocity width for ejected systems.

A second major characteristic, used in the classification of the narrow line systems, is the presence or absence of absorption due to heavy elements. The 1 % of systems which do show heavy elements are known as the metal line systems, while the remaining 99 % are the Ly-α systems. These two types of systems do differ in several observational respects, but the reasons for these differences are not clearly understood. Whereas it is widely considered that they are two totally separate populations, I wish to stress that much can be gained by examining the differences and similarities. We will see that the questions that arise when one examines the proposition that they are, in fact, closely related are central to any thorough understanding of their nature. The comparative study of the system properties which follows shows that the observational parameters which are used to classify systems are far from ideal. It also highlights the importance of seeking a general framework in which the properties of all narrow lined systems can be understood.

2. ABUNDANCES

Metal abundances have been estimated for less than a dozen absorption line systems. The main impediment is the scarcity of systems which show enough lines from enough ions to allow simultaneous estimates of both ionization and abundances. Difficulties are also encountered with inadequate spectral resolution, line blending, the existence of multiple subcomponents, and slight differences in the velocity distributions of the different ions in a given system.

Published abundance estimates range from near solar, to about 0.01 of solar abundances. These estimates are sufficiently accurate to show that abundances do differ, but they do not allow a determination of the full range or distribution. Moreover they are unlikely to be fully representative of all metal line systems because estimates are usually made for only those systems which show the largest number of metal lines, a bias which will lead to above average abundances.

Metal line systems are sufficiently common that only a few percent can occur within the Holmberg radii of galaxies. We should expect that abundances will usually be at levels anticipated for the outer regions of galactic haloes at generally high redshifts. Values around 0.01 solar might be appropriate, depending on the history of enrichment. An evaluation of the distribution of abundances as a function of epoch can be expected in the coming decade. These results will be of extreme importance in the determination of the sequence of metal enrichment during and prior to galaxy formation. For example, if it is found that metal systems with high abundances are common at the largest redshifts ($z \simeq 4$), then enrichment by Population III stars would be indicated for theories which have galaxies collapsing at later epochs.

Attempts to set upper limits on the abundances of the Ly-α systems are thwarted by our lack of precise knowledge of their level of ionization, and hence their total column densities. Two approaches have been followed.

Ly-α systems typically have neutral hydrogen column densities of N(HI)=10^{14} cm^{-2} (Sargent et al. 1980; Carswell et al. 1987). Even if these systems have very high ionizations corresponding to total column densities of $\log N = 19$, it is observationally difficult to set abundance limits which are substantially below those found in the metal line systems. Norris, Hartwick and Peterson (1983) attempted to overcome this problem by summing portions of the spectra of two QSOs, shifted to align the expected positions of metal lines at the redshifts of some 65 Ly-α systems. They did not detect CIV or NV, but OVI was marginally present. Searches in higher quality spectra of other QSOs by Sargent and Boksenberg (1983) and Norris and Peterson (1986) have failed to confirm the OVI detection, leading to typical upper limits on the abundances of metals of about 0.03 solar, a limit which is sensitive to the assumed level of ionization.

Impressive upper limits on the metal abundances have also been determined for two Ly-α systems which have logN(HI) = 17, amongst the largest N(HI) known for Ly-α systems. In both cases the limits are about 0.001 solar for an assumed level of ionization (Sargent and Boksenberg 1983; Chaffee et al. 1986). The basic assumption is that these rare systems with logN(HI) = 17 should have about the same level of ionization as the typical systems with logN(HI) = 14. The large N(HI) then corresponds to large total column densities needed to justify the low abundances. Chaffee et al. quantify this argument by assuming that the Ly-α systems are in pressure equilibrium with a hot low density intergalactic medium. All clouds at a given redshift should then have about the same density and ionization. It follows that the logN(HI) = 17 systems will be about 1000 times the size of the typical Ly-α systems, even though they are only about 1 % as common. Since detection probability is proportional to cross-sectional area, the logN(HI) = 17 systems must then represent only 1 in 10^8 of Ly-α absorbing clouds, and they would have 10^9 times the mass of a typical cloud. It is then far from obvious that these should be considered typical clouds.

Nevertheless these results are extremely important because they show that some systems probably do have low abundances. This follows from the observation that some Ly-α and some metal line systems do have sizes in excess of 1 kpc, and it is reasonable to assume that the same applies to those systems with logN(HI) = 17, which are intermediate in N(HI).

In summary, there is no direct observational evidence showing that the typical Ly-α system has a metal abundance which is well below the range shown by metal line systems, but there is at least a 3 order of magnitude range in metal abundances for the narrow line systems.

3. COLUMN DENSITIES

Ly-α and metal line systems are clearly distinguished in terms of N(HI). Since HI is the only species detected in the Ly-α systems, and it is ubiquitous in metal line systems, N(HI) is the only column density that can be used for a comparative study of the different narrow lined systems. As we saw above, this is unfortunate since it would be much more useful to classify systems in terms of their total column densities, or better still, their masses.

The first determination of the distribution of N(HI) in the Ly-α systems by Carswell et al. (1984) revealed a roughly power law distribution. This has now been extended to include a hopefully representative small sample of all narrow lined systems (Tytler 1987a).

The most impressive feature of the column density distribution is the fact that the narrow lined systems cover an enormous range of logN(HI), from below 12.8 for the weakest Ly-α systems, up to 21.9 for metal systems (Smith et al. 1986). This 9 order of magnitude spread is comparable to the difference in luminosity between the white dwarf stars and the supergiants. Here then is the clearest evidence for diversity amongst the absorption systems, a diversity which encourages one to search for subclasses of systems.

It then came as a surprise to find that to first order the column density distribution could be represented by a single power law. It appears that the majority of systems with logN(HI) ≤ 17 are Ly-α systems, while those with larger N(HI) are predominantly metal line systems. If the power law fit to the Ly-α systems is extrapolated up to the larger column densities of the metal line systems, one obtains a good estimate of the frequency of occurrence of the latter. These preliminary results suggest that there is some commonality between the Ly-α and metal line systems, as has been noted by Bergeron and Boissé (1984).

New data now show that the power law slope flattens at both low column densities below logN(HI) $\simeq 14.35$ (Carswell et al. 1987), and also above logN(HI) $\simeq 20$ (Smith et al. 1986). In addition, the distribution remains very poorly constrained at other N(HI). However a number of interesting questions are suggested. Why are there so few Ly-α systems with logN(HI) ≥ 17? Why are metal line systems with logN(HI) ≤ 17 extremely rare, and could possible connections between the Ly-α and metal line systems account for these observations?

The simplest response is that the Ly-α are really metal line systems which have too low a total column density to show metal lines. For all reasonable levels of ionization, metal lines would be observed if total column densities exceed $10^{18.3}$ for abundances greater than 0.01 solar (Chaffee et al. 1986). For a typical logN(HI) $= 14$, this limits ionization to $n(H^+)/n(H) \leq 10^{4.3}$. Those Ly-$\alpha$ systems which have larger N(HI) must have lower ionization or lower metal abundances.

4. IONIZATION

Since the time of their discovery it has been recognized that the Ly-α systems should be photoionized by the intergalactic radiation field (Arons 1972). Although direct spectroscopic measurements of the level of ionization in the Ly-α systems will have to wait for far-ultraviolet observations with the HST, estimates are already available for the metal line systems. Bergeron and Stasinska (1987) have found that line strengths observed in metal systems indicate a very narrow range for the ionization parameter $U = n_\gamma/n_e$. High ionization systems have $1.5 \times 10^{-3} \leq U \leq 2 \times 10^{-2}$, and low ionization systems have $2.5 \times 10^{-4} \leq U \leq 2.5 \times 10^{-3}$.

Let us postulate that additional systems exist with the same range of gas densities, but sufficiently small line of sight columns that they are, like the Ly-α systems, optically thin in the Lyman continuum. Since these systems would experience the same background ionizing radiation, and have the same range of values for the ionization parameter, we deduce that they should have $1.9 \leq \log n(H^+)/n(H) \leq 3.8$. This shows that most metal line systems do have ionizations such that metal lines would not be detected if their N(HI) were as low as the values found in typical Ly-α systems. Note that this level of ionization is not low enough to hide the metals in the exceptional Ly-α systems with logN(HI) $= 17$. These systems must have abundances of under 0.003 solar if they have gas densities like the metal line systems.

The gas densities in Ly-α systems have also been deduced from limits on the sizes of the systems. Sargent et al. (1982) failed to find any correlation between the Ly-α systems in the spectra of a close pair of QSOs, leading to a maximum size of about 1 Mpc. However a strong correlation was found by Foltz et al. (1984) in two images of the gravitationally lensed QSO 2345+007A,B. This observation leads to a minimum size of about 5 kpc for Ly-α systems with both low and moderate equivalent widths.

The implications are most interesting for the low equivalent width systems. At least 5 of 7 systems with Ly-α rest frame equivalent widths in the range 0.12-0.23 Å were seen in both lensed images. These systems will have $13.5 \leq \log N(HI) \leq 14.6$, maximum total gas densities of $n \leq 10^{-3.5} cm^{-3}$, and total column densities of $\log N \geq 18.2$ assuming that they are spherical. This alone is sufficient to limit abundances to under 0.01 solar. The level of ionization is high with $\log n(H^+)/n(H) \geq 4.7$, about an order of magnitude above the maximum level of ionization found for common metal line systems. Were this same ionization to apply to systems with larger N(HI), even lower abundances would apply.

Persuasive as these arguments are, they are not unique, as a number of investigators have pointed out. The Ly-α systems need not be pressure supported, in which case they could have a variety of levels of ionization. It is also possible that the Ly-α lines seen in the lensed QSO images arise from the passage of light through highly flattened rather than spherical clouds, in which case the level of ionization would be much lower than assumed.

We are then left with two possible explanations for the scarcity of metal line systems with the low N(HI) typical of the Ly-α systems. The first is that, at these low column densities, metals do not exist, or are much less abundant than in the observed metal line systems. The second is that such systems do occur, and they have the same gas densities and low levels of ionization as the observed metal line systems. The metal lines would not be observed at the N(HI) of typical Ly-α systems.

Further evidence that Ly-α systems may have a low level of ionization comes from the observation of unexpectedly narrow Ly-α lines in a few systems. Chaffee *et al.* (1983) reported the observation of a line which, if Ly-α, must have a temperature of under 16,700°K to limit thermal line broadening. Other similar systems have been observed, although there is disagreement about their frequency of occurrence. These systems must either be of high density and hence low ionization, or they could be cooled by metals or H_2.

Barcons and Fabian (1987) have stressed that an intergalactic medium which is capable of producing the X-ray background in the energy range 3-300 keV is only compatible with Ly-α clouds if the latter are of high density and low ioinization. A medium with a current density $n = 10^{-6} cm^{-3}$ and $T = 3 \times 10^8$ °K would crush Ly-α clouds of $T \simeq 10^4$ °K unless their densities were $n \simeq 1 - 10 \ cm^{-3}$.

Explanations for the lack of Ly-α systems with $\log N(HI) \gg 17$ are then that the processes which make the clouds do not operate at these column densities, that metals form in Ly-α clouds whenever column densities are this large, or that metals exist for a wide range of column densities and are only observed when column densities exceed this value.

5. SHEETS OR SPHERES?

If the ionization of the Ly-α clouds is low enough to permit substantial metal abundances, then the observations of Foltz *et al.* (1984) demand that they must be thin sheet like structures. Barcons and Fabian (1987) have discussed this possibility and concluded that one might at the same time account naturally for the distribution of column densities if such thin Ly-α clouds were flat, rather than curved, so that large column densities could arise in clouds observed edge on. Rees (1980) noted that conduction will lead to the rapid evaporation of such clouds unless magnetic fields are present. But magnetic fields are desirable in this scenario, since they would tend to produce the filamentary or sheet like structures, elongated along the local field direction.

While it is not necessary that the sheets be fully contiguous, differences in the velocities of any component clouds must be limited to ≤ 25 km s^{-1}, both to conform with the Foltz et al. observations, and the lack of clustering of Ly-α systems in general.

It is interesting to note that a metal line system is also seen in the lensed QSO spectra obtained by Foltz et al. Striking differences are noted between the two lines of sight, implying that two separate clouds are being observed, despite the similarity in redshift. Indeed one of the clouds shows strong lines of low ionization ions which must arise in gas of relatively high density, and consequently the thickness of this cloud must be much less that the separation of the two lines-of-sight.

An even more interesting case is the metal line system at $z_{abs} = 1.1249$ in the spectra of both components of the original lensed QSO 0957+561A,B. Young et al. (1981) measured a velocity difference of only 8 ± 11 km s^{-1} along the two lines-of-sight which are separated by 8 kpc. Noting that the observation of absorption lines of similar strengths and velocities is extremely unlikely if the two absorbing clouds belong to a galaxy halo, the authors discuss an origin in a face-on galaxy. It seems that this may be an example of a metal line system which has the sheet like structure suggested for the Ly-α clouds.

6. EVOLUTION

The most recent analyses of the distribution in redshift of the Ly-α systems (Hunstead et al. 1987; Tytler 1987b) show that the number of systems observed per unit redshift is best represented by $N(z) \propto (1+z)^\gamma$ with $\gamma = 2.3 \pm 0.4$ for $1.5 \leq z \leq 3.8$.

The data on metal line systems are less consistent. A substantial fraction of metal line systems are optically thick in the Lyman continuum. Over the redshift range $0.4 \leq z \leq 3.5$ these systems yield $\gamma = 1.1 \pm 0.5$ (Bechtold et al. 1984). Lanzetta et al. (1987) find that the lower ionization systems which are characterized by strong MgII lines have $\gamma = 2.4 \pm 0.8$ for $0.2 \leq z \leq 2.1$. Interesting limits are not yet published for the high ionization systems dominated by CIV but Sargent on page 6 of this volume presents results of his new CIV survey. One further result can be deduced for the metal line systems which have extremely strong damped Ly-α lines.

Wolfe et al. (1986), Smith et al. (1987) and Tytler (1987a) have concluded that these systems occur at $z \simeq 2.7$ with about 5-6 times the frequency expected from 21 cm observation of local galaxies. If the association with galaxies is correct then these systems should have $\gamma \simeq 2.3$.

In summary it seems that with the exception of the systems which are optically thick in the Lyman continuum, the Ly-α and metal line systems could have similar rates of evolution. It is important to check this possibility. If confirmed, one should be suspicious of models in which the Ly-α systems and metal line systems are totally unrelated entities, and particularly if the Ly-α are pressure supported by the intergalactic medium while the metal line systems are not.

7. CLUSTERING

There exist impressively large differences in the clustering properties of the metal line and the Ly-α systems. It was shown by Sargent et al. (1980) that the metal line systems cluster strongly on velocity scales of a few hundred km s^{-1}, while the Ly-α systems cluster much less strongly, if at all. Sargent et al. (1980) suggested that the metal line systems arise in galaxies and thus cluster like galaxies, while the Ly-α systems are a separate population with a widespread intergalactic distribution.

Considering the absorption system clustering from a comparative perspective in the context of cold dark matter dominated models, Salmon and Hogan (1986) arrive at two possible interpretations. If $\Omega = 0.2$, the metal systems can be associated with galaxies which are an unbiased sample of the mass distribution. Some unknown mechanism must stop the Ly-α systems from following the gravitational potential of the galaxies.

In an $\Omega = 1$ model the correlation function is very weak at the redshifts of the absorbers. The Ly-α clouds could be an unbiased sample of the mass, but the metal line systems would have to be more strongly clustered than the mass, either because they are biased like galaxies, or because non-gravitationally induced velocities are causing the clustering. A particularly interesting result of this study was that the velocity correlation was found to be only weakly, if at all, dependent on the number density, diameter or mass of the absorption system cloud, implying that gravitational clustering alone can not account for the differences between the Ly-α and the metal line systems.

This is consistent with the suggestion of York et al. (1986) that much of the clustering of the metal line systems might be ascribed to the dynamics of explosions in dwarf galaxies which are actively forming stars.

It will be possible to test these ideas by evaluating the variation of the absorption system clustering with epoch, since the growth of clustering of the Ly-α systems will be particularly dramatic in an $\Omega = 1$ universe, while hydrodynamic effects arising during galaxy formation should die out.

8. INTERPRETATIONS

Ignorance about the relative importance of events during early epochs currently allows for a rich variety of interpretations of the QSO absorption line systems. A consensus has arisen that most progress can be made when the formation of the absorption systems is treated as an intergral part of a general scenario for galaxy formation.

Among the most detailed of such descriptions is the work of Ostriker, Ikeuchi and their co-workers. Ikeuchi and Ostriker (1986) present a unified picture for the formation of ordinary galaxies, young low mass blue compact galaxies and Ly-α clouds. They hypothesize the occurrence of explosions (pregalactic objects or expanding voids) at redshifts 5 – 20. The shocks produced by these explosions ionize the intergalactic medium and create shells which are unnstable to fragmentation. The large mass fragments collapse to form galaxies, while those with low masses survive as Ly-α clouds confined by the pressure of the ambient intergalactic medium.

In this picture there is thus a natural continuity between the Ly-α clouds and the dwarf galaxies which might account for the metal line systems.

The chief weakness of this model is that it invokes hypothetical energetic explosions. There is no consensus that these explosions are required to be strong enough to ionize the intergalactic medium (e.g. Donahue and Shull 1987). Rees (1986) notes that the pressure of the intergalactic medium is thus determined in an ad hoc manner, a point which is also made by Vishniac and Bust (1987) who find that the clouds may be confined by ram pressure rather than by the pressure of the ambient intergalactic medium. If the sites of the explosions are young galaxies, then Visniac and Bust argue that one must either postulate the distruction of clouds in dense environments, or following Salmon and Hogan (1986), require that galaxy clustering be minimal at redshifts $\simeq 2.5$.

Shocks also play a role in many other models. Hogan (1987) has recently discussed a possible origin for the Ly-α clouds in the context of the Fall and Rees (1985) theory for the collapse of protogalaxies. Hogan notes that if the Ly-α clouds are of relatively low ionization, their contribution to the mass density of the universe is insignificant compared to galaxies (Tytler 1987a). They could then arise in a very minor gas phase occurring during the formation of galaxies. Shocks produced as protogalaxies are assembled from subunits would be observed as Ly-α clouds with total column densities in excess of 10^{17} cm^{-2}, with lower N(HI) occurring in clouds which are moderately ionized. Since the protogalaxies would have sizes of $\simeq 300$ kpc prior to collapse, coherent gas flows might produce Ly-α lines which are closely correlated in velocity over lines of sight separated by tens of kpc, as required by the observations of Foltz et al. (1984). This model avoids the need to postulate a medium to confine the Ly-α clouds, but it is then unclear whether the redshift evolution of the clouds will match the observations, since this will depend on the unknown details of the evolution of protogalaxies.

Attempts to model long lived Ly-α clouds without using pressure support center on the use of cold dark matter. Rees (1986) notes that low mass 'minihaloes' are predicted to exist at the present epoch in the cold dark matter theory for the formation of galaxies. Such minihaloes may be able to trap and stably confine gas. Radii of order 8 kpc are indicated, while the gas density would be low enough to allow a high level of ionization. Ikeuchi and Norman (1987) present further details of this type of model. In common with Rees, they stress that there need be no sharp demarcation between the Ly-α and the metal line systems, since collapse and star formation are inevitable in those haloes which have somewhat higher masses than the Ly-α clouds.

It will be important to test how well these models can preserve the distinction between the Ly-α and metal line system in terms of clustering. If both types of system come from the low mass end of the same spectrum of perturbations, then it may be that the clustering of the metal systems is largely due to hydrodynamic velocities of clouds which arise in the same events which produce the metal enrichment.

Prospects for observational discrimination between these interpretations seem excellent. The ionization of the Ly-α clouds should be revealed by the comparison of H and He lines. The frequency of occurrence of low temperature Ly-α clouds may be estimated from the numbers of lines with extremely low velocity dispersions. Observations of low redshift systems will be especially exciting. Here we can see whether metal or Ly-α systems are associated with dwarf galaxies, we can search for the expected changes in clustering, and extend constraints on the evolution of the clouds up to about 10 billion years.

It is appropriate to end with an analogy drawn from the recent history of extragalactic astronomy. Following their discovery, QSOs were widely regarded as a totally unique new phenomena. We are now aware that in spite of great differences in luminosity, appearance, and space density, the Seyfert galaxies display much of the same physics. A great deal has been learnt from this recognition of underlying similarity. It is suggested that a similar situation applies to the Ly-α and the metal line systems seen in QSO spectra.

REFERENCES

Arons, M. 1972, *Ap. J.*, **172**, 553.
Bechtold, J. A., Green, R. F., Weymann, R. J., Schmidt, M., Estabrook, F. B., Sherman, R. D., Wahlquist, H. D. and Heckman, T. M. 1984, *Ap. J.*, **281**, 76.
Bergeron, J. and Boissé 1984, *Astr. Ap.*, **133**, 374.
Bergeron, J. and Stasinska, G. 1987, *Astr. Ap.*, in press.
Carswell, R. F., Morton, D. C., Smith, M. G., Stockton, A. N., Turnshek, D. A., and Weymann, R. J. 1984, *Ap. J.*, **278**, 486.
Carswell, R. F., Webb, J. K., Baldwin, J. A. and Atwood, B. 1987, *Ap. J.*, in press.
Chaffee, F. H., Jr., Foltz, C. B., Bechtold, J. and Weymann, R. J. 1986, *Ap. J.*, **301**, 116.
Chaffee, H. F.,Jr., Weymann, R. J., Latham, D. W. and Strittmatter, P. A., 1983, *Ap. J.*, **267**, 12.
Donahue, M. and Shull, M. J. 1987, preprint.
Fall, S. M. and Rees, M. J. 1985, **298**, 18.
Foltz, C. B., Weymann, R. J., Röser, H.-J. and Chaffee, F. H. 1984, *Ap. J. (Letters)*, **281**, L1.
Hunstead, R. W., Murdoch, H. S., Pettini, M. and Blades, J. C. 1987, *Ap. J. (Letters)*, submitted.
Ikeuchi, S. and Norman, C. A. 1987, *Ap. J.*, in press.
Ikeuchi, S. and Ostriker, J. P. 1986, *Ap. J.*, **301**, 522.
Lanzetta, K. M., Turnshek, D. A., Wolfe, A. M. 1987, *Ap. J.*, **322**, 739.
Norris, J., Hatrwick, F. D. A. and Peterson, B. A. 1983, *Ap. J.*, **273**, 450.
Norris, J. and Peterson, B. A. 1986, in preparation.
Rees, M. J., 1980, *In* Physical Cosmology, ed. R. Balian, J. Audouze, D. N. Schramm, pp. 615-660. Amsterdam: North-Holland Publishing Co.
_____. 1986, *M.N.R.A.S.*, **218**, 25P.
Salmon, J. and Hogan, C. 1986, *M.N.R.A.S.*, **221**, 93.
Sargent, W. L. W. and Boksenberg, A. 1983, *In* Quasars and Gravitational Lenses, 24th Liège International Astrophysical Colloquium, pp. 518-537.
Sargent, W. L. W., Young, P. J., Boksenberg, A. and Tytler, D. 1980, *Ap. J. Suppl.*, **42**, 41.
Sargent, W. L. W., Young, P. J. and Schneider, D. 1982, *Ap. J.*, **256**, 374.
Smith, H. E., Cohen, R. D. and Bradley, S. E., 1986, *Ap. J.*, **310**, 583.
Tytler, D. 1987a, *Ap. J.*, **321**, 49.
_____. 1987b, *Ap. J.*, **321**, 69.
Wolfe, A. M., Turnshek, D. A., Smith, H. E. and Cohen, R. D., 1986, *Ap. J. Suppl.*, **61**, 249.
York, D. G., Dopita, M., Green, R. and Bechtold, J. 1986, *Ap. J.*, **311**, 610.
Young, P., Sargent, W. L. W., Boksenberg, A. and Oke, J. B. 1981, *Ap. J.*, **249**, 415.

DISCUSSION

Jenkins: I'd like to propose a simple test of your small covering factor hypothesis. Since the line to continuum ratio of QSOs varies as you go down the Lyman series, I would think you get anomylous Ly-β, either too strong or too weak if your idea is right. It seems to me this would be a very interesting thing to check in the spectra.

Tytler: Yes. There are also tests associated with that because Ly-α absorption lines in the Ly-β emission line region should be very far from the quasar in the intervening hypothesis. There should be no complications. It's going to take a while to measure this.

Norman: The size of the broad line region very roughly goes as the square root of the luminosity, so there should be a real luminosity effect there.

Ostriker: A comment and a question. The comment is: I wonder what difference it makes what you call these things, whether you call them metal line systems or Ly-α cloud systems or what. If you know their properties, if you know whether or not they're clustered, if you know their metal abundance, and if you know their sizes, it matters not the slightest. It's just something we can hang our hat on, that's all.

Tytler: It matters to me in terms of jumping to hasty conclusions and I think there has been rather a lot of that. Obviously, we want to know the properties.

Ostriker: Yes. I think its only the properties that matter. The second is a question. If the inverse effect turns out to be a simple proximity effect due to ionization by the quasars and that can be shown on a quantitative basis to be just what you'd expect, would that rule out the possibility that the clouds are small?

Tytler: No. The preprint you recently circulated shows that it's going to a fairly complicated affair. You're just looking for a slight change in the number of absorbers as you go near to the quasar. I don't really think we're going to get a lot out of it in terms of being able to say its all caused by this effect or its all caused by that effect.

Ostriker: So do you say that there might be two effects which do the same thing?

Tytler: Yes. I don't see how those data, as I currently see them, will lead to anything very decisive. I can't yet see that.

Ostriker: Well, it doesn't pay to study that then.

Tytler: Yes. (Laughter)

Rees: Can you say a bit more about the importance of helium? You said that the helium line would be important, but isn't it the case that if the ionization level is high, then essentially all you need to do is soak up the UV continuum and if the UV continuum is like a power law, then the helium would be much more highly ionized than the hydrogen?

Tytler: For clouds that are close to neutral I would not expect to see much HeII.

Rees: So do you say this is an important test of whether the clouds are very neutral or not?

Tytler: Yes. But only if they're very nearly neutral. For moderate to high ionization $N(\text{He II})/N(\text{H I})$ is nearly constant as Chaffee et al. (*Ap. J.*, **301**, 116, 1986) showed.

Buss: In close pairs of quasars you could distinguish between the two cases that Ostriker was talking about, that is, you see at the other quasar's redshift a lack of lines. That would not depend on the covering factor.

Tytler: I presume by "covering factor" you mean cloud size.

Bergeron: Of all the recent data of the neutral class, there do exist a few Ly-α systems with column densities of 10^{17} cm^{-2} in each one or more. And I think then, OI should have been detected by now easily. I could think of the one in 2126 which is exactly 10^{17} cm^{-2} and OI falls outside the Ly-α forest region for this one; I could think also of 2206–199 which has the huge HI column density of about 10^{20} cm^{-2} in which CIV has been discovered and therefore its not entirely neutral. Can you comment about that?

Tytler: What I expect to happen is that most of the current Ly-α systems with column densities in regions of 10^{16}–10^{17} cm^{-2} won't show metal lines. This won't tell me very much, because it won't tell me the distribution of ionization, which is what I really want to know. I want to know the distribution of the ionization for systems that are optically thin in the Lyman continuum. I want to measure that somewhere in the low column density range. As soon as I know that, then I'll know the size distribution. Just knowing that some of them show metals doesn't tell me what the distribution is.

Bergeron: Yes, but if you see OI and not CIV then you can prove your case, but if you see CIV and not OI it would disprove your theory by saying its highly ionized.

Tytler: I agree with that in the region of 10^{17} cm^{-2}, where it starts to be optically thick.

Bajtlik: I wanted to comment on your point about shocks and then sheets. The poster upstairs, in which we look at proximity effects, shows that if you're more than about 10 Mpc from the quasar, you are into the ambient radiation field, you are not dominated by the local quasar. You find that in order to survive 10^8 years or so, one of these Ly-α clouds must be fairly dense, and/or large. Since we have constraints on the size, I suggest that maybe they are either shock compressed sheets or thermally unstable sheets and the densities are higher.

Balbus: Do I understand your getting nT in the clouds of about 10^4 °K? If so, I'm wondering about the confinement.

Tytler: I'm not getting any nT values.

Balbus: You can assume a density of ~ 1 cm^{-3}.

Tytler: I can assume a lower temperature.

Balbus: How low do you want it to be?

Tytler: It must be below about 2×10^4 °K if they are going to be almost neutral. We have to be more careful about this though. If I'm actually pinning down a value for nT, then I have specified the level of ionization. Since a variety of ionization values are possible for individual Ly-α clouds, a range of nT values should be considered. The observational limits on this range become less restrictive as the H I column density and metal abundance drop. It is only for the largest column densities that a narrow range is required to conceal high metal abundances.

Balbus: Well, depending upon how comfortable one is with pressure confinement, it seems to me that puts real constraints on small dense clouds.

Tytler: So you'd expect them to collapse?

Balbus: Well, I'd expect them either to evaporate or, if you want to pressure confine them with an intergalactic medium, I think if you want to have reasonable densities then it makes the intergalactic medium very hot.

Norman: What says you have to pressure confine them?

Balbus: Well, there not gravitationally confined.

Norman: Why not? That's the possibility Martin raised in his talk.

Green: I have two questions. One is, in deriving your distribution of column densities, how did you treat the multiplicity of components for systems where the column density was derived from the total Lyman limit optical depth? The other question is, if you have very thin sheets, how does the distribution of aspect ratios jive with the observations.

Tytler: Yes, these questions have been thought about. The first question concerns the velocity resolution. In the case of Ly-α clouds, lines separated perhaps more than about 60 km s^{-1} could be counted as an individual entity, whereas in the case of a Lyman continuum absorber with a very large column density anything within at least a thousand km s^{-1} will be lumped together. When you separate out the components in the high column density clouds you're going to see more components, for a start, so you're presumably going to increase the number of low column density clouds and somewhat decrease the number with larger columns. The column density distribution will then become steeper for the metal line systems which are resolved into components. Apart from the change of slope, this effect is less important than it might seem. I am not interested in the small scale velocity distribution of the gas because I do not want to distinguish between, for example, an unresolved gas cloud with a given velocity width, and a second cloud which has the same total column density and velocity spread but which is resolved into several clumps. The number of components observed could also depend on the ion that was observed. The distribution of column densities is probably most useful if one ignores the number of components in systems.

Your question about aspect ratios has been discussed by Barcons and Fabian (*MNRAS*, **224**, 675, 1987), who show that power law distributions can be generated provided that the thin sheets are very flat.

Jenkins: One worries about the reliability of column densities for very strongly saturated Ly-α lines, when you're getting up to $10^{16}-10^{17}$ cm^{-2} in using the standard curve of growth analysis technique. Has anyone used both the standard curve of growth technique and the Lyman limit absorption on the same system to verify that you're getting the right answer.

Tytler: Yes, that was done by Carswell early on, and you get different numbers. It might move the points by factors of 5 or so. The factor of 5 is half of a column density bin when the whole range is 9 orders of magnitude. I can be crude if I'm talking about these things.

J. Bahcall: I'd like to ask a crude question, too. (Laughter) It concerns the matter of how much the field has progressed in 7 or 8 years. Wal referred to a proposal that I made maybe 7 or 8 years ago that was just an extreme proposal that both the Ly-α lines and the metal lines came from the same big halos, with the Ly-α halos being bigger than the inner halos where the metal lines came from. On that basis you can now understand the result, particularly with the picture that Martin presented yesterday, that the Ly-α's evolved more rapidly because they were the outer part that fell in perhaps first. However, I haven't followed the field very carefully recently. I wonder, is there something that can kill that sort of a maximal proposal; that everything comes from galaxies, very big halos and not so big halos. Is there something that can kill that idea that's been developed in the interim?

Tytler: Well, I actually rather favor that idea. If I want to have metals all over the place, well outside the inner regions of galaxies, and I already know that there are metals out to about 100 kiloparsecs anyway, then I would favor an environment in which there are large structures and star formation. I like that basic picture. I'm not aware of any detailed problems. I assume there's a test that will probably come from the clustering, which should be continuous between the two.

Wolfe: Shouldn't they be clustered in that situation?

Tytler: Yes. Also, consider the distribution in the metal abundances, somehow the frequency of occurrence should relate to how far out you are, and that should somehow fit in with the abundances in globular clusters and stars. Those things must eventually fit together.

Shull: Do we know enough about the frequency of cooling flows to rule that out? Those could be Mpc sizes.

Tytler: If you wanted them just in the very richest clusters, then I think you would not have enough.

Shull: There's evidence now that a lot of ellipticals have them.

Tytler: You need more than just a lot. You need all of them to have them to account for the frequency of occurrence.

Shull: No, these are large cross-sections.

Tytler: 100 kpc?

Shull: They're quite a bit larger.

Bergman: No, it isn't Mike. The x-ray surface brightness of the ellipticals have been well observed. Your probably thinking of sizes from M87 which is more like a cluster size.

Sancisi: I feel that many people are worried about the size of the galaxy's halo in HI, and so on. I point out that there is a population of objects that we know of, they are very frequent, and have very large cross-sections, of the order of 100–300 kpc—they are the interacting galaxies. It seems to me that for isolated galaxies, we can trace them no more than 1 or 2 or 3 optical radii. But there is a population of 2, 3, 4 galaxies interacting, that have spread almost everywhere. So if you can find a way of making them much more frequent at high redshift (I heard yesterday that possible mergers and encounters may be much more frequent at that time), then you have got a kind of population that perhaps gives you the cross-section and the amount of intersections that you need.

Tytler: In general terms you want to consider any process that distributes gas that has been enriched, so anything that will spread the gas back out again is marvelous. So that process is of interest. I don't see the numbers working at present. You also have to explain the number of absorption systems at low redshifts, too. They're not that rare.

J. Bahcall: Renzo, isn't it true that if every galaxy had a Mpc halo with 10^{15} cm^{-2} neutral hydrogen you'd never see it?

Sancisi: Yes that's right, but first is the question of how far you want to go out, and then you will probably have tidal disruptions in any case. They would not be very distant from one another.

Rees: I want to go back to the question of the very small clouds and specifically how they could arise. You've got to have a very hot intergalactic medium, then you've got to have a magnetic field presumably to supress conductivity, and, so presumably the natural expectation would be that they would be filaments or sheets. Do you think that significantly eases your arguments concerning the effects of clouds covering the emission line region.

Tytler: Slightly, but I still might want to consider whether they have metals in them as well. You haven't gotten very far just by arriving at sheets.

Rees: So do you say they are confined by high pressure medium which is a medium in a protogalaxy?

Tytler: The difficulty with fitting what I'm talking about into any particular theory is that I'm saying quite a lot of things, most of which are unconventional, so you have to choose some subset of the possibilities, and build something around them. I'm not comfortable with any particular area of this. I don't know how much of this is right.

Kronberg: If the Ly-α clouds were confined by a hot medium then at some point we would begin to see Faraday rotation from the medium. I've looked for that in a very few systems, 3 or 4, and there's no excess in the Faraday rotation in the Ly-α systems that I've seen. There is, however, regularly Faraday rotation that sticks out like a sore thumb in all the metal line systems that have substantial column depth.

Tytler: Then do you say the Faraday rotation doesn't increase with redshift?

Ostriker: Can you define your remark in terms of some parameter?

Kronberg: Yes. For the strong absorbing systems, with the best estimates that Judith Perry and I were able to make for column density a few years ago, the typical magnetic field strengths are of the order of a few to a few tens of microGauss. Now, that's in the really strong systems, of course.

J. Bahcall: That's for the metal systems.

Kronberg: Yes, that's right.

J. Bahcall: But I think Jerry's question referred to the Ly-α systems where it's more relevant.

Kronberg: Okay, in that case, we don't detect anything.

Ostriker: Is it a limit?

Kronberg: Its a limit on field strength times column density. If I take all the strong absorbing systems out of the universe which have excess variable rotation and look at the rest of the QSOs which have only weak absorption, and I ask what is their excess Faraday rotation, the answer is I can't measure any. So the only thing I can talk about then is some widespread medium for which I have an estimate which was about a few times 10^{-9} Gauss locally, but of course its scaled by a number of other parameters like H_o and q_o, and so on.

Rees: Phil's data is perhaps important, providing the Ly-α systems may not be the same, but as far as conductivity is concerned, my point was that in order to allow these sheets to survive you need a magnetic field. The field strength needed is far lower than Phil's limit. Something like 10^{-12} Gauss is sufficient to suppress conductivity. So it's consistent to have a field which can affect thermal conductivity and yet be well below the limit to detect Faraday rotation.

P. Shapiro: I have a comment on having neutral clouds with a density of 1 cm^{-3} with the help of shocks. If you want to make those shocks in the intergalactic medium, you have to make them at very high redshift in order to get the pressure as high as claimed and still have the gas cool behind the shock. So if you're talking about shocks at redshifts less than 10, you cannot make them at that high a pressure.

THE PROPERTIES OF THE GASEOUS GALACTIC CORONA

Blair D. Savage
Department of Astronomy
University of Wisconsin-Madison

1. INTRODUCTION

In recent years significant progress has been made in obtaining information about the physical conditions of the gaseous matter in the halo of the Milky Way. Progress has also been achieved in obtaining the beginning of a theoretical understanding of what ionizes and what supports gas found at large distances from the galactic plane. Theorists have begun to realize the important role the gaseous halo may play in providing a place for the gas in the underlying disk of the Milky Way to vent its energy.

The relationship between the gas of the Milky Way halo and the narrow absorption lines seen in the spectra of QSOs is only weakly established. While many QSO observers generally believe that a subset of the QSO absorption line phenomena likely is related to the absorption produced in the halos and/or extended disks of intervening galaxies as first proposed by Bahcall and Spitzer (1969), it is clear that an enormous amount of work remains to establish this connection with certainty and to determine what fraction of the observed features actually have their origin in gaseous halos.

Fortunately the Milky Way gaseous halo can be studied in many different ways (see Table 1) and a large body of observational and theoretical work has begun to provide important new insights about its properties. Since we know so little about the gaseous halos surrounding other galaxies, it seems appropriate to use the gaseous halo around our own galaxy as a basis for comparison with some of the phenomena seen in QSO spectra which might represent absorption by the halos and/or disks of other intervening spiral galaxies. It is not the intention of this review to emphasize the relation between what we see in our own halo and what is commonly seen in QSO spectra. Instead the emphasis will be on a discussion of what is currently known about the Milky Way's gaseous halo.

Many observers of galactic interstellar matter consider halo gas to be that gas having a distance away from the galactic plane, $|z|$, which exceeds 0.5 kpc. Furthermore many of the galactic studies of this gas are for matter at large $|z|$ but in the solar region of the galaxy. In contrast, the QSO absorption line observers will most often only sample galaxies through absorption sight lines which pass through the outermost edges of intervening galaxies. Such sight lines may pass through disk gas and halo gas. Understanding the modifications caused by this major difference in sampling represents a very important first step in interrelating what we see locally in our own

Table 1. Observational techniques used to study Milky Way halo gas

Technique	Species or Gas Phase Sampled	References*
21 cm emission	HI	1, 2, 3
optical absorption	CaII, TiII, NaI, KI	4, 5, 6
ultraviolet absorption	many ions	7, 8, 9, 10
optical emission	Hα, OIII	11
ultraviolet emission	CIV, OIII	12
pulsar dispersion	e's	13
free-free absorption and emission	e's	14
X-ray emission	hot gas (10^6 °K)	15, 16
non-thermal radio emission	high energy e's and B field	17
Faraday rotation	e's and B field	18
γ Ray emission	high energy e's and photon field	19

*References (This list identifies recent reviews and/or research papers which provide a survey of the relevant literature).
1. Lockman (1984), 2. Lockman (1986), 3. Giovanelli (1986), 4. Albert (1983), 5. Jenkins (1986), 6. Morton and Blades (1986), 7. Savage and deBoer (1981), 8. Savage and Massa (1987), 9. Pettini and West (1982), 10. Savage and Jeske (1980), 11. Reynolds (1986), 12. Martin and Bowyer (1987), 13. Harding and Harding (1982), 14. Spitzer(1978 and references therein), 15. Marshall and Clark (1984), 16. Nousek et al. (1982), 17. Beuermann, Kanbach and Berkhuijsen (1985), 18. Sofue, Fujimoto and Wielebinski (1986), 19. Stecker (1979)

halo to the absorption we might expect to see at high redshift in the spectra of QSOs.

2. DISTRIBUTION OF THE GAS

Many of the observing techniques identified in Table 1 have provided information about the galactic distribution of Milky Way halo gas. Halo gas spans a wide range of physical conditions. It is important to study its distribution using diagnostic methods which can adequately sample the many phases of gas believed to be present. For example, the extensive 21 cm HI measures permit detailed studies of the neutral component of halo gas while ultraviolet absorption line studies reveal the neutral, moderately ionized and highly ionized gas of the halo. Additional information about ionized gas in the halo can be inferred from pulsar dispersion studies, radio free-free absorption studies and diffuse Hα emission studies. The very hot (10^6 °K) gas of the halo can in principle be studied through its X-ray emission, and non-thermal radio emission measures can be used to trace the presence of high energy cosmic ray electrons and the magnetic field in the halo. Below we briefly review the results of these different approaches to the study of Milky Way halo gas.

2.1. HI 21 cm Emission Line Studies

An extensive body of information about halo gas and gas in the outer Milky Way exists in the radio astronomy HI 21 cm literature. For a short overview see Lockman (1986). A significant problem associated with 21 cm emission line measurements of halo gas is the presence of emission by bright HI situated in the radio antenna side lobes. Such emission can often be confused with the emission by low column density halo gas and great care must be exercised to avoid or eliminate the problem. Unfortunately, the vast majority of the papers reporting 21 cm results totally ignore this problem even though it is well documented that the errors can be large (vanWoerden, Takakubo and Braes 1962; Lockman, Jahoda and McCammon 1986).

The z extent of galactic HI varies with radial position in the galaxy. The z distribution for the inner galaxy and solar neighborhood has been studied by Lockman (1984) and Lockman, Hobbs and Shull (1986). Studies of the outer galaxy are found in Knapp, Tremaine and Gunn (1978), Henderson, Jackson and Kerr (1982), Kulkarni, Blitz and Heiles (1982) and Burton and Lintel-Hekkert (1986).

For galactocentric distances $R < 3$ kpc, the HI layer is narrow and well described by a single Gaussian with FWHM < 0.2 kpc. The inner galaxy does not seem to have HI at large z distances. Starting at about $R = 4$ kpc the z distribution of the neutral gas layer becomes more complex (see Figure 1) and can be roughly described as a combination of components with exponential scale heights of about 0.12 kpc and 0.5 kpc (see Table 2). The more extended component while recognized in earlier 21 cm studies (Oort 1962; Shane 1971) has, until the work of Lockman and his collaborators, been largely ignored by galactic radio astronomers.

In the solar neighborhood, a method for evaluating the z extent of the neutral halo gas is by intercomparing ultraviolet measures of HI Ly-α absorption to halo stars at known z distances and HI 21 cm emission data for the same direction. Larger column densities from the 21 cm measurements would be expected if gas exists in the halo beyond the star. From such a program involving 25 stars Lockman, Hobbs and Shull (1986) have shown for high latitude lines of sight that 5×10^{19} atoms cm^{-2} or 15 percent of the HI emitting gas exists in the halo with $|z| > 1$ kpc. The data are consistent with a two component z distribution consisting of equal amounts of gas in a narrow component and an extended component which have scale heights of 0.135 and 0.5 kpc, respectively. The extended component of HI was also apparent in the results of the Copernicus satellite HI Ly-α and H_2 survey for those sight lines through the intercloud medium which showed small H_2 column densities (Bohlin, Savage and Drake 1978).

Beyond the solar circle the thickness of the HI layer is observed to increase nearly linearly and at large galactocentric distances (> 20 kpc) the scale height of the gas approaches 1.5 kpc (Kulkarni, Blitz and Heiles 1982; Henderson, Jackson and Kerr 1982). In addition, the gas warps to positive z for galactic latitudes between approximately 60 and 120 degrees and to negative z for galactic latitudes between approximately 150 and 300 degrees. Figure 2 which is from Burton and Lintel-Hekkert (1986) illustrates quite well the large z extension and the warping of gas at a galactocentric distance of 17 kpc.

The actual radial extent of the Milky Way as detected in neutral hydrogen is difficult to estimate reliably because of uncertainties in the galactic rotation curve at large R. Also it is not clear what effect the warp of the galaxy will have on the dynamics of the hydrogen at large galactocentric distances. In summarizing this uncertain subject, Lockman (1986) concludes that it is likely that our Galaxy has a normal sized HI disk with a radius to the neutral hydrogen column density level of 10^{19} cm^{-2} of about 30 kpc which is about a factor of 1.5 of the Holmberg radius.

Certain aspects of the intermediate and high velocity cloud phenomena detected in the HI 21 cm emission line relates to galactic halo gas and/or very high z extensions

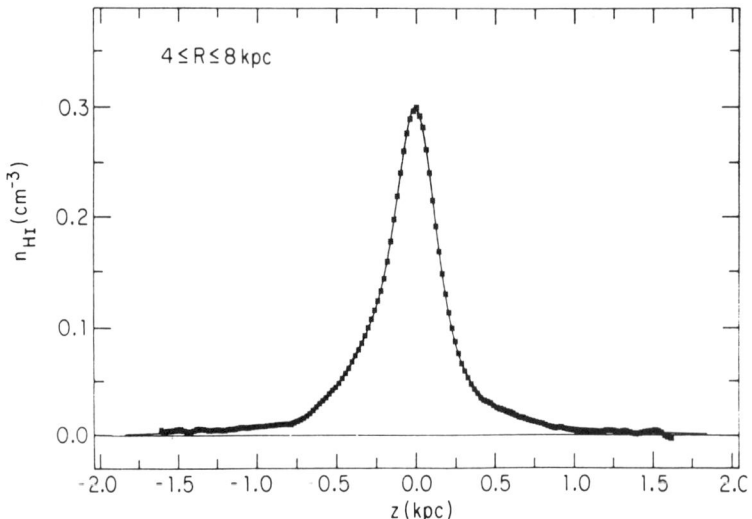

Figure 1. The HI volume density versus distance from the galactic plane from Lockman (1984). These results are based on 21 cm emission line data and are averaged over galactocentric distances betweeen 4 and 8 kpc. The data clearly show the extended component of HI in addition to the more familiar disk component of the neutral gas. The scale heights and midplane densities for the various components are listed in Table 2. At small $|z|$ the indicated values of $n(HI)$ (cm^{-3}) underestimate the true values because of optical depth effects. Although the component of HI which extends to high z was first seen by Oort (1962) and Shane (1971), it has been largely ignored until the important recent work of Lockman.

of distant parts of the Milky Way (for a review see Giovanelli 1986). The literature on this subject is very extensive but important to study because the dynamical phenomena which inject gas into the halo may drive a circulation which produces many of the intermediate and high velocity neutral HI clouds (Bregman 1980; Kaelble, deBoer and Grewing 1985).

The very interesting studies of Heiles (1984) of HI bubbles, superbubbles and 'worms' provide evidence for the injection of gas into the halo by various events occurring in the galactic plane. The existence of such structures seems compatible with the galactic fountain and galactic convection ideas for the origin of halo gas discussed in §6.

2.2. Optical and Ultraviolet Absorption Line Studies

Recent optical and ultraviolet absorption line studies have provided important information about the distribution, abundances (see §4) and kinematics (see §5) of

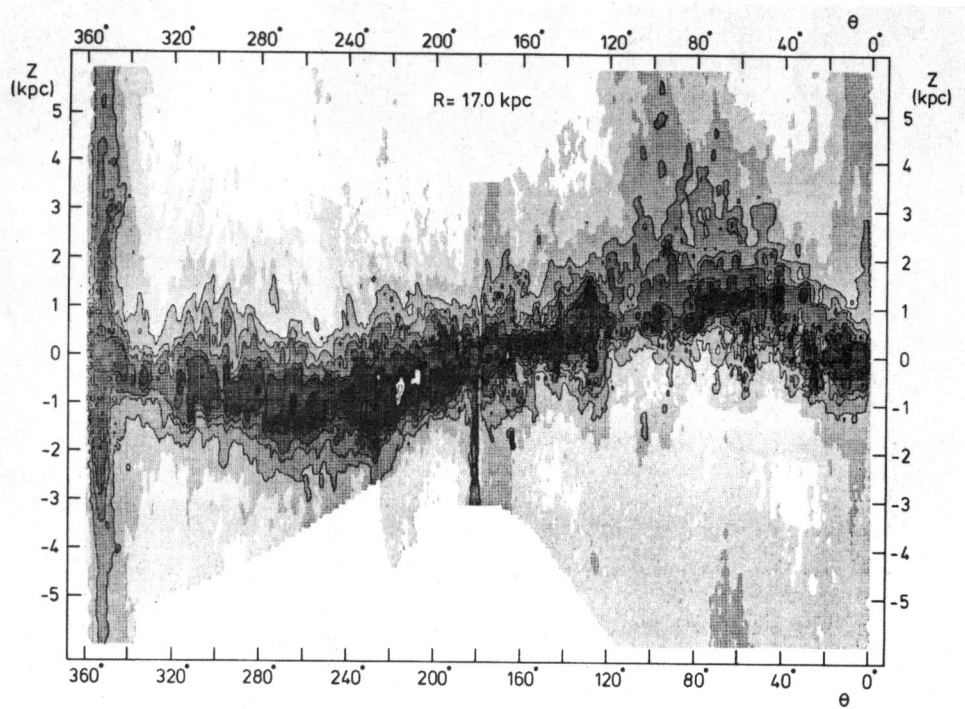

Figure 2. *The z extension of HI and the galactic warp at a galactocentric distance of 17 kpc is illustrated in this figure from Burton and Lintel-Hekkert (1986). The contour divisions occur at values of n(HI) of 0.01, 0.02, 0.03, 0.05, 0.07, 0.16, 0.30, 0.45, 0.6, 0.9, and 1.2 atoms cm^{-3}. Grey scale divisions occur at levels of 0.001, 0.004, 0.01, 0.02, 0.03, 0.05, 0.07, 0.16, 0.30, 0.45, and 0.60 atoms cm^{-3}. In this presentation the dependence of density on galactocentric cylindrical coordinates (z, θ, and R) is illustrated for R = 17 kpc. Near $\theta = 0°$ and $180°$, confusion of HI 21 cm emission from distant and local gas invalidates the velocity-distance assignment used to convert the observed profiles into gas density as a function of distance. A flat rotation curve with Θ (R > 10 kpc) = 250 km s^{-1} was assumed. The galactic warp which extends from about +1 kpc to about −1 kpc is clearly seen. While the faintest contours are very uncertain because antenna side lobe contamination was ignored in the analysis, the data do suggest a very large z extension for the HI in the outer galaxy. For example, at $\theta = 260°$, the first contour which represents n(HI) = 0.01 cm^{-3} extends from z = +0.5 to −2.5 kpc with a peak density occurring at z = −1.0 kpc.*

Table 2. Milky Way gas scale heights for the solar neighborhood

Ion	n_o (cm^{-3})	H (kpc)	Technique	Ref.
HI	0.1	0.8	uv absorption	1
SiIV	2×10^{-9}	3	uv absorption	1
CIV	7×10^{-9}	3	uv absorption	1
NV	3×10^{-9}	2	uv absorption	1
OVI	3×10^{-8}	0.3(+0.2,−0.15)	uv absorption	2
HI	0.15	0.37	uv absorption	3
SII	8.4×10^{-7}	0.32	uv absorption	3
ZnII	4.0×10^{-9}	0.35	uv absorption	3
SiII	8.0×10^{-7}	0.35	uv absorption	3
MnII	7.0×10^{-9}	0.25	uv absorption	3
FeII	8.0×10^{-8}	1.1	uv absorption	3
CaII	7.5×10^{-10}	1.0	optical absorption	4
HI	0.0053	0.48	21 cm emission	5
HI	0.09	0.25	21 cm emission	5
HI	0.16	0.11	21 cm emission	5
e's	0.03	0.8	pulsar dispersion and free-free absorption	6, 7
e's		0.6	Hα emission	8

References:

1. Savage and Massa (1987); exponential scale heights derived from IUE data for 90 stars whose sight lines mostly sample interarm and intercloud gas.

2. Jenkins (1978); Copernicus OVI survey. Unfortunately only a few stars with $|z| > 1$ kpc were observed. The indicated scale height is very uncertain.

3. van Steenberg and Shull (1987b); Gaussian scale heights derived from IUE measurements of 261 stars.

4. Morton and Blades (1986); comparison measures of CaII with HI are not available although it may be possible to reconstruct the comparison from HI measures in the literature.

5. Lockman (1984); values listed represent fits of 3 components to the observed z distribution for galactocentric distances between 4 and 8 kpc.

6. Harding and Harding (1982),

7. Readhead and Duffett-Smith (1975), and

8. Reynolds (1986)

Milky Way halo gas. Many of the recent optical studies have produced interstellar line profiles with higher resolution and higher signal to noise than the ultraviolet studies. The optical data are therefore better suited for detailed studies of individual velocity components in the absorbing gas. Unfortunately, with the exception of TiII which has nearly the same ionization potential as HI, the species available to galactic optical observers are mostly trace ionization stages of elements of relatively low abundance (e.g. CaII, NaI and KI). In contrast, the ultraviolet observations, which have mostly been obtained with the International Ultraviolet Explorer (IUE) satellite, sample a wide range of ionization states of relatively abundant elements. Since the ultraviolet data of galactic halo gas provides the closest connection to the QSO absorption line literature those data will be emphasized in this discussion.

The first survey of the optical CaII lines toward stars at high galactic latitudes by Munch (1956) and Munch and Zirin (1961) revealed that the mean number of intermediate and high velocity CaII components increased with z distance to the background star. From these observations it was concluded that interstellar clouds exist at z distances as large as about 1 kpc. These results motivated Spitzer's (1956) paper entitled *On a Possible Interstellar Galactic Corona* in which it was postulated that the neutral clouds found at large distances from the galactic plane require the pressure support of an exterior medium and this medium was most likely a hot gaseous corona. The pioneering observations of Munch and Zirin have been supplemented through important recent optical measurements of Albert (1983), Keenan et al. (1983), and Morton and Blades (1986). For a recent review of these and other optical results see Jenkins (1986).

The first studies in the ultraviolet with the IUE satellite of Milky Way halo gas involved measures of the interstellar absorption lines in the spectra of hot stars situated in the Large and Small Magellanic Cloud (Savage and deBoer 1979, 1981; Prevot et al. 1980; Gondhalekar et al. 1980). Other studies utilized Milky Way halo stars as background sources (Pettini and West 1982; deBoer and Savage 1983, 1984; Savage and Massa 1986, 1987). Extragalactic supernovae, the brightest QSOs and Seyfert galaxies have also been used as probes (Pettini et al. 1984; York et al. 1983, 1984). All these data reveal an absorbing gaseous region with a wide range of physical conditions. Because of the richness of the ultraviolet region of the spectrum the data are able to reveal absorption produced by a range of ionization states from CI to CIV, NI to NV, SiII to SiIV, etc.

The occurrence of the brightest supernova in 384 years in the LMC has provided a new sight line to the LMC for which high quality ultraviolet and optical absorption line data are now available to probe Milky Way halo gas (deBoer et al. 1987; Dupree et al. 1987; Vidal-Madjar et al. 1987). Examples of these spectra are shown in Figures 3 and 4. The strong absorption extending away from zero LSR velocity in weakly and highly ionized features originates in Milky Way disk and halo gas. At high velocities (250 km s^{-1}) absorption associated with the LMC is apparent.

A subject of considerable controversy in recent years has been the possible origin of absorption at intermediate velocities toward the LMC. Songaila et al. (1986) have proposed an LMC and/or Magellanic Stream origin for this gas while Savage and deBoer (1979, 1981) attributed the absorption to the Milky Way. This absorption is apparent in most LMC star spectra near 60 and 120 km s^{-1} (Savage and deBoer 1981) and is associated with gas having HI column densities of about $2 - 5 \times 10^{18}$ atoms cm^{-2} (McGee and Newton 1986). If this absorption is associated with the Milky Way and if the gas in the halo corotates with the gas in the disk, then the absorbing clouds would have z distances of -7 and -15 kpc, respectively (Savage and deBoer 1981). However, there is evidence that the corotation assumption is not valid for halo gas at $|z| > 1$ kpc. In fact, the rotational motion of high z gas seems to lag behind the rotation of the disk gas (see §5). For halo gas rotating more slowly than disk gas, the absorption line velocities for gas toward the LMC would imply clouds closer to the disk than the numbers given above.

Because of the uncertainties of the motions of halo gas, it is unreliable at this time to use the observed velocities of the gas to estimate its distance. A better proceedure for estimating the z extent of halo gas is to obtain measures of absorption toward stars at different z distances. With such data, the stratification of the gas away from the galactic plane can be extablished by examining at what z distance the measured column densities projected onto the z axis (N $|\sin b|$) no longer increase. Measures of this type were obtained as part of the important halo gas survey of Pettini and West (1982). In that work it was concluded that the density distribution of SiIV and CIV peaked at a z distance of about 2 to 3 kpc. Savage and Massa (1986, 1987) have recently completed work on a new survey of SiIV, CIV and NV in galactic halo and disk gas. They report absorption line measurements toward 40 B stars at a wide variety of z distances. When these data are combined with the earlier Pettini and West (1982) data, the results shown in Figure 5 are obtained. In that figure N(ion)$|\sin b|$ is plotted against $|z|$ for HI, SiIV and CIV. Figure 5d shows the expected behavior of N(ion)$|\sin b|$ versus $|z|$ if the gas has a simple exponential distribution with scale heights of 0.3, 1.0, 3.0, and 10 kpc. The data for SiIV and CIV are roughly consistent with a scale height of 3 kpc while the HI data for the same star sample yields a smaller scale height of about 0.8 kpc (for a summary of scale height estimates see Table 2). The curves for SiIV and CIV show evidence for an enhancement in the density distribution of these ions over the simple exponental curve for z near 1 to 2 kpc. Such an enhancement is predicted by various models of photoionized halos (see §6 and Figure 5d).

The work of Savage and Massa (1985, 1987) included sight lines extending over distances as large as 9 kpc through the low halo ($|z| < 2$ kpc). Such measurements permitted the clear detection of NV absorption in halo gas for 10 lines of sight. NV is of great importance for the theories of halo gas because of the ions accessible to the IUE, it is the most difficult to create by photoionization (see §6). The data shown in Figure 6 also reveal that the character of halo gas in the inner galaxy (galactocentric distances between 4 and 6 kpc) is not substantially different than that of halo gas in the vicinity of the sun. This is interesting because there is a three fold enhancement in the number of HII regions and supernova remnants between the solar neighborhood and the inner galaxy.

An exceedingly important ion for studying hot galactic halo gas is OVI since oxygen is abundant and the conversion of OV to OVI requires 113 eV. In collisionally ionized gas OVI peaks in abundance near 3×10^5 °K. Unfortunately, the information about OVI in the galactic halo is sparse because the OVI doublet near 1035 Å required the far ultraviolet capability of the Copernicus satellite which was unable to guide on stars fainter than about seventh magnitude and was limited to only a few of the brighter halo stars. From an OVI survey which included several stars with $|z| \sim 1$ kpc, Jenkins (1978) concluded that OVI has a scale height of 0.3 (+0.2,−0.15) kpc. However, this estimate is very uncertain.

A search for [FeX] $\lambda 6375$ absorption toward the LMC and several very distant stars situated in the lower galactic halo by Pettini and d'Odorico (1986) produced 3σ

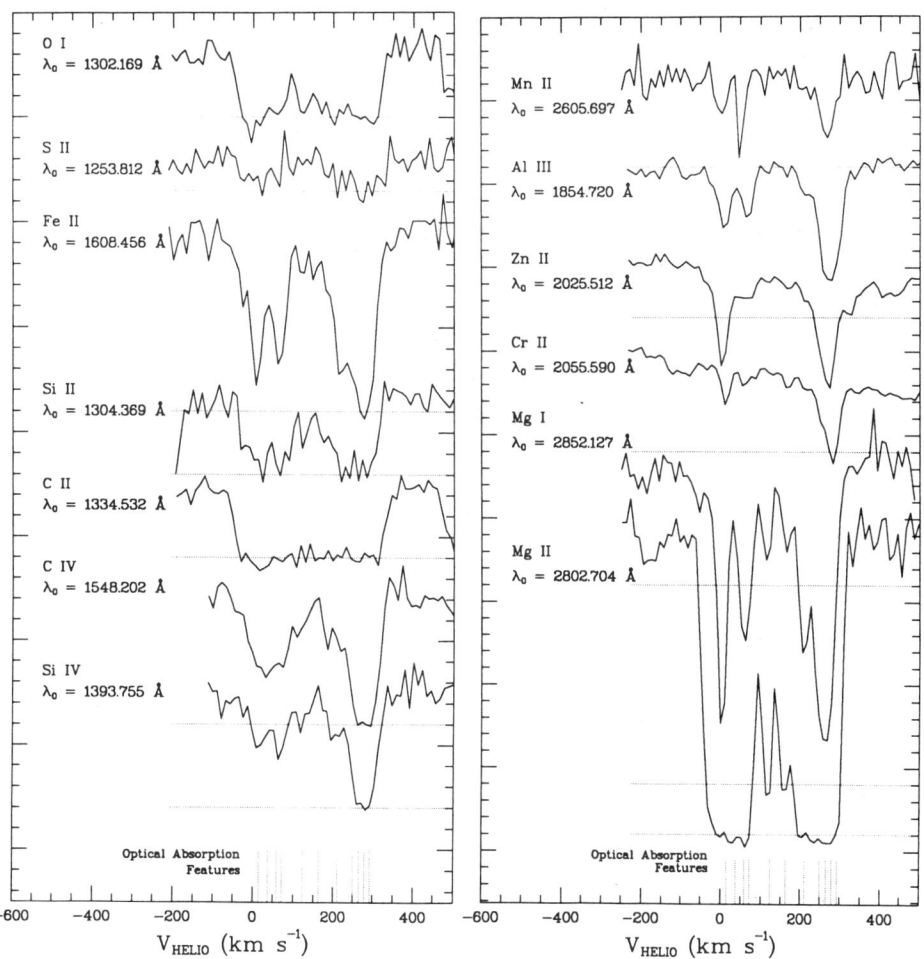

Figure 3. *Selected interstellar absorption lines in the ultraviolet spectrum of supernova 1987A. Intensity is plotted versus heliocentric velocity with the zero level of intensity indicated by the horizontal lines. The velocity of CaII absorption lines as determined from high quality optical spectra (see Figure 4) are marked on the velocity axis. The absorption extending away from 0 km s^{-1} is produced in the Milky Way. Absorption centered near 270 km s^{-1} occurs in the LMC. Absorption at the intermediate velocities of 60 and 120 km s^{-1} may be produced in the distant parts of the halo of the Milky Way, although Songaila et al. (1986) have proposed that these features are more likely associated with the LMC. In either case, these features may be representative of zero redshift absorption by gas in the outer parts of galaxies. The absorption between 0 and 100 km s^{-1} can be characterized as mixed ionization absorption with the strong low ionization lines of CII, SiII and MgII about three times stronger than the high ionization lines of CIV and SiIV. This IUE spectrum which has a resolution of 25 km s^{-1} is from Dupree et al. (1987). Similar examples of ultraviolet absorption toward the LMC are found in Savage and deBoer (1981) and deBoer et al. (1987).*

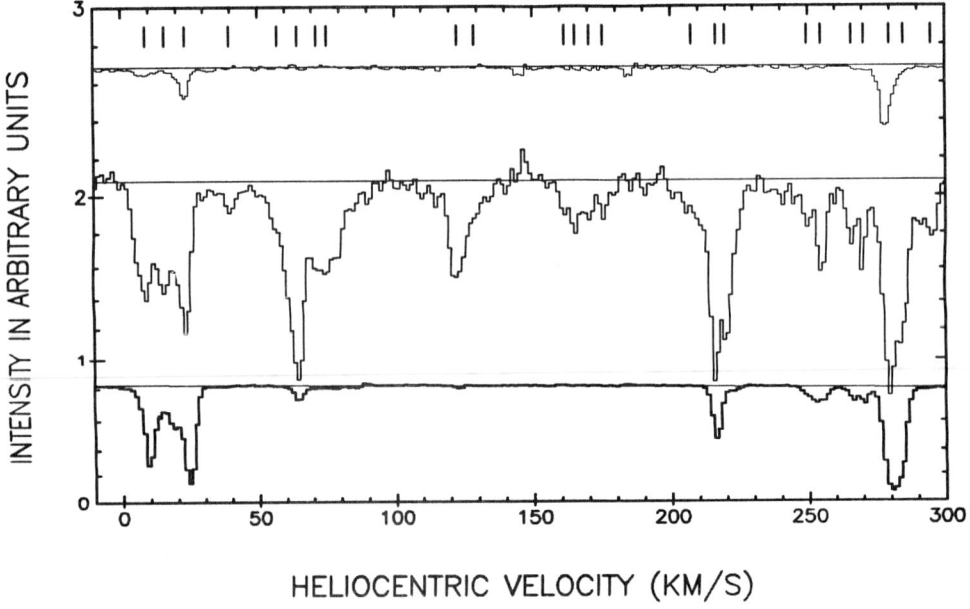

Figure 4. Optical interstellar absorption lines of KI, CaII and NaI in the spectrum of SN 1987A. At the top, tick marks show the locations of a 'CaII forest' of 24 interstellar absorption components clearly detected in the H and K lines. A comparison of the CaII and NaI absorption in the components near 10 and 25 km s^{-1} versus that in the components near 65 and 125 km s^{-1} provides an excellent example of the effect first noted by Routly and Spitzer (1952) wherein the strength of the CaII absorption increases relative to that of NaI for clouds moving at intermediate and high velocities. The most likely explanation for this effect is the destruction of dust grains containing Ca in high velocity clouds either through shock processing of grains or through the processing of grains in the gas/dust environment which may have existed before the formation of the high velocity clouds. The data shown here were obtained at a resolution of about 3 km s^{-1} which is approximately eight times better than for the IUE measurements of Figure 3. A comparison of this figure with Figure 3 shows that, at the lower resolution of the IUE data, multiple velocity components are severely blended. A curve of growth analysis that does not properly allow for this blending will likely produce very uncertain column densities. The data shown were obtained by Vidal-Madjar et al. (1987) using the high resolution instrumentation of the European Southern Observatory.

upper limits as low as 4 mÅ which corresponds to $N(FeX) \leq 6 \times 10^{16}$ cm^{-2}. In the most favorable case (HD156359) the upper limit to the column density of 10^6 °K gas implied by these data exceeds by about 3000 times the column density of gas at about 2×10^5 °K detected through NV absorption. Since the column density of 10^6 °K gas which might produce [FeX] is expected to be about 50 to 150 times higher than that of the cooler phase, it would appear that another 10 to 20 times increase in detection sensitivity is probably needed to record FeX absorption toward a star like HD156359.

Although most of the attention has focused on the highly ionized gas in the halo as traced by SiIV, CIV and NV, the neutral and weakly ionized halo gas is actually about 10 times more abundant (Savage and deBoer 1981). The neutral and weakly ionized gas which is recorded in the ultraviolet spectra of the IUE satellite has been studied by van Steenberg and Shull (1987a,b) for a group of 261 stars. From these data, the estimates of the exponential scale heights of HI, SII, ZnII, MnII, SiII and FeII listed in Table 2 were obtained. The group of stars contains a large percentage which are situated in relatively dense regions of the galactic disk, thus the derived value of the HI scale height (0.37 kpc) from these measurements is smaller than for the data shown in Figure 5a (0.8 kpc). This difference illustrates how the nature of the object sample can influence the results of these scale height estimates. Thus, comparing HI scale heights with that of other ions *for the same sample of stars* is the most informative exercise. In such a comparison we see that based on the van Steenberg and Shull data the scale heights of SII, ZnII, SiII and MnII are roughly comparable to that for HI while the scale height for FeII (1.1 kpc) exceeds by a factor of about 3 that of HI. For SII, ZnII, MnII, SiII and FeII, the ionization stage listed is the dominant ionization state of the element in the neutral interstellar gas. The more extended z distribution for FeII compared to other heavy elements was recognized in an earlier analysis of IUE data by Jenkins (1983).

A large scale height of 1.0 kpc has also been found for CaII based on the optical absorption line data of Morton and Blades (1986) who included results from Hobbs (1974), Albert (1983) and Keenan *et al.* (1983). In the case of this result, the scale height comparison with HI was not available although it might be possible to construct such a comparison using the extensive HI Ly-α survey results of Shull and van Steenberg (1985) or Bohlin, Savage and Drake (1978).

In the interstellar gas of the galactic disk which contains dust, elements are often depleted by large factors and the amount of depletion increases with line of sight density (for a recent review see Jenkins 1987). For $\langle n_H \rangle = 3$ atoms cm^{-3}, Zn, Si, Mn, Fe and Ca are depleted by factors of 3, 40, 20, 100, 4000, respectively (see Figure 5 in Jenkins 1987) while for $\langle n_H \rangle = 0.03$ cm^{-3} the depletions are factors of 2, 4, 6, 16 and 16. It is noteworthy that the two elements with the greatest depletion (Fe and Ca) are the ones having scale heights for the neutral gas phase which greatly exceed the HI scale height. A plausible explanation for this result is that the gas to dust ratio increases with $|z|$ and the effect is most easily recognized in data for the most highly depleted species. A shift in the gas to dust ratio with $|z|$ could be caused by a dust destruction processes which preferentially destroys dust in low density environments or by a halo injection process which primarily injects matter containing a greater gas to dust ratio than is found in the matter of the galactic disk. However, the selective production of gaseous Fe by Type I supernovae occurring in the galactic halo is an alternate possibility which can not be ruled out (see Jenkins 1983, 1986; van Steenberg and Shull 1987a,b).

2.3. Soft X-ray Studies

Hot plasmas emit X-ray continuum and line radiation. The continuum is from Bremsstrahlung, radiative recombination and two photon emission from one and two electron atoms. The line emission is from collisionally excited lines and recombination followed by radiative cascades. Various broad band X-ray instruments have been used to study the diffuse X-ray background from the interstellar medium. For example, the Wisconsin soft X-ray measurements of McCammon et al. (1983) and Bloch et al. (1986) employ detectors with the bands: Be(0.078-0.111 keV), B(0.130-1.88 keV), C(0.160-0.284 keV) and M_1(0.440-0.930 keV). The interpretation of the data requires a good understanding of the geometry of the hot emitting gas and the cooler absorbing gas. To reach optical depth one in the absorbing medium requires larger column densities as the X-ray band energy increases. For the Be, B, C and M_1 bands listed above an X-ray absorption optical depth of one is obtained for hydrogen column densities of 1×10^{19}, 6×10^{19}, 1.1×10^{20} and 1.1×10^{21} atoms cm^{-2}.

Figure 5 *(Opposite page).* These figures, which are based on results from the IUE satellite, provide estimates of the z extension of interstellar HI, SiIV and CIV. In (a), (b), and (c), $N(ion)|sinb|$ for HI, SiIV and CIV is plotted against $|z(kpc)|$ for approximately 60 stars located at various distances away from the galactic plane. The figure is from Savage and Massa (1985) and includes data for approximately 30 B stars from their survey plotted as open and closed circles. The results from the Pettini and West (1982) halo gas survey are shown as squares. Extragalactic measurements from Savage and deBoer (1981) and York et al. (1983) are shown as triangles. In (d) the solid lines illustrate the expected behavior of $N(ion)|sinb|$ versus $|z|$ if the gas has a simple exponential density distribution with scale heights of 0.3, 1.0, 3.0, and 10.0 kpc. These curves are most easily compared to the data by making a transparency of the figure and placing it over the data points. The data points for HI in (a) are well represented by an exponential distribution with a scale height of about 0.8 kpc. The choice of the object sample has a significant effect on the appearance of such plots. In the case of the data shown above, the objects at small z are mostly situated along sight lines with relatively low gas densities, thus the inferred exponential scale height for the HI matches that generally associated with the component of HI that extends to high z (see Table 2). The data for the highly ionized component of the gas can be compared to the data for HI in order to compare the z extension of the neutral and highly ionized gas. In the case of the SiIV and CIV measurements of (b) and (c) the inferred scale heights of roughly 3 kpc are substantially greater than for the HI. However, the latter estimates are strongly influenced by the few extragalactic data points (LMC, SMC, and 3C273). The SiIV and CIV measurements suggest the existence of an abrupt (factor of 2 to 3) increase near $|z| \approx 1$ kpc in $N(ion)|sinb|$ compared to the simple exponential curve. This increase could be the result of the photoproduction of SiIV and CIV by the extragalactic EUV radiation. The dashed and dot-dashed lines in (d) show the model results for the photoionized halo calculations of Chevalier and Fransson (1984) for SiIV and CIV, respectively.

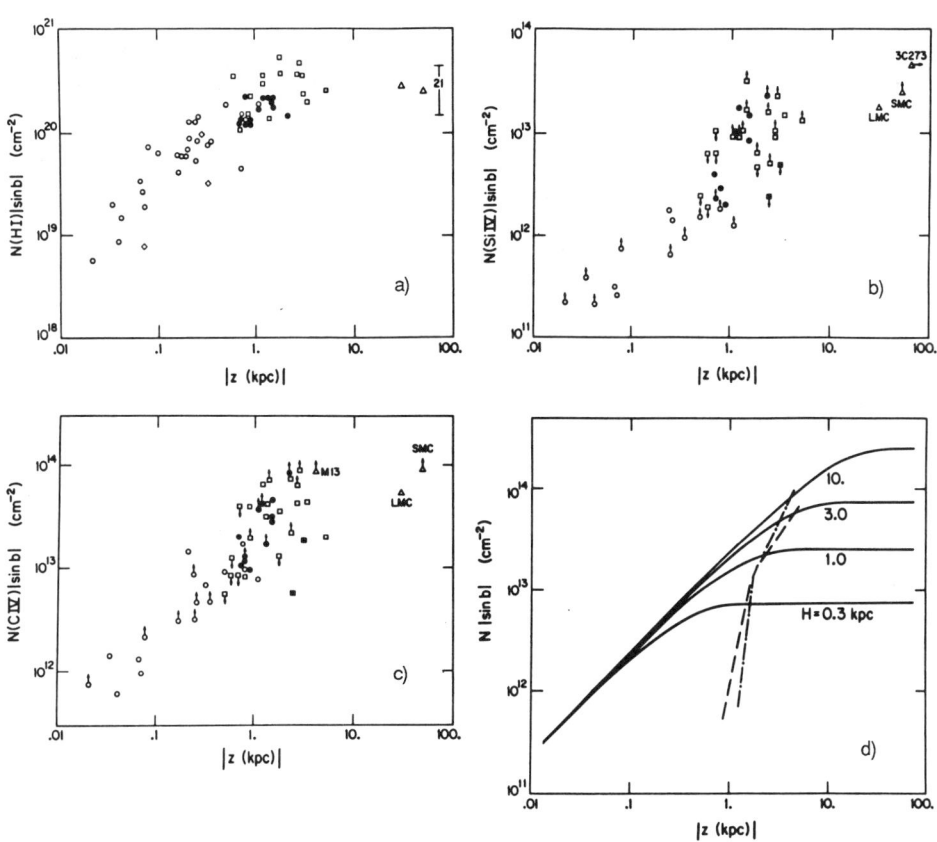

The relatively small absorption mean free path for Be and B band X-rays implies that emission observed in these bands must arise from relatively nearby gas (i.e. $r < 100$ pc; Bloch et al. 1986). For the higher energy bands the situation is less clear and has been the subject of a debate which has extended over more than a decade. Several groups have claimed the detection of X-rays from Milky Way halo gas while other groups working with similar data advocate a local origin for the X-rays. The data which have been cited as favoring the detection of a hot halo component are the MI band (0.5 to 1.0 keV) data of Nousek et al. (1982) and the SAS-3 C band data of Marshall and Clark (1984). In both cases the measurements are compatible with the existence of hot halo gas with an emission measure perpendicular to the galactic plane of about 0.004 cm^{-6} pc and a temperature of about $2-3 \times 10^6$ °K. If the X-ray emitting gas has a z extension of about 3 kpc which is the value inferred from the IUE measures of CIV and NV, its density would be about 10^{-3} cm^{-3}. However, the X-ray data are also compatible with a more local origin for the emission.

2.4. Optical and Ultraviolet Emission Line Studies

Observations of faint optical emission lines from the diffuse interstellar medium can be used to study the warm ionized gas of the Milky Way disk and halo. Studies of Hα relating to gas at high z distances have been reviewed by Reynolds (1986). Faint Hα emission from high $|z|$ gas in the Perseus spiral arm can be traced to 20 degrees away from the galactic plane which implies the detection of ionized gas at $|z| \approx 1$ kpc. The scale height estimated for this ionized gas is about 0.6 kpc. An independent estimate of the ionized gas scale height can be obtained from the variation in the Hα intensity distribution from the galactic pole to the galactic plane. The result is also 0.6 kpc. The Hα intensity at high galactic latitudes implies that toward the galactic pole the emission measure of the warm ionized gas is $\int n_e^2 dx = 2$ cm^{-6} pc. When this number is compared with the electron column densities (from pulsar dispersion measures) inferred for gas located in the direction of the galactic pole, i.e. $\int n_e dx = 8 \times 10^{19}$ cm^{-2} (see §2.5), it is possible to evaluate the amount of clumping of the ionized gas. The result is a volume filling factor of about 0.2 for gas parcels having an electron density of about 0.2 cm^{-3}.

A very important recent discovery is the definite detection of ultraviolet emission from CIV λ1550 and OIII] λ1663 in several directions at high galactic latitude by Martin and Bowyer (1987). These data appear to convincingly establish that the origin of much of the CIV found in Milky Way halo gas is through processes occuring in hot collisionally ionized gas near 10^5 °K rather than in warm (10^4 °K) photoionized gas (see §6). This result is compatible with the direct detection of NV absorption in galactic halo gas by Savage and Massa (1985). These ultraviolet results suggest that very interesting information about 10^5 °K gas in the halo might be obtained from ground based studies of emission from diffuse [OIII] λ5007.

2.5. Radio Studies of the Ionized Gas

A number of radio astronomical techniques have provided information about the ionized gas of the low halo (i.e. $|z| < 1$ kpc). These include pulsar dispersion measures which yield values of $\int n_e dx$ to a large number of pulsars (see Harding and Harding 1982) and free-free absorption measures toward extragalactic radio sources which provide $\int n_e^2 dx$ to the edge of the galaxy [Spitzer (1978) reviews these and other radio measurements of the ionized gas]. The radio measurements are compatible with the existence of a diffuse but patchy layer of ionized gas with an exponential scale height of about 0.8 kpc, an average mid-plane density $\langle n_e \rangle = 0.03$ cm^{-3}, an average mid-plane square density $\langle n_e^2 \rangle = 0.009$ cm^{-6}, and a column density to the galactic pole of $\int n_e dx = 8 \times 10^{19}$ electrons cm^{-2}. The 21 cm emission line studies and the Ly-α absorption studies toward the galactic pole yield neutral hydrogen column densities averaging about 2.5×10^{20} atoms cm^{-2}. Thus about 25 % of the gas in the direction of the galactic poles is ionized.

2.6. Nonthermal Radio Emission, Magnetic Fields and the Cosmic-Ray Halo

A full understanding of the physical nature of the gaseous galactic halo must consider the inter-relationships among the thermal gas phases, the magnetic field and the cosmic-rays. The magnetic field and cosmic-rays may have a role in supporting the thermal gas at the large z distances it is found. Through its pressure the magnetic field may influence the motions of halo gas. The field also provides a way of propagating waves into the halo which might dissipate and heat the gas (Hartquist 1983). The scale heights of the galactic magnetic field and cosmic rays are uncertain but the existing data all point toward relatively large scale heights (> 1 kpc).

Observations of cosmic-ray radio-active isotopes with long half-lives such as ^{10}Be (Garcia-Munoz, Simpson, and Wefel 1981; Guzik et al. 1985) imply that the cosmic-rays spend much of their life moving in galactic regions of relatively low gas density and that they therefore probably spend a significant portion of their lives at large distances away from the galactic plane. Studies of the non-thermal radiation from the Milky Way provide information about cosmic ray electrons interacting with the galactic magnetic field. The non-thermal radiation is estimated to have a half intensity z extension which ranges from 1 to 3 kpc (Beuermann, Kanbach and Berkhuijsen 1985 and references therein). However, it is difficult to separately derive from such data the cosmic ray scale height and the magnetic field scale height. The results for the γ-ray emission which might be attributed to Compton interactions between a cosmic ray electron halo and various background photon fields also suggest a thickness of the high energy electron halo of about 3 kpc (Stecker 1979).

Information about magnetic fields at large z distances follows from a statistical comparison of the Faraday rotation measures of extragalactic radio sources and pulsars. The extragalactic sources show a substantial excess of rotation measure compared to the pulsars. Since the pulsars are mostly situated at $|z| < 0.5$ kpc the excess can be attributed to contributions from electrons located in magnetic fields at $|z| > 0.5$ kpc (Sofue et al. 1979; Sofu, Fujimoto and Wielebinski 1986). However, such a comparison will be influenced by a component of Faraday rotation which is associated with the extragalactic source.

3. IONIZATION CHARACTERISTICS OF MILKY WAY HALO GAS

Gas in the Milky Way halo spans a wide range of physical conditions. Neutral condensations exist which produce strong absorption lines of such species as HI, CII, OI, NI, NII, MgII, SiII, FeII, AlII etc. In addition, lines from more weakly ionized gas are often found including such species as CI and MgI, NaI and CaII. The neutral gas may have temperatures ranging from 10^2 °K to more than 10^3 °K and densities ranging up to a few atoms cm^{-3}. The neutral halo clouds likely exist in a much hotter medium (the hot ISM) with a temperature of order 10^6 °K and density of order 10^{-3} to 10^{-4} cm^{-3}. This hot medium which can be studied through its X-ray emission or OVI absorption has only been adequately studied at small z distances. This is because the X-ray emission from high z gas is difficult to discriminate from X-ray emission produced by nearby low z gas (see §2.3), and because interstellar OVI absorption studies with the Copernicus satellite by Jenkins (1978) were limited to relatively bright and therefore nearby stars. Highly ionized halo gas as traced by NV, CIV, SiIV, SiIII and AlIII has been extensively studied with the IUE satellite (see §2.2). The NV and some of the CIV absorption is probably produced in collisionally ionized cooling gas with temperatures near 10^5 °K. Photoionization in 10^4 °K gas may be responsible for the AlIII, SiIII, SiIV and some of the CIV absorption (see §6).

For a galactic pole to pole path through the halo and disk perpendicular to the galactic plane at the solar position in the galaxy, the Milky Way would produce the absorption line strengths listed in Table 4. These values were estimated with reference to the IUE UV data for the objects listed at the end of the table. Comments to the table indicate if the equivalent width of a particular feature is dominated by disk absorption (i.e. for $|z| < 0.5$ kpc) or by halo gas absorption (i.e. for $|z| > 0.5$ kpc). As can be seen, even within those absorption lines formed by particular ions (e.g. Si II), some lines are mostly produced in the disk (i.e. the weak lines) while others owe their great strength to absorption by low column density high velocity halo clouds. A QSO observer recording a spectrum similar to that listed in Table 4 would refer to the system as 'mixed ionization' and would note the strong damped Ly-α absorption line. The observer would also remark that the strongest low ionization lines of CII and SiII have equivalent widths which exceed by a factor of 2 to 3 the high ion counterparts (e.g. CIV and SiIV). Most of the lines listed in Table 4 would not be detected because many of the QSO absorption line surveys have minimum detectable equivalent widths in the rest frame which exceed 0.2Å.

The only fine structure excited line listed in Table 4 is CII* $\lambda 1335$. Most of this feature arises in the higher density gas of the galactic disk although for some sight lines through halo gas, moderate density halo clouds would be capable of producing measurable but weak absorption in this excited state. For example, CII* absorption which has $W_\lambda \approx 0.1$ Å is seen in the -60 km s^{-1} cloud toward the halo star HD93521 in the direction $l = 183°$ and $b = 62°$ at a distance $r = 0.8$ kpc (Caldwell 1979). This cloud is estimated by Danly (1987) to have a z distance exceeding 0.27 kpc.

Two ions of very great importance for tracing hot gas through the Milky Way halo are NV and OVI. The equivalent widths of the NV lines produced in the Milky Way halo are relatively small and would be difficult to detect in most QSO spectra, particularly if the lines are blended with Ly-α forest lines. However, theoretical calculations suggest a pole to pole column density of about 10^{15} atoms cm^{-2} for OVI (see §6). This column density would produce a strongly saturated OVI doublet at 1032 and 1038 Å. For a velocity spread parameter $b = 30$ km s^{-1}, the two components of the doublet would have equivalent widths of 0.32 and 0.25 Å, respectively. The OVI equivalent widths would be two or three times larger if the gas producing the OVI absorption shares the multicomponent absorbing characteristics of the weakly ionized gas. Even in the presence of the Ly-α forest, lines of this strength might be detected if they exist in QSO spectra. Greater efforts should be made to establish

4. ELEMENTAL ABUNDANCES IN MILKY WAY HALO GAS

Information about elemental abundances in halo gas can provide important clues about the origin of the gas. For a review of the abundances in neutral halo gas see Jenkins (1986). If the halo gas is supplied from the disk, its abundances should approximate those found in disk gas provided allowances are made for the effects of heavy-element depletion onto dust grains. If the halo gas is supplied from intergalactic space, its abundances may reflect those found in relatively unprocessed matter. Type I supernova explosions occurring in the halo might provide for the selective enrichment of particular elements in the gas. The amount of mixing between disk and halo gas will influence the elemental abundances in halo gas. This mixing involves not only the vertical mixing, i.e. mixing in the z direction, but also the radial mixing (if it occurs) since galaxies are observed to have radial abundance gradients. The possible presence of dust introduces a major complication in interpreting abundance data for elements measured in the gas phase. For elements such as Fe, Ti, Ca and Al, the gas phase depletion for normal diffuse cloud matter which contains dust ranges from factors of 50 to 1000. Therefore, even a very modest amount of grain destruction can have an enormous effect on the measured gas phase abundances.

TiII is the best ion available to ground based optical telescopes for the study of the abundance of a normally highly depleted element. This is because the TiII lines are usually weak and therefore suitable for deriving accurate column densities provided the lines can be detected. Furthermore, TiII is the dominant ionization state of Ti in neutral hydrogen clouds and therefore the gas phase abundance relative to HI follows directly from a comparison of the column density ratio N(TiII)/N(HI). The TiII studies of Stokes (1978) and Albert (1983) reveal that although the gas phase TiII abundance is typically 10^{-2} solar in low velocity neutral disk gas, its abundance often equals and sometimes exceeds 10^{-1} solar in high velocity clouds which are found either in the galactic disk or in the galactic halo. Similar studies with the CaII lines also reveals a shift toward more gaseous CaII in high velocity and/or halo clouds [see §2.2, Albert (1983) and Figure 4]. However, it is more difficult to convert the CaII data to abundance relative to HI because of uncertainties associated with the ionization equilibrium between CaII and CaIII.

The study of elemental abundances in halo gas in the ultraviolet with the IUE satellite has been limited by the low signal to noise of typical data and by the low resolution of the IUE spectrograph for interstellar studies. For the intermediate velocity gas toward the LMC (which may or may not be associated with the Milky Way) Savage and de Boer (1981) performed a curve of growth study and obtained column density estimates or limits for neutral O and singly ionized Mg, Al, Si and Fe. When these results are combined with high sensitivity 21 cm radio data of HI emission at similar velocities from McGee, Newton and Morton (1983), the data indicate that the composition is within a factor of three of solar. The most detailed halo gas abundance study based on ultraviolet data has been for the high velocity gas toward the bright halo star HD93521 observed by the Copernicus satellite (Caldwell 1979). Although this star is in the low halo ($|z| \approx 0.7$ kpc), the elemental abundance results are informative. In the clouds at $v_{LSR} < -10$ km s^{-1}, the abundances of Fe, Si, Ar, and P are found to approach those found in the sun. This result for HD93521 is also confirmed for Ti in the ground based TiII measurements of Albert (1983).

Collectively, the available gas phase abundance information suggests that the gas in the low and perhaps distant halo has abundances which in some cases approach those found in the sun. This is in marked contrast to the highly depleted gas phase abundances usually found for disk gas. The most likely explanation for these results is that the neutral gas in the halo exists in regions relatively devoid of solid matter. The process(es) which inject matter into the halo therefore seem capable of destroying dust or of preventing dust from forming. Alternately it may be that dust destruction proceeds much more rapidly in the lower density gas of the halo. Finally, because of the additional complications involved with understanding the effects of the gas/dust interactions the existing data do not rule out the possibility that certain elements (e.g. Fe) are selectively enriched in the halo through Type I supernova explosions (Jenkins 1983).

5. KINEMATICS OF MILKY WAY HALO GAS

The observed motions of galactic halo gas may yield insights about the origin of the gas. For example, in the galactic fountain model as described by Bregman (1980), hot (10^6 °K) supernova heated disk gas moves upward (toward larger $|z|$) and radially outward (toward larger R) before suffering a thermal instability and forming into clouds which fall toward their point of origin. The upward motion is a consequence of the gas being hot and buoyant and/or greatly elevated in pressure due to a recent explosion or the collective effects of stellar winds. The radial outflow is caused by an assumed radial pressure gradient in the halo of the galaxy. If the radially outflowing gas maintains its angular momentum, its rotational velocity must change, i.e. it will not co-rotate with the gas in the disk below. After the outflowing gas cools and condenses into clouds, the clouds are expected to return on ballistic orbits to a region near the point of origin.

The galactic fountain model provides specific predictions about the kinematics of halo gas. The details of the predictions depend on the gas phase being probed. The very hot (10^6 °K) gas which can not be studied easily would be expected to exhibit motions away from the plane and away from the galactic center. In contrast, the cooler condensed gas should be observed to be returning to the galactic disk with a component of motion toward the galactic center. Because of conservation of angular momentum, this returning gas might also be expected to rotate more slowly than gas in the disk below.

In a general study of the motions of the HI 21 cm high velocity clouds (which may be related to the returning matter in a galactic fountain flow) Kaelble, de Boer and Grewing (1985) concluded that the motion pattern is consistent with the gas having a z velocity between 0 and -100 km s^{-1}, a galactocentric velocity between 0 and -100 km s^{-1} and a slower rotation than for disk gas. Thus, this analysis seems compatible with the basic idea of a fountain, a result also reached by Bregman (1980) in an analysis of a smaller subset of 21 cm data.

Optical and ultraviolet absorption line data provides an important complement to the 21 cm data because a wider span of ionization can be probed and because the origin of the absorption is constrained to the space between the sun and star whose distance can often be estimated. From an analysis of ultraviolet data, there now is strong evidence for infall of the weakly ionized gas toward the galactic plane with velocities in the z direction of up to 100 km s^{-1} (deBoer and Savage 1984;

Danly 1987). However, there are pronounced differences between the north and south galactic polar regions with the north galactic zone showing significantly more gas moving at large negative velocities (Danly 1986). The infalling gas is visible in the 21 cm emission line profiles for the galactic polar regions presented by Kulkarni and Fich (1985) and is discussed in detail in the work of Wessilius and Fejes (1973) and Danly (1987). The HI data reveal that toward the north galactic pole ($75 < b < 90°$) half the HI is falling toward the plane with v_z between -20 and -100 km s^{-1}. In the case of CIV and SiIV, the measures of Danly (1987) for stars at high north galactic latitudes do not show convincing evidence for infall or outflow while the measurements toward the south galactic pole suggest outflow.

From an analysis of Copernicus satellite data for stars with a typical $|z|$ of 0.24 kpc, Jenkins (1978) was unable to find a systematic flow associated with OVI absorption in excess of 10 km s^{-1} away from the galactic plane. From this he inferred that the rate of escape of hot material from $z \sim 0.24$ kpc is about 10 times smaller than estimates of the rate of infall of cool gas or of 10^5 °K gas (see §6). It would be very valuable to obtain such OVI measurements in the future for stars having substantially larger $|z|$ distances with the Lyman (previously known as FUSE) spacecraft.

The analysis of the velocities of CIV for halo stars at large distances in the inner galaxy by Savage and Massa (1987) shows that the observations are compatible with the notion that the highly ionized halo gas as traced by CIV rotates more slowly than disk gas, although there is no evidence for radial inflow or outflow.

An important topic concerns the expected velocity structure of halo gas absorption when viewing through the entire halo at different view angles. Blades and Morton (1983) and York (1982) have argued, based on optical CaII measurements that the velocity structure of the absorption seen along extended paths through the Milky Way halo is usually confined to 1 or 2 components over a velocity range of less than 50 km s^{-1} centered on zero LSR velocity. d'Odorico, Pettini and Ponz (1985) disagree with this claim and point out that multiple CaII profiles are in fact almost invariably seen when the sight lines through the halo are in directions where galactic rotation produces a spread of velocities and when spectra with adequate signal to noise and resolution are obtained. Since the CaII doublet is not a sensitive tracer of the low column density halo gas, it is important to look at the existing ultraviolet data. Danly (1987) has shown that the strong ultraviolet lines of SiII, CII and MgII exhibit full widths at half intensity ranging from about 100 to 140 km s^{-1} toward distant stars at the north galactic pole and from 70 to 110 km s^{-1} toward distant stars at the south galactic pole. These numbers for the north galactic polar region are considerably larger than values shown for NII by Cowie and York (1978) in their Figure 2b. The difference involves the fact that the Danly sample includes stars at much greater z distances and there appears to be a substantial change in the kinematics of the gas at about $|z| = 1$ kpc. Figure 6 (from Danly, Savage and Lockman 1987) illustrates ultraviolet line profiles toward HD100340 and HD18100. HD100340 is a star in the north galactic polar region at about $z = 5.2$ kpc in the direction $l = 259°$ and $b = 61°$ while HD18100 is in the south galactic polar region at $z = -3.4$ kpc in the direction $l = 218°$ and $b = -63°$. The substantial difference between absorption in the north and south galactic polar regions is well illustrated in these spectra. For HD100340 the FWHM of the strong low ionization lines of SiII, CII and MgII is about 140 km s^{-1} even though the object does not lie in the direction of the pronounced intermediate velocity cloud overlying the north galactic pole. A similar result was obtained for the ultraviolet bright star, VZ 1128, which is situated in the globular cluster M3 at $z = 10$

kpc in the direction $l = 42°$ and $b = 79°$ (deBoer and Savage 1984). For HD18100 the strong low ionization lines have FWHM of about 90 km s^{-1}.

The observations of weaker ultraviolet lines toward the very distant stars near the north galactic pole reveals the multicomponent absorbing nature of the gas (e.g. see the SiII $\lambda 1526$ line for HD100340 in Figure 6). Thus, the multicomponent absorbing character of the medium even exists in those directions relatively unaffected by galactic rotation. For a number of the sight lines observed by Savage and Massa (1985) toward the inner galaxy the strong ultraviolet lines of SiII, CII and MgII sometimes attain full widths at half intensity as large as 200 km s^{-1}.

The data shown in Figure 6 imply that a pole to pole observation through the Milky Way halo at the solar position in the galaxy would produce strong ultraviolet absorption lines of CII, SiII and MgII with full velocity widths at half intensity of about 140 km s^{-1}. Inclined sight lines would produce wider lines because of the consequences of galactic rotation. These line widths are about 2 to 3 times wider than those commonly predicted based on observations of the CaII line.

6. THE ORIGIN OF GALACTIC HALO GAS

Theories for the origin of the gaseous galactic halo must be able to explain the support and ionization of the gas. Two competing models for the support of the gas are the 'galactic fountain model' (Shapiro and Field 1976; Bregman 1980; also see Chevalier and Oegerle 1979 and Habe and Ikeuchi 1980) and the cosmic ray supported halo models (Hartquist, Pettini and Tallant 1984; Chevalier and Fransson 1984). In the galactic fountain model it is noted that gas will flow into the halo as a consequence of supernova explosions which heat and elevate the pressure of gaseous regions in the galactic disk. The regions of elevated pressure may breakout from the galactic plane and provide an injection of gas into the halo where the gas may cool and return to the disk in a flow pattern resembling a fountain. Calculations of the hydrodynamics of the breakout process produced by correlated supernovae in OB associations are found in Mac Low and McCray (1987) while the physical effects of falling clouds on the galactic disk are considered by Tenorico-Tagle et al. (1987). A process that will occur on a more gradual basis is 'galactic convection'. Here it is recognized from far ultraviolet absorption line studies of OVI (Jenkins 1978) and soft diffuse X-ray studies (McCammon et al. 1983) that there exists in the galactic disk a hot ($T \approx 0.3$ to 2×10^6 °K) low density ($n \approx 10^{-3}$ cm^{-3}) phase of the interstellar medium which may fill much of the volume of the galactic disk. Hot (10^6 °K) gas in the solar region of the galaxy is buoyant. Its thermal scale height is about 7 kpc. The gas will therefore tend to flow outward away from the galactic plane into the halo where it may cool and return to the disk as cooler clouds in a convective like flow. The spatial characteristics observed for the the cool gas as traced through its HI 21 cm emission illustrated in Figure 3 is suggestive of a fountain like or convective like flow. In the cosmic ray supported halo models it is assumed that the gas is supported at its large z distances by the pressure of outwardly streaming cosmic rays.

For the ionization of the gas, the possibilities for the production of the highly ionized species (SiIV, CIV and NV) include electron collisional ionization in 0.8 to 3×10^5 °K gas and photoionization by hot white dwarf stars (Dupree and Raymond 1983), by normal Population I stars, and by the extragalactic EUV background. A number of recent calculations have concentrated on determining the production of the highly ionized gas by photoionization (Hartquist, Pettini and Tallant 1984; Chevalier

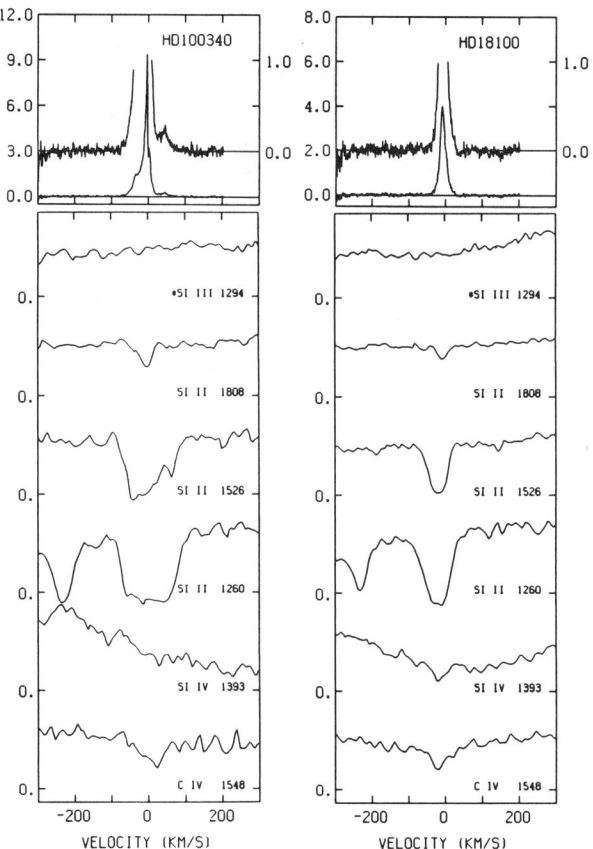

Figure 6. *Sample UV interstellar absorption line data for HD100340 ($l = 259°$, $b = 61°$, $z \approx 5.2$ kpc) and HD18100 ($l = 218°$, $b = -63°$, $z \approx -3.4$ kpc). The various absorption lines are plotted on an LSR velocity basis with tick marks indicating the zero level of intensity. The upper most absorption line is a stellar photospheric feature which provides information about the possibility of stellar blending. HD100340 and HD18100 have relatively large $v \sin(i)$ and stellar blending is not a problem. The three interstellar lines of SiII illustrate the changing appearance of the absorption produced by a single ion with the f-value of the transition. The lines of SiII $\lambda 1808$, $\lambda 1526$, and $\lambda 1260$ have oscillator strengths of 0.0055, 0.23 and 0.959, respectively. The strongest SiII line near 1260 Å clearly reveals the low column density high velocity gas of the halo. For HD100340 the strong low ionization lines of OI, CII, SiII and MgII have full velocity widths at half intensity of 140 km s^{-1} while for HD18100 the widths are about 90 km s^{-1}. At the top of each figure, 21 cm emission line data obtained with the NRAO 140 ft telescope using a technique designed to reduce the effects of antenna side lobe contamination are also shown. A comparison of the 21 cm data with the SiII $\lambda 1260$ data reveals the extreme sensitivity of the strong UV absorption lines to trace amounts of gas. This figure is from Danly, Savage and Lockman (1987). The data are representative of ultraviolet absorption along very distant sight lines toward the north and south galactic polar regions.*

and Fransson 1984; Fransson and Chevalier 1985; Bregman and Harrington 1986). The predictions of Chevalier and Fransson (1984) are shown in Figure 5d. From this work it appears possible to understand the observed amounts of SiIV and CIV and in particular the sudden rise in N(ion)|sin b| near $z = 1$ kpc from photoionization by the EUV background. However, the various calculations have difficulty producing the observed amount of NV (Savage and Massa 1985,1987). NV is an important ion since among those ions accessible to the IUE, it requires the greatest amount of energy for its production (77 eV). Most hot stars containing He have strong He^+ edges at 54 eV. Therefore, the only stellar sources that might be capable of converting NIV into NV are the very hot hydrogen white dwarfs. In order to explain the observed NV, Edgar and Chevalier (1986) calculated the amount of SiIV, CIV, NV and OVI produced in cooling gas in a galactic fountain flow. For a galactic fountain mass flow rate of 4 $M_\odot$ yr^{-1} on each side of the galactic plane, they predict the column densities perpendicular to the galactic plane listed in Table 3. We see that in the cooling fountain flow enough NV is produced to explain the observations. Their calculation also predicts CIV and OIII] emission line strengths which are compatible with the recent diffuse background measurements of Martin and Bowyer (1987). The theory predicts a large OVI column density [N(OVI) = 5.8–6.0 $\times 10^{14}$ cm^{-2}] while the measurements of Jenkins (1978) yield a column density of approximately 3×10^{13} cm^{-2} out to a $|z|$ of about 1 kpc. If the theory is correct such a large difference will require the presence of substantial quantities of OVI in the halo for $|z| > 1$ kpc. Future measures with the proposed Lyman spacecraft will provide a crucial test of the cooling fountain gas theory.

Table 3. Column densities in cooling fountain gas

Ion	Predicted[a] N(cm^{-2})	Observed[b] N(cm^{-2})
SiIV	(3.3–6.4) $\times 10^{12}$	$\approx 2 \times 10^{13}$
CIV	(4.3–7.9) $\times 10^{13}$	$\approx 1 \times 10^{14}$
NV	(2.8–3.6) $\times 10^{13}$	$\approx 3 \times 10^{13}$
OVI	(5.8–6.0) $\times 10^{14}$	$> 3 \times 10^{13}$

[a] Predicted values assume a fountain mass flow rate of 4 $M_\odot$ yr^{-1} to each side of the galactic plane. The values listed are column densities through the halo (N Sin |b|) on one side of the galaxy.
[b] Observations are from Savage and Massa (1987) for SiIV, CIV and NV and from Jenkins (1978) for OVI. The value for OVI is listed as a lower limit because the measures only extend to $|z| \approx 1$ kpc.

Savage and Massa (1987) found that the explanation for the support and ionization of halo gas requires a blending of the ideas from the galactic fountain models and the photoionized halo models. In this situation SiIV would mostly be produced by photoionization while NV would mostly be produced by collisional ionization in cooling fountain gas. CIV therefore, probably represents an intermediate situation with important contributions from collisional ionization and photoionization.

If a galactic fountain flow actually exists, it seems reasonable to interpret the inflowing neutral gas at large |b| as gas associated with that flow. This interpretation is consistent with the abundance studies of Caldwell (1979) for HD93521 and Albert (1983) for TiII toward many stars. Hot (10^6 °K) upflowing disk gas would be expected to be devoid of grains because of rapid destruction during the heating of that gas. When this hot gas cools and returns to the plane, it is unlikely that conditions will be favorable for grain formation. Therefore, the returning gas should have nearly solar abundances in the gas phase as is observed. In a galactic fountain flow the relative amounts of gas residing in high ionization states (e.g. SiIV, CIV and NV) versus gas in low ionization states (e.g. SiII, CII and NII) will be influenced by the cooling rates of the gas through the various temperature regimes. The observation that the column density of weakly ionized gas exceeds that of the highly ionized gas by about a factor of ten seems compatible with the fact that gas rapidly cools from about 3×10^5 to 10^4 °K.

The gaseous galactic halo is a region of our galaxy that is now ideally suited for innovative observational and/or theoretical studies. In a brief period of time (about 10 years) the subject of the gaseous halo has moved from one where even the very existence of such a region was doubted by many to one which is now considered of vital importance for the overall regulation of the interstellar medium of the galactic disk (e.g. see Cox 1981 and Cowie 1987).

7. ABSORPTION BY MILKY WAY HALO GAS AND THE ABSORPTION LINE SYSTEMS SEEN IN THE SPECTRA OF QSOS

Table 4 lists the approximate absorption line strengths expected for a sight line through the Milky Way disk and halo at the position of the sun. The measurements are appropriate for a sight line which is inclined at 60° from the galactic disk and which passes through halo and disk gas with positive and negative z. The resulting absorption line system would be called a damped Ly-α system with mixed ionization. The HI Ly-α line would mostly arise in disk gas while the highly ionized lines of NV, CIV and SiIV would mostly arise in halo gas. The lines of lower ionization would be produced in disk and halo gas. The absorption line system would exhibit multiple components and have a velocity spread as recorded in the strongest lines of CII, SiII and MgII of about 140 km s^{-1}. As shown by Savage and Jeske (1980) absorption line systems with the basic characteristics of the system listed in Table 4 are sometimes found in the spectra of QSOs. The frequency of occurrence of the subset of QSO systems having the characteristics of the Milky Way 'solar region' system listed in Table 4 is small because of the presence of the strong damped Ly-α line from disk gas having a HI column density of about 6×10^{20} cm^{-2}. Even if only the halo part of the absorption is considered, the fact that the strong low ionization lines of SiII and CII are about 2× stronger than the CIV and SiIV absorption appears to place the absorption system listed in Table 4 in a category which is different from the most common QSO mixed ionization systems having absorption redshifts between approximately 1 and 2 (Wolfe 1986; Danly, Blades and Norman 1987).

There are many reasons why we might expect the character of the Milky Way halo absorption line system found for gas in the solar neighborhood and those systems most commonly found in QSO spectra to appear different. First, the frequency of occurrence of the most common systems implies that if they are associated with galaxy halos then the absorption likely occurs many Holmberg radii from the center of the absorbing galaxy. Thus we need to consider how the appearance of the Milky Way halo system might change as the sight line passes further and further from the galactic center. Second, the character of the extragalactic background radiation which illuminates and ionizes the galaxy halo gas will change with redshift. Third, the character of the halo gas will change as the vigor of the fountain changes. Thus the halo gas characteristics may be sensitive to galactic evolution and in particular to the massive star production rate.

The appearance of a galaxy disk-halo absorption line system will change with galactocentric distance because of a number of factors. Some of these include:

1. changes in the vigor of the galactic fountain flow with galactocentric distance because of changes in the production rate of massive stars;
2. changes in the gravitational acceleration in the z direction, g_z, with galactocentric distance. g_z determines the degree of heating necessary to have gas reach large z distances. The observed increase of the HI scale height with galactocentric distance (see §2) may be associated with a decrease in g_z;
3. galaxy radial abundance gradients which will influence the detectability of halo gas observed via metal absorption lines;
4. changes in the relative importance of photoionization to collisional ionization with galactocentric distance which will modify the ionization state of the halo and disk gas;
5. changes in the physical density and extreme ultraviolet optical depth of the halo gas and associated disk gas with galactocentric radius which will modify the ionization state of the gas; and
6. changes in the kinematical properties of the halo gas with galactocentric distance which will almost certainly occur because of changes in the vigor of the fountain and changes in the gravitational acceleration.

It would be interesting to consider the consequences of the six effects listed above and then also evaluate the consequences of changing the extragalactic background and the stellar content of the underlying galaxy. Unfortunately such a task is full of uncertainties.

The best way to establish what the Milky Way looks like at large galactocentric distances is to look in those directions where galactic rotation permits a velocity separation of distant matter from local matter, i.e. in or near the directions $l \approx 90°$ and $l \approx 270°$ and at low enough galactic latitudes to permit the viewing of very distant gas at large $|z|$ distances. Some objects which have been observed at optical and/or ultraviolet wavelengths which may meet these criteria are:

1. SN1980K in NGC6946 at $l = 96°$ and $b = +12°$ (Pettini et al. 1982). The published data for SN1980K reveal strong absorption in MgII $\lambda 2800$, MgI $\lambda 2852$ and FeII $\lambda 2599$ to high negative velocities of about -150 km s^{-1} which can be attributed to absorption by high $|z|$ gas situated in the outer parts of the Milky Way. Unfortunately, the spectra obtained were too noisy to study absorption by highly ionized gas;
2. SN1983N in M83 at $l = 315°$ and $b = +32°$ (Jenkins et al. 1984; d'Odorico, Pettini and Ponz 1985). The published optical data for SN1983N reveal multicomponent CaII absorption associated with the Milky Way at velocities of -45, -8 and $+43$ km s^{-1}. In addition a component of unknown origin is found at $+248$ km s^{-1}. Although IUE ultraviolet data were obtained for this object, they have not been published;
3. the nuclear region of the galaxy NGC3783 in the direction $l = 287°$ and $b = +23°$ shows Milky Way absorption in the CaII H and K lines at $v_{LSR} = -39, -21, -1, +40$, and $+60$ km s^{-1}. In addition there is a CaII feature of unknown origin at 241 km s^{-1} (West et al. 1985); and

Table 4. Milky Way halo and disk gas absorption characteristics[a]

species	λ(Å)[b]	log [fλ][c]	W_λ (Å)[d]	species	λ(Å)[b]	log [fλ][c]	W_λ (Å)[d]
HI	1215.67	2.70	18.	SII	1259.52	1.30	0.35
CI	1656.93	2.35	0.07	SII	1253.81	1.13	0.25
CI	1560.31	2.10	0.06	SII	1250.59	0.83	0.14
CI	1277.21	2.30	0.08	MnII	2576.11	2.87	0.16
CII	1334.53	2.20	0.70	MnII	2593.73	2.76	0.14
CII*	1335.71	2.20	0.12	MnII	2605.70	2.62	0.12
CIV	1548.20	2.48	0.34	FeII	2599.40	2.77	0.80
CIV	1550.77	2.18	0.18	FeII	2585.88	2.25	0.75
NI	1200.71	1.72	0.25	FeII	2382.03	3.04	0.80
NI	1200.22	2.03	0.25	FeII	2373.73	2.08	0.40
NI	1199.55	2.20	0.30	FeII	2343.49	2.70	0.80
NV	1242.80	1.97	0.06	FeII	1608.46	2.19	0.30
NV	1238.81	2.28	0.12	NiII	1741.56	2.07	0.08
OI	1302.17	1.81	0.60	CrII	2055.59	2.54	0.07
MgI	2852.13	3.75	0.30	ZnII	2062.02	2.62	0.06
MgI	2026.48	2.35	0.06	ZnII	2025.51	2.92	0.12
MgII	2802.70	2.92	1.25				
MgII	2795.53	3.22	1.35				
AlII	1670.79	3.50	0.50				
AlIII	1862.79	2.70	0.13				
AlIII	1854.72	3.00	0.25				
SiII	1808.01	0.83	0.13				
SiII	1526.72	2.55	0.45				
SiII	1304.37	2.28	0.47				
SiII	1260.42	3.08	0.60				
SiII	1193.28	2.78	0.50				
SiIII	1206.51	3.30	0.55				
SiIV	1402.77	2.57	0.12				
SiIV	1393.76	2.87	0.23				

[a] For a galactic sight line at the position of the sun extending through halo and disk gas to each side of the galactic disk at an inclination of approximately 60° from the disk. The list is incomplete for λ(rest) < 1200 Å, which is near the short λ limit of the IUE spectrograph.
[b] Wavelengths from Morton and Smith (1973) are vacuum for λ < 2000 Å and air for λ > 2000 Å.
[c] The f values are from the recent compilation by Pwa and Pottasch (1986).
[d] The equivalent widths listed are inferred from IUE absorption measurements toward HD100340 ($l = 250°$, $b = 61°$, $z = 5.2$ kpc) and HD18100 ($l = 218°$, $b = -63°$, and $z = -3.4$ kpc). The absorption toward these two stars seems representative of absorption toward the north and south galactic poles, respectively. For highly saturated lines the entry is entirely determined by the measurements for HD100340 since the strongly saturated lines toward this star have FWHM $\approx$ 140 km s^{-1} compared to about 90 km s^{-1} for HD18100. Samples of the spectra are found in Figure 6. For unsaturated lines the equivalent widths listed are a straight addition of the measurements toward HD100340 and HD18100. In some cases the data have been supplemented with measurements based on other directions. Numerous weaker lines are not listed. For a reasonably complete line list for the 1200 to 3000 Å wavelength region of all the interstellar lines detected toward ζ Oph see Pwa and Pottasch (1986). The absorption toward HD100340 resembles that seen toward the LMC at velocities less than 140 km s^{-1}, while that toward HD18100 resembles the absorption seen toward the SMC at velocities less than 80 km s^{-1}. Data for the LMC and SMC sight lines were compared to QSO absorption line measurements in Savage and Jeske (1981). The absorption line equivalent widths listed above have contributions from disk and halo gas. Those lines which primarily are produced in the disk include all the weak low ion lines and the HI Ly-α line. The strong low ionization lines and the lines of intermediate and high ionization are mostly produced in halo gas.

4. stars and other sources in the LMC near $l = 280°$ and $b = -33°$ (Savage and deBoer 1981; Dupree et al. 1987; deBoer et al. 1987). The most extensive data currently available which may provide clues about what the Milky Way looks like in absorption at large galactocentric distances is the extensive collection of ultraviolet and optical data for sight lines to the LMC. If Savage and deBoer (1981) are correct in assigning the absorption features seen near 60 and 120 km s^{-1} to absorption in distant parts of the Milky Way, then those features provide the best existing information about what the outer parts of the Milky Way looks like in the resonance lines of abundant elements (see Figure 3 and 4 and the many spectra shown in Savage and deBoer 1981). Even if some or all of the 60 and 120 km s^{-1} absorption is instead associated with sheets of matter in the vicinity of the LMC as advocated by Songaila et al. (1986), it is possible that the absorption characteristics of this gas may be representative of the absorption produced in the outlying regions of galaxies at redshifts of zero. It is noteworthy that these absorption features have mixed ionization characteristics with the strong low ionization lines of CII and SiII being two to three times stronger than the higher ionization lines of CIV and SiIV (see Savage and deBoer 1981; Savage and Jeske 1981). It seems reasonable to propose that such a behavior is a general characteristic of gas in the outer regions of spiral galaxies at small redshift.

I am grateful to my colleagues at the University of Wisconsin-Madison for their many helpful discussions relating to the properties of Milky Way halo gas. I'm appreciative of support for this work through NASA grant NAG 5-186.

REFERENCES

Albert, C. E. 1983, *Ap. J.*, **272**, 509.
Bahcall, J. N., and Spitzer, L. 1969, *Ap. J. (Letters)*, **156**, L63.
Beuermann, K., Kanbach, G., and Berkhuijsen, E. M. 1985, *Astr. Ap.*, **153**, 17.
Bloch, J. J., Jahoda, K., Juda, M., McCammon, D., Sanders, W. T., and Snowden, S. L. 1986, *Ap. J. (Letters)*, **308**, L59.
Bohlin, R. C., Savage, B. D., and Drake, J. F. 1978, *Ap. J.*, **224**, 132.
Bregman, J. N. 1980, *Ap. J.*, **236**, 577.
Bregman, J. N., and Harrington, P. J. 1986, *Ap. J.*, **309**, 833.
Burton, W. B., and Lintel Hekkert, P. 1986, *Astr. Ap. Suppl.*, **65**, 427.
Caldwell, J. A. R. 1979, Ph.D. Thesis, Princeton University.
Chevalier, R. A., and Fransson, C. 1984, *Ap. J. (Letters)*, **279**, L43.
Chevalier, R. A., and Oegerle, W. R. 1979, *Ap. J.*, **227**, 398.
Cowie, L. L. and York, D. G. 1978, *Ap. J.*, **220**, 129.
Cowie, L. L. 1987, in *Interstellar Processes*, eds. D. J. Hollenbach and H. A. Thronson, Jr. (Dordrecht: D. Reidel Pub. Co.), p. 245.
Cox, D. P. 1981, *Ap. J.*, **245**, 534.
Danly, L. 1986, in *Proceedings of the NRAO Conference on Gaseous Galactic Halos*, eds. J. N. Bregman and F. J. Lockman, (Greenbank: NRAO SP), p. 45.
_____. 1987, Ph.D. Thesis, University of Wisconsin-Madison.
Danly, L., Blades, J. C. and Norman, C. 1987, in *QSO Absorption Lines: Probing the Universe, A Collection of Poster Papers*, eds., J. C. Blades, C. A. Norman and D. A. Turnshek, (ST ScI Publication), p. 88.
Danly, L., Savage, B. D., and Lockman, F. J. 1987, *Ap. J. Suppl.*, in preparation.

deBoer, K. S. and Savage, B. D. 1983, *Ap. J.*, **265**, 210.
⎯⎯⎯⎯⎯⎯⎯⎯⎯. 1984, *Astr. Ap.*, **136**, L7.
deBoer, K. S., Grewing, M., Richtler, T., Wamsteker, W., Gry, C., and Panagia, N. 1987, *Astr. Ap.*, **177**, L37.
d'Odorico, S., Pettini, M., and Ponz, D. 1985, *Ap. J.*, **299**, 852.
Dupree, A. K., Kirshner, R. P., Nassiopoulos, G. E., Raymond, J. C., and Sonneborn, G. 1987, *Ap. J.*, **320**, 597.
Dupree, A. K., and Raymond, J. C. 1983, *Ap. J. (Letters)*, **275**, L71.
Edgar, R. J., and Chevalier, R. A. 1986, *Ap. J. (Letters)*, **310**, L27.
Fransson, C., and Chevalier, R. A. 1985, *Ap. J.*, **296**, 35.
Garcia-Munoz, M., Simpson, J. A., and Wefel, J. P. 1981, *Proc. 17'th Inter. Cosmic Ray Conf.*, (Paris), **2**, 72.
Giovanelli, R. 1986, in *Proceedings of the NRAO Conference on Gaseous Galactic Halos*, eds. J. N. Bregman and F. J. Lockman, (Greenbank: NRAO SP), p. 63.
Gondhalekar, P. M., Willis, A. J., Morgan, D. H., and Nandy, K. 1980, *M.N.R.A.S.*, **193**, 875.
Guzik, T. G., Wefel, J. P., Garcia-Munoz, R., and Simpson, J. A. 1985, Proc. 19'th Inter. Cosmic Ray Conf., (La Jolla), **2**, 76.
Harding, D. S., and Harding, A. K. 1982, *Ap. J.*, **257**, 603.
Habe, A., and Ikeuchi, S. 1980, *Prog. Theor. Phys.*, **64**, 1995.
Hartquist, T. W., Pettini, M., and Tallant, A. 1984, *Ap. J.*, **276**, 519.
Hartquist, T. W. 1983, *M.N.R.A.S.*, **203**, 117.
Heiles, C. 1984, *Ap. J. Suppl.*, **55**, 585.
Henderson, A. P., Jackson, P. D., and Kerr, F. J. 1982, *Ap. J.*, **263**, 116.
Hobbs, L. M. 1974, *Ap. J.*, **191**, 381.
Jenkins, E. B. 1978, *Ap. J.*, **220**, 107.
⎯⎯⎯⎯⎯⎯⎯⎯⎯. 1983, in *Kinematics, Dynamics and Structure of the Milky Way*, ed. W. L. H. Shuter, (Dordrecht: Reidel), p. 21.
Jenkins, E. B., Rodgers, A. W., Harding P., Morton, D. C., and York, D. G. 1984, *Ap. J.*, **281**, 585.
Jenkins, E. B. 1986, in *Proceedings of the NRAO Conference on Gaseous Galactic Halos*, eds. J. N. Bregman and F. J. Lockman, (Greenbank: NRAO SP), p. 1.
⎯⎯⎯⎯⎯⎯⎯⎯⎯. 1987, in *Interstellar Processes*, eds. D. J. Hollenbach and H. A. Thronson, Jr., (Dordrecht: D. Reidel Pub. Co.), p. 533.
Kaelble, A., deBoer, K. S., and Grewing, M. 1985, *Astr. Ap.*, **143**, 408.
Knapp, G. R., Tremaine, S. D., and Gunn, J. E. 1978, *A. J.*, **83**, 1585.
Keenan, F. P., Dufton, P. L., McKeith, C. D., and Blades, J. C. 1983, *M.N.R.A.S.*, **203**, 963.
Kulkarni, S. R., Blitz, L. and Heiles, C. 1982, *Ap. J. (Letters)*, **259**, L63.
Kulkarni, S. R., and Fitch, M. 1985, *Ap. J.*, **289**, 792.
Lockman, F. J. 1984, *Ap. J.*, **283**, 90.
⎯⎯⎯⎯⎯⎯⎯⎯⎯. 1986, in *Proceedings of the NRAO Conference on Gaseous Galactic Halos*, eds. J. N. Bregman and F. J. Lockman, (Greenbank: NRAO SP), p. 63.
Lockman, F. J., Hobbs, L. M., and Shull, M. 1986, *Ap. J.*, **301**, 380.
Lockman, F. J., Jahoda, K., and McCammon, D. 1986, *Ap. J.*, **302**, 432.
Mac Low, M.-M., and McCray, R. C. 1987, *Ap. J.*, submitted.
Martin, C. and Bowyer, S. 1987, *B.A.A.S.*, **18**, 1036.

Marshall, F. J., and Clark, G. W. 1984, *Ap. J.*, **287**, 633.
McCammon, D., Burrows, D. N, Sanders, W. T., and Kraushaar, W. L. 1983, *Ap. J.*, **269**, 107.
McGee, R. X., and Newton, L. M. 1986, *Proc. Astr. Soc of Australia*, **6**, 358.
McGee, R. X., Newton, L. M., and Morton, D. C. 1983, *M.N.R.A.S.*, **205**, 1191.
Morton, D. C., and Smith, A. 1973, *Ap. J. Suppl.*, **26**, 333.
Morton, D. C., and Blades, J. C. 1986, *M.N.R.A.S.*, **220**, 927.
Munch, G. 1956, "in preparation" referenced by Spitzer (1956).
Munch, G., and Zirin, H. 1961, *Ap. J.*, **133**, 11.
Nousek, J. A., Fried, P. M., Sanders, W. T., and Kraushaar, W. L. 1982, *Ap. J.*, **258**, 83.
Oort, J. H. 1962, in *The Distribution and Motion of Interstellar Matter in Galaxies*, ed. L. Woltjer (New York: Benjamin), p. 71.
Pwa, T. H. and Pottasch, S. R. 1986, *Astr. Ap.*, **164**, 116.
Pettini, M., and West, K. A. 1982, *Ap. J.*, **260**, 561.
Pettini, M., and d'Odorico, S. 1986, *Ap. J.*, **310**, 700.
Pettini, M. *et al.* 1982, *M.N.R.A.S.*, **199**, 409.
Prevot, L. *et al.* 1980, *Astr. Ap.*, **90**, L13.
Reynolds, R. J. 1986, in *Proceedings of the NRAO Conference on Gaseous Galactic Halos*, eds. J. N. Bregman and F. J. Lockman (Greenbank: NRAO SP), p. 53.
Readhead, A. C. S., and Duffett-Smith, P. J. 1975, *Astr. Ap.*, **42**, 151.
Routly, P. M., and Spitzer, L. 1952, *Ap. J.*, **115**, 227.
Savage, B. D. and Jeske, N. 1981, *Ap. J.*, **244**, 768.
Savage, B. D., and deBoer, K. S. 1979, *Ap. J. (Letters)*, **230**, L77.
_____. 1981, *Ap. J.*, **243**, 460.
Savage, B. D., and Massa, D. 1985, *Ap. J. (Letters)*, **295**, L9.
_____. 1987, *Ap. J.*, **314**, 380.
Shane, W. W. 1971, *Astr. Ap. Suppl.*, **4**, 315.
Shapiro, P. R., and Field, G. B. 1976, *Ap. J.*, **205**, 762.
Shull, M., and van Steenberg, M. E. 1985, *Ap. J.*, **294**, 599.
Sofue, Y., Fujimoto, M., and Kawabata, K. 1979, *Pub. Astr. Soc. Japan.*, **13**, 125.
Sofue, Y., Fujimoto, M., and Wielebinski, R. 1986, *Ann. Rev. A. A.*, **24**, 459.
Songaila, A., Blades, J. C., Hu, E. M., and Cowie, L. L. 1986, *Ap. J.*, **303**, 198.
Spitzer, L. 1956, *Ap. J.*, **124**, 20.
_____ 1978, in *Physical Processes in the Interstellar Medium*, (New York: John Wiley and Sons).
Stecker, F. W. 1979, in *The Large-Scale Characteristics of the Galaxy*, ed. W. B. Burton, (Dordrecht: Reidel), p. 475.
Stokes, G. M. 1978, *Ap. J. Suppl.*, **36**, 115.
Tenorico-Tagle, G., Franco, J., Bodenheimer, P., and Rozyczka, M. 1987, *Astr. Ap.*, **179**, 219.
Tomisaka, K., and Ikeuchi, S. 1986, *Pub. Astr. Inst. Japan*, **38**, 697.
van Steenberg, M., and Shull, J. M. 1987a, *Ap. J.*, submitted.
_____. 1987b, *Ap. J.*, submitted.
van Woerden, H., Takakubo, L., and Braes, L. L. E. 1962, *Bull. Astr. Inst. Netherlands*, **16**, 321.
Wessilius, P. R., and Fejes, I. 1973, *Astr. Ap.*, **24**, 15.

West, K. A., Pettini, M., Penston, M. V., Blades, J. C., and Morton, D. C. 1985, *M.N.R.A.S.*, **215**, 481.

Wolfe, A. M. 1986 in *Proceedings of the NRAO Conference on Gaseous Galactic Halos*, eds. J. N. Bregman and F. J. Lockman (Greenbank: NRAO SP), p. 259.

Vidal-Madjar, A., Andreani, P., Cristiani, S., Ferlet, R., Lanz, T., and Vladilo, G. 1987, *Astr. Ap.*, **177**, L17.

York, D. G. 1982, *Ann. Rev. Astr. Ap.*, 20, 221.

York, D. G., Wu, C. C., Ratcliff, S., Blades, J. C., Cowie, L. L., and Morton, D. C. 1983, *Ap. J.*, **274**, 136.

York, D. G., Ratcliff, S., Blades, J. C., Cowie, L. L., Morton, D. C., and Wu, C. C. 1984, *Ap. J.*, **276**, 92.

DISCUSSION

Shull: If it is the last scenario you describe, that is making metal line systems, you would naturally expect that there would not be dust grains because they would get destroyed in the shocks. Is that consistent, though, with the low metallicities that we're hearing about?

Savage: I presented a picture for what I feel is happening here in the Milky Way. I did not attempt to make the transition from these data to what one sees in quasar systems with z of 1 or 2. In reference to the Milky Way, if the input of gas to the halo is from a fountain flow of hot gas, one would expect the rapid destruction of dust grains and therefore expect to measure solar gas phase abundances in the returning cooling gas of the fountain. This expectation seems consistent with the observations of such species as TiII in high z gas.

Shull: If we do make this leap, which I guess I make but you are not prepared to make, can you mix the fountain gas with pristine gas and still make the model work?

Savage: If you want to mix fountain gas having solar abundances with pristine gas having low metal abundances you will naturally get a gas mixture having lower than solar metal abundances. However, another leap I'm not ready to take is that of accepting the elemental abundance information that has so far been derived from quasar absorption line data. In most cases there are large and unknown uncertainties associated with the simple curve-of-growth methods which have been applied to the quasar measurements to obtain column densities.

Balbus: What was the scale height for the convective layer in the Galaxy?

Savage: The scale height for CIV and SiIV is of the order of 3 kpc, although when you look at the plots that produced this estimate, the first thing you would conclude is that the z distribution is not easily described by a simple exponential scale height. In fact, there is some evidence that a factor of 2 to 3 enhancement of density occurs near a z of about 1 kpc. Such an enhancement may be produced by photoionization by extragalactic EUV radiation.

Balbus: I also assume that the 3 kpc is about the break-out velocity times the cooling time.

Morton: With this picture of the fountain, how do the velocities work out? Are the velocities consistent with the expectations of the model?

Savage: Inflow is seen for weakly ionized gas from the north galactic pole with velocities up to about -100 km s^{-1}. Much of this inflow is associated with a large sheet of intermediate velocity gas which has a radial velocity ranging from -50 to -70 km s^{-1}. This inflow is certainly consistent with the predictions of the fountain model which predicts inflow for cool gas and outflow for the very hot gas (which can not be spectroscopically studied). However, measurements toward the south galactic pole do not reveal such an infall and one wonders to what extent our view toward the north and/or south is being biased by the existence of local phenomena.

Morton: What is the angular size of the sheet? Is it too big for that kind of pattern?

Savage: Statements about what size clumps are expected to be produced is beyond what I'm willing to make at this stage. Laura Danly has studied this feature extensively and can comment on its extent.

Danly: It occupies all of the high latitude sky from galactic longitude of 80 to 200 degrees. So, it occupies maybe a third of the sky over the northern galactic hemisphere.

Wolfe: In comparing your Milky Way halo work to the quasar absorption data, what interests me the most, of all the subsystems of the metal-line systems that we've heard about, are the MgII systems with weak CIV. These systems are the most similar to absorption seen in the Milky Way halo.

Savage: Those MgII and weak CIV systems, and also those showing damped Ly-α, appear to show the greatest similarity to what we see locally.

Wolfe: If you're just talking about lines-of-sight that avoid the disk, that just happen to go through a halo region, the subset of systems that this resembles the most is the MgII systems with weak CIV.

Ozernoy: What Doppler parameters would be observed by an external observer looking through our galaxy?

Savage: That same question was going through my mind when I was hearing the discussions of Doppler parameters over the past day and a half. The result will certainly depend on the resolution of the data. For low resolution measurements (say 100 km s^{-1} in the rest frame), the multi-component nature of the absorption would produce artifically large values of b for the low ionization lines of say SiII, MgII and FeII. I would guess values of 30 to 40 km s^{-1} might be obtained. At higher resolution (say 30 km s^{-1}), if individual components are analyzed, the b values would be smaller and perhaps of order 10 to 20 km s^{-1}. In the case of the weaker high ionization lines the inferred b values would be about 30 to 40 km s^{-1} at low and at higher resolution.

Boksenberg: Going back to the fountain picture, you said that the cool gas coming back is consistent with what you're seeing, do you see anything in the direction of outflow which might represent swept-up but outflowing gas?

Savage: Occasionally positive velocity gas is seen, but negative velocity gas appears more common.

Norman: In terms of comparing sight-lines through our Galaxy and quasar absorption lines, Danly, Blades and I have made a detailed statistical analysis of the best data available on sight-lines through our Galaxy from stars at high galactic distances and quasar absorption line data. We really find that doing Kolmogorov-Smirnoff tests on the data, using standard parameters, these things are different at the 99.99 % level. That is very similar to work Art Wolfe did some time ago, but we have a bigger sample.

Wolfe: Were the quasar data selected on the basis of CIV measurements?

Norman: Yes.

Savage: You have a high ion selected survey and you're going to have a survey biased towards strong high ionization systems. This potential bias has to be incorporated into any statements of frequency of occurrence of systems of different types. It would be interesting to do exactly the same analysis with a MgII selected survey.

Norman: Yes, we have looked at that.

Savage: And what's the answer?

Norman: The effect is still there.

Wolfe: In the survey that Lanzetta, Turnshek and I did, we employed a MgII selected sample without any regard to CIV and there is a subset that looks very much like the Milky Way halo. It's about 30 to 40 percent of the total.

Savage: How does the MgII system frequency of occurence compare to that of the high ionization systems?

Wolfe: That subset is lower, about one third; so its about a third of a third.

Norman: Yes, that's about right.

Wolfe: The CIV surveys that have been done have very low thresholds. Just the fact you're doing a CIV selected sample doesn't automatically rule out the idea that you're going to get CIV dominated systems. You can still see very weak CIV.

P. Shapiro: Just a simple argument that might suggest a way of segregating Ly-α metal-poor clouds from metal line clouds along the lines of the galactic fountain. If I take metals out of the gas, I think the radiative cooling time for this gas, that might have been supernovae heated, is ten times larger. Therefore travelling at the sound speed, as this gas does, it gets 10 times further out. So this would be a natural mechanism for segregating the metal-poor clouds that were formed in a galactic fountain-like way from the much smaller radius metal line clouds, giving a much larger scale by a factor of 10 for the Ly-α metal-poor clouds. I am thinking of higher redshift protogalactic environments where you're forming metals, and some gas clouds have metals and some do not, and the natural segregation of clouds that may arise.

Savage: What you have said is also interrelated with ideas of dust destruction rates in hot gas since many of the heavy elements in normal disk gas are situated in the grains.

Morton: Let me ask P. Shapiro a question. How much metal contamination can you tolerate for that effect to be still important?

P. Shapiro: Since you get a factor of 10 enhancement in cooling, I would guess that if you had less than a tenth solar abundance in the metals, that you lose that extra factor of 10 in cooling. You need to have a nearly solar abundance in order to get the full factor of 10 difference in scale height.

Morton: So you can have a modest contamination and still get the effect?

P. Shapiro: Yes.

Blades: I had a question for Art Wolfe, concerning the presence of CII excited (CII*). In the sub-sample of MgII systems do you see CII*? I don't find one convincing case of a real CII* detection in a quasar system.

Wolfe: I don't think we do. Am I right?

Turnshek: In our MgII selected sample there are none. Of course, there are some in the damped Ly-α systems: PKS0458-020 (Wolfe et al. 1985, Ap.J. (Letters), **294**, L67) is the best example. Also, there are some proposed detections in objects that Sargent and collaborators have studied, but they are in the Ly-α forest and there is always a possibility of a chance coincidence.

Wolfe: Yes that's right, we do see some in the damped Ly-α systems, at in at least two.

Savage: I didn't say that you find CII* in high-latitude gas that I listed is actually associated with disk gas that you're penetrating in order to view high z objects. CII* seems not to be generally detected in high z gas. However, the extensive intermediate velocity feature we discussed earlier which is situated toward the north galactic pole at a z distance of perhaps 500 pc does have CII*.

ANALOGIES AT LOW REDSHIFT FOR QSO ABSORPTION LINE REGIONS

D. G. York
Astronomy and Astrophysics Centre
University of Chicago

Abstract. Recent observations have added some precision to our knowledge of QSO absorption line regions. These new data suggest inadequacies in the standard picture that halos of normal galaxies are sufficient to explain many of the absorption line systems. Rather, extended active star formation around galaxies may be the best way to describe the absorbing regions.

1. INTRODUCTION

In 1969, Bahcall and Spitzer suggested that halos of normal galaxies might give rise to QSO absorption lines. The existence of extended gas, out to 50 – 100 kpc, could not be ruled out with other data at that time. Spitzer (1956) showed that an extended region of OVI, NV, CIV, and SiIV might exist around our own Galaxy. Later observations by the Copernicus satellite (Rogerson, Spitzer et al. 1973; Rogerson, York et al. 1973) and by IUE (Savage and Massa 1985) showed that these ions exist throughout our disk. Evidence that they exist very far out is lacking, except where other galaxies, such as the LMC and SMC, provide the background sources, in which case the absorption seen may be intrinsic to those galaxies, not the Milky Way halo.

In this short contribution, I review new data on the nature of the absorbers (§2), review recent searches for an extended halo (§3), and show that little evidence exists for gaseous halos 50 – 100 kpc in size around normal galaxies and that star forming regions provide a better description of a region capable of producing QSO absorption lines (§4). This view is subject to observational tests (§5) and has implications for models of galaxy formation (§6). This article is mainly a summary of the work of my colleagues and I and is not written as an exhaustive review.

2. DESCRIPTION OF ABSORBERS

The key new results on the QSO absorption line systems are the high resolution data recently obtained at 10 - 20 km s^{-1} resolution (Bechtold et al. 1987; Blades et al. 1985). Systems near $z \sim 1 - 2$ are shown to split into many (> 10) components of narrow absorption lines. The components are often unresolved and first estimates of their Doppler parameters, b, range from 4 - 10 km s^{-1} for species such as FeII and CIV. For these species, many of the individual components are not too saturated, allowing the doublet ratio method to be used in deriving b (Spitzer 1978). The result implies a turbulent region of rather normal clouds ($T < 2 \times 10^4$ K in several cases).

A second property that comes from the high resolution data is the non-uniformity of ionization. Bechtold et al. show that the ratio of optical depths, τ ($\tau \propto Nf\lambda$, where N is the column density, f the oscillator strength, and λ the wavelength), for SiII and SiIV varies from component-to-component by over a factor of 100. To me, this result implies ionization sources spread throughout a turbulent gas, with some regions shielded from UV ionizing radiation for SiII, but all observed regions effectively unshielded from radiation at wavelengths > 912Å. This result needs to be confirmed and tested in a number of other systems.

A third recent result is the several instances in which authors have suggested local ionization sources well above the intensity of the cosmic UV background for particular systems (Bechtold et al. 1987; Meyer and York 1987). This result could be much more general. Two obvious sources of ionization in the absorbers are stellar UV photons and the cosmic UV background. The latter produces low photon densities and requires low electron densities ($\sim 10^{-3}$ cm^{-3}) in the absorbing clouds. The former can produce high photon densities and allows higher electron densities. Observationally, electron densities in the range 1 - 10 cannot be ruled out in most cases simply because absorption lines from the collisionally excited levels are not well enough observed. The possible transitions are CII*λ1335, blended with very strong CII λ1334 in most cases, or, when resolvable, producing an equivalent width, W_λ, < 100 mÅ at $T = 10^4$ K, $n_e/n_H = 1$, $n_e < 10$ cm^{-3}; SiII* $\lambda\lambda$1190, 1193, 1260, 1309, subject to a more severe W_λ constraint than CII*, but, more seriously, usually lost in the Ly-α forest; and most promising, SiII* λ1533, weak but typically unblended. For the last transition, Meyer and York (1987) give a 5σ detection of $W_\lambda = 60$ mÅ (rest frame), in a system at $z = 2.81$. As better data are obtained, the line may become more commonly reported.

Of course, actual densities can only be derived when N(SiII), N(SiII*), and T are well known. The results quoted above are only indicative that future work along these lines can pay off. For instance, N(SiII*) can be found from λ1533; N(SiII) from λ1808; T from line width measurements of SiII λ1808 and FeII λ2385, or other weak FeII lines, thermal and turbulent effects being separated by requiring the thermal component to satisfy b_T (SiII) $\sim \sqrt{2} b_T$ (FeII).

Additionally, I point out the importance of MgI observations. While N(MgII) is difficult to determine from the often saturated λ2798Å doublet, all indications are that N(MgI)/N(MgII)< 10^{-2} (York and Morton 1988). In the high n_e regime with a cosmic UV background, most magnesium would be in MgI; as shown by all collisional ionization calculations, (e.g., Chaffee et al. 1986). The importance of MgI is that it is detectable at 1×10^4 °K, owing to strong dielectronic recombination, whereas species such as CI, SI, etc. yield only upper limits to column densities because their recombination is by much slower radiative processes. The detection of MgI in many MgII systems thus implies high radiation density, i.e., local ionization sources, if n_e is high.

It is clearly crucial to measure N(SiII*), N(SiII), N(MgII), N(MgI), b(SiII), b(FeII), and b(MgI) in the same systems. All can be measured using ground based telescopes and CCDs or photon counters for systems with $z > 1$ (so SiII* λ1533 is at $\lambda > 3100$Å) and $z < 2.5$ (so MgI 2852 is redshifted to $\lambda < 10,000$Å).

Finally, there has been recent success in actually detecting galaxies within 200 kpc (projected) of absorption line systems, e.g., see the discussion by Bergeron (1988) in this volume. Detailed data for six cases are given by Yanny et al. (1987). The available absorption-line selected sample ($z = 0.3 - 0.9$) is luminous galaxies ($M_B < -20$) and most have strong emission lines. The prototype is Q0235+164, where an intense emission patch is situated 2 arcsec from the QSO with $z = 0.525$, identical with that of a strong MgII absorber seen in the QSO (Smith et al. 1977; Wolfe et al. 1982; Cohen et al. 1987).

3. CONSTRAINTS ON HALOS

The several detections of galaxies near QSO absorbers confirms that material associated with galaxies produces the absorbers. This result is reassuring in the sense that stars are the only known site for heavy element nucleosynthesis, stars exist mainly in galaxies, and QSO systems contain heavy elements. The problem continues to be one of specifying what 'material associated' means, and what implications that meaning may have for the study of the history of galaxies.

The conventional view is that normal galaxies have extended halos. This view was strongly reinforced by Savage and de Boer (1981) and Savage and Jeske (1981) who found complex absorption in spectra of LMC and SMC stars, attributed part of it to the Milky Way halo, and pointed out the similarity of their spectra with QSO absorption line systems. Wolfe (1983) has questioned the significance of this similarity. I take the view that the LMC line-of-sight has some significant similarity to a QSO absorption line system, but that the absorption is caused not by gas in our halo, but by the gas in the Magellanic Clouds.

This view derives from several observations. First, a histogram of the number of calcium components as a function of velocity for LMC/SMC lines-of-sight is extended to high positive LSR velocities (up to > 250 km s^{-1}), whereas the similar histogram for all other extragalactic sight-lines (to QSOs and Seyferts, for example) is not much different from the same histogram for calcium in stars within 1000 pc of the sun: the last two histograms are well confined to within ± 30 km s^{-1} of the local zero velocity. In this regard, the LMC/SMC sight-lines are unique (York 1982). Songaila et al. (1986) argue from regional similarities of profiles within the clouds that the uniqueness derives from the fact that the CaII absorption at high velocity resides within the clouds, not in the halo of the Galaxy.

Second, searches for extended high velocity gas in Mkn509, Akn120, and 3C273 from the Milky Way halo have failed (York et al. 1984), most notable in 3C273 which has the same Galactic longitude as, and a complementary latitude to, the LMC. While the ions seen in QSO absorbers exist in our Galaxy, they do not unambiguously extend to great distances in any know direction.

Third, searches for CaII in QSOs behind known galaxies do not typically show absorption (Morton et al. 1986), even though a few remarkable cases are known (Boksenberg and Sargent 1978; Blades et al. 1981; Carilli et al. 1987). A detailed review of searches for CaII is given elsewhere in this volume (Blades 1988). Penston (private communication) appears to be confirming this result. On the other hand, CaII does show up in QSO absorbers (Robertson et al. 1988) and is common in LMC/SMC spectra of gas similar in other regards to QSO absorbers, as noted above.

On the theoretical side, Bregman (1981) and others (e.g., Shapiro and Field 1976) have pointed out the plausibility of a Galactic fountain that could be driven by supernovae and result in wide spread ejecta in halo regions of galaxies. According to Bregman, the complex systems at postulated 50–100 kpc distances for QSO absorbers could only be produced in very active star burst galaxies, not by normal galaxies.

Finally, the recent data revealing complex systems of many clouds with a wide range of ionization seems unlikely to arise in the halo of our Galaxy. Only very few clouds are thought to exist at large distances from the plane (the Oort clouds), and most ionizing sources of any power (i.e., O stars) are restricted to the plane.

A popular approach to extending the halo hypothesis is to assume there are enough dense clusters of galaxies that halos, producing moderately simple absorption, overlap in projection; then the velocity dispersion of galaxies within the cluster leads to broader lines. Pettini et al. (1983) attempted to explain two systems in the BL Lac 0215+015 in this way, and concluded that all the clusters expected to be dense enough to explain the complex systems (1 or 2) had to be in that line-of-sight. Thus, to explain all the known systems would require clusters to have properties at high z quite different from those at low z.

All of these arguments will be the basis of further research, and all may be shown to be wrong. However, the weight of the evidence is against the hypothesis that halos of normal galaxies, at high redshift, explain the QSO absorbers. It is not yet proven that normal galaxies exist at high redshift or that normal galaxies have halos.

4. STAR FORMING REGIONS

York et al. (1986) suggested that star forming regions might better account for the absorbers.

The available data on star forming regions in resonance absorption lines is sparse. Gas with velocities of $+30$ to -120 km s^{-1} is seen toward stars within 200 pc of the Orion belt, with ions up to (weak)CIV (Cowie et al. 1979). Stars near η Carina show velocities from -350 to $+350$ km s^{-1} in CaII as well as CIV and several intermediate ions are observed (Walborn and Hesser 1975; Walborn 1982; Walborn and Hesser 1982). A search in several other galactic HII regions failed to show high velocity gas in CIV or SiIV, but showed that at low velocity these ions are seen if stars earlier than O6 are present (Cowie et al. 1981).

Gas rich dwarf galaxies show strong interstellar absorption lines, up to 5 Å, in all the ions seen in QSO systems. The prime example may be the LMC, with the caveat that the association of the wide spread velocity components with the clouds themselves is based on the arguments in this paper. (Ideally, we will someday be able to use stars at a wide range of distances between the Sun and the LMC to probe interstellar gas and find out where the strong absorption arises, directly). A second example is found in the spectrum of the UV-bright clump of stars in NGC 1705 (Lamb et al. 1985). Ample tertiary examples are found in the UV atlas of gas rich dwarf galaxies spectra of Rosa et al. (1984) as pointed out by York, Caulet et al. (1988).

To be specific, large star forming regions (~ 1 kpc) contain many interstellar components (< 10) spread over $500 - 1000$ km s^{-1} (producing W[MgII] ~ 1.7 to 4 Å if the components are saturated) with a wide range of ions detectable (CI – CIV), and with varying ionization from component to component. The same words describe what is known of QSO absorption lines.

Of course, to explain the high frequency of QSO absorption lines requires the postulated star formation to cover a large area of the sky, implying galaxies much more actively forming stars than our own Galaxy. (York 1982 estimated $< 5\%$ chance

5. OBSERVATIONAL TESTS

5.1. General Comments

Evidently, the QSO absorbers must still be regarded as having several possible origins.

About 10% of the known absorbers could be formed in normal galaxies, statistically speaking, without postulating halos. Other papers in this volume discuss the possibility that large 21 cm extensions of a small percentage of galaxies may cause some absorbers, or that the detritus of radio sources and their jets may account for some systems, particularly those absorbers with $z_{abs} \sim z_{em}$. However, all the suggestions I know of require the postulate that an otherwise unverified change occur as one looks back in time if all the absorbers are the be explained. With regard to the several possibilities mentioned here, clustering properties must change; the average level of star formation must change; galaxies must typically have larger halos than can be verified today; or double-lobed radio galaxies must be more common, i.e., occur in a larger fraction of galaxies than they do today.

Of course, we may eventually find the answer to be 'all of the above.' QSO absorption lines may not represent a homogeneous population of objects, but may represent different physical phenomena—overlapping halos of normal galaxies, star forming regions, large disks, radio lobes and associated material, and general galaxy ejecta now unbound.

For observational tests, it is easiest to think of extreme hypotheses, realizing that when the data to support these tests are available, a continuum of possibilities may result.

Therefore, I simply place in opposition the extreme cases that:
1. all the absorbers arise in star forming regions; and
2. all absorbers arise in halos of normal galaxies (with galaxy clustering as needed).

5.2. Star Formation Hypothesis

For the first case, star forming regions with absorption lines, CI-CIV, will produce emission.

The emission can be sought by tuning a Fabry Perot to the redshifted position of the emission line using the known absorber redshift, then imaging the field of the QSO. Interference filters or dense-packed fiber bundles feeding a spectrograph can also be used (Yanny et al. 1987; York, Yanny et al. 1988; Yanny et al. 1988).

York et al. 1986 make the case to search for [OII] or [OIII], the former being at shorter wavelengths ($\lambda\lambda 3726, 3729$) and observable over a higher redshift range with low noise detectors ($z = 0$ to 1.5). We argued that star forming regions would have dust and that Ly-α, the most obvious line for high z searches, would be heavily absorbed as a result. Robertson et al. (1988) have found that CaII is quite underabundant compared to FeII and MgII in three low z MgII systems, suggesting depletion onto grains, as in the interstellar gas of our Galaxy. They point out that a high radiation field could also explain the apparent underabundance. On the other hand, Meyer and York (1987) find NiII to be in roughly solar abundance compared to S and Si, in a system at $z = 2.81$, though all are found at 1/30 solar abundance compared to HI. Finally, McCarthy et al. (1987) have detected Ly-α emission from a 100 kpc patch at $z \sim 1.9$, which also contains lines of CII] and CIII]. Heavy-element containing regions at high z do not necessarily have dust. Thus Ly-α emission may be a good search line for $z > 1.6$ systems. (It has not, however, shown up in many cases searched at $z_{abs} = z_{em}$ (Djorgovski et al. 1987; Hu and Cowie 1987).

As noted earlier, emission lines of [OII] are common in galaxies near absorbers, for the small sample so far observed. From our own work, it appears that emission at 5×10^{17} ergs cm^{-2} s^{-1} directly on top of the QSO (within ~ 4 arcsec) is not common, implying that massive star bursts do not accompany the absorbers. Many absorbers would be explained by regions like 30 Dor or the Carina nebula, however, and these giant HII regions will be difficult to observe at high redshift, largely because they subtend only a few tenths of arcseconds. It seems that detection levels of 5×10^{-19} ergs arcsec^{-2} must be reached before giant HII regions can be ruled out as the absorbers. This limit is within reach in excellent seeing with large telescopes. Presumably, these absorbers must be so common that they will show up all around the QSO; the one actually illuminated by the QSO will be difficult to detect because of the QSO continuum radiation.

A second test of this hypothesis is to search for SiII* and MgI, as described earlier. Regular detection would force one to postulate local sources of UV radiation as the ionizing flux. (The cosmic UV flux is inadequate to explain the observed ionization levels when $n_e > 0.1$ cm^{-3}.) The obvious source for such high radiation fields is hot stars just formed. Other consequences (e.g., variable ionization) already discussed can also be searched for.

5.3. Normal Galaxy Halo Hypothesis

The normal halo hypothesis assumes low densities, hence produces no detectable emission. One can continue to search for absorption in the outer parts of normal known galaxies (Monk et al. 1986), and, when found, compare the profiles with QSO absorption line profiles. Proving that normal galaxies exist at high z and that the halos have similar properties is a much larger task. The former will depend on large surveys (probably designed for other reasons). The latter will depend on finding higher z galaxies with QSOs behind them, and a comparison of the halo absorption at low z and high z. This is probably most simply done by searching for galaxies near QSOs known to have absorption lines, then correlating the number of galaxies per arcsec^2 at the absorber redshift with the width or strength of the absorber. Providing clustering of galaxies explains the large QSO absorber strengths, a positive correlation (with scatter) should be found. One must reach deep into the luminosity function of galaxies for this to be conclusive.

5.4. Comment on Statistical Tests

It should be evident that statistical studies of properties of the absorbers must currently be interpreted with two possibilites in mind. The strong lines may contain many widely spread components in the same galaxy, the dispersion arising from mechanical energy of star formation and destruction. Thus a strong or a weak system may equally indicate the detection of only one galaxy. Even two components separated by 1000 km s^{-1}, a not uncommon situation, may arise in one galaxy. Alternatively, a strong system may represent components from several (3-5) galaxies projected on top of each other, the dispersion in velocity arising from gravitational motion of the galaxies with respect to each other. Then weak systems (few components) represent single galaxies and strong systems or widely separated components represent several galaxies. Until the issues raised here are resolved, statistical studies based on system strengths or line widths cannot be directly compared with galaxy clustering statistics. Likewise, statistics of clustering of metal line systems cannot be easily compared with Ly-α—only clustering studies. The latter have strong HI lines but low mass, and may be widely dispersed by mechanical forces of galaxy formation. Again, it is difficult to say, a priori, which velocity spreads are due to mechanical, gravitational, or (in this case) cosmological effects.

6. GALAXY FORMATION

The possibility that the QSO absorbers resemble giant HII regions in their properties, and the frequency of QSO absorbers, imply that, in projection, large areas around a galaxy may well be forming stars. The converse is perhaps more immediately useful: if large areas around potential wells of galaxies are forming stars, they should be detected in absorption against QSOs. Thus, specific models of galaxy formation may possibly be ruled out. Regions at $z = 2$ with 400 kpc extent filled with star formation are trivially ruled out, for instance. If regions over a volume of radius > 100 kpc involve star formation at $z \sim 2$, the regions must be far apart.

This discussion raises the possiblity that a large amount of information is available on galaxy formation (thousands of absorbers are known) if we can only figure out what to do with it.

Spinrad and collaborators (previously referenced, and Spinrad et al. 1985) have pointed out possible regions of galaxy formation associated with 3CR radio sources, at z up to ~ 2. These regions, in absorption, would presumably look like QSO absorption line regions of the most extreme strength. The emission line widths are as large as 1000 km s^{-1}, a wide range of ions is encountered (HI-CIII]), and the area on the sky covered is $\lesssim 100$ kpc in emission. Presumably, the discrepancy between line width and ionization temperature is accounted for by a cloudy medium that produces components each < 20 km s^{-1} in width, unresolvable in emission profiles. They are ~ 1000 times more luminous in emission than the objects listed by Yanny et al. (1987), but have low specific star formation rates (in $M_\odot$ yr^{-1} pc^{-2}), comparable to limits set by York, Yanny et al. (1988) for the absorbers. Finally, some have obviously associated galaxies, while others do not (McCarthy et al. 1987). Similarly, the spectra of galaxies near QSOs have a wide range of [OII] equivalent widths, some implying few associated stars, some implying very luminous (and cool) stellar components.

The understanding of these two growing data sets will almost certainly add to our empirical knowledge of galaxy formation.

I acknowledge many discussions with my colleagues, whose names are associated with mine in the bibliography.

REFERENCES

Bahcall, J., and Spitzer, L. 1969, *Ap. J. (Letters)*, **156**, L63.
Bechtold, J., Green, R., and York, D. G. 1987, *Ap. J.*, **312**, 50.
Bergeron, J. 1988, *This volume*.
Blades, J. C., Hunstead, R. W., and Murdoch, H. S. 1981, *M.N.R.A.S.*, **194**, 669.
Blades, J. C., Hunstead, R. W., Murdoch, H. S., and Pettini, M. 1985, *Ap. J.*, **288**, 580.
Blades, J. C. 1988, this volume.
Boksenberg, A., and Sargent, W. L. W. 1978, *Ap. J.*, **220**, 42.
Bregman, J. N. 1981, *Ap. J.*, **250**, 7.
Carilli, C., and van Gorkom, J. H. 1987, *Ap. J.*, **319**, 683.
Chaffee, F. H., Foltz, C. B., Bechtold, J., and Weymann, R. J. 1986, *Ap. J.*, **301**, 116.
Cohen, R. D., Smith, H. E., Junkkarinen, V. T., and Burbidge, E. M. 1987, *Ap. J.*, **318**, 557.
Cowie, L. L., Songaila, A., and York, D. G. 1979, *Ap. J.*, **230**, 469.
Cowie, L. L., Hu, E. M., Taylor, William and York, D. G. 1981, *Ap. J. (Letters)*, **250**, L25.
Djorgovski, S., Spinrad, H., Pedelty, J., Rudnick, L., and Stockton, A. 1987, *A. J.*, **93**, 1307.
Hu, E., and Cowie, L. L. 1987, *Ap. J. (Letters)*, **317**, L7.
Lamb, S., Gallagher, J. S., Hjellming, M. S., and Hunter, D. A. 1985, *Ap. J.*, **291**, 63.
McCarthy, P. J., Spinrad, H., Djorgovski, S., Strauss, M. A., van Breugel, W., and Liebert, J. 1987, *Ap. J. (Letters)*, **319**, L39.
Meyer, D., and York, D. G. 1987, *Ap. J. (Letters)*, **319**, L45.
Monk, A. S., Penston, M. V., Pettini, M., Blades, J. C. 1986, *M.N.R.A.S.*, **222**, 787.
Morton, D. C., York, D. G., and Jenkins, E. B. 1986, *Ap. J.*, **302**, 272.
Pettini, M., Hunstead, R. W., Murdoch, H. S., and Blades, J. C. 1983, *Ap. J.*, **273**, 436.
Robertson, J. G., Morton, D. C., Blades, J. C., York, D. G., and Meyer, D. 1988, *Ap. J.*, in press.
Rogerson, J. B., Spitzer, L., Drake, J. F., Dressler, K., Jenkins, E. B., Morton, D. C., and York, D. G. 1973, *Ap. J. (Letters)*, **181**, L97.
Rogerson, J. G., York, D. G., Drake, J. F., Jenkins, E. B., Morton, D. C., and Spitzer, L. 1973, *Ap. J. (Letters)*, **181**, L110.
Rosa, M., Joubert, M., and Benvenuti, P. 1984, *Astr. Ap. Suppl.*, **57**, 361.
Savage, B. D., and deBoer, K. 1981, *Ap. J.*, **243**, 460.
Savage, B. D., and Massa, D. 1985, *Ap. J. (Letters)*, **295**, L9.
Savage, B. D., and Jeske, N. A. 1981, *Ap. J.*, **244**, 768.
Shapiro, P. R. and Field, G. B. 1976, *Ap. J.*, **205**, 762.
Smith, H. E., Burbidge, E. M., and Junkkarinen, V. T. 1977, *Ap. J.*, **218**, 611.
Songaila, A., Blades, J. C., Hu, E. M., and Cowie, L. L. 1986, *Ap. J.*, **303**, 198.
Spinrad, H. S., Filippenko, A. V., Wyckoff, S., Stocke, J. T., Wagner, R. M., and Lawrie, D. G., 1985, *Ap. J. (Letters)*, **299**, L7.
Spitzer, L. 1956, *Ap. J.*, **124**, 20.
Spitzer, L. 1978, *Physical Processes in the Interstellar Medium* (New York: Wiley).

Walborn, N. R. and Hesser, J. E. 1975, *Ap. J.*, **199**, 535.
Walborn, N. R. 1982, *Ap. J. Suppl.*, **48**, 145.
Walborn, N. R. and Hesser, J. 1982, *Ap. J.*, **252**, 156.
Wolfe, A. M. 1983, *Ap. J. (Letters)*, **268**, L1.
Wolfe, A. M., Davis, M. M., and Briggs, F. H. 1982, *Ap. J.*, **259**, 495.
Yanny, B., Hamilton, D., Schommer, R., Williams, T. B. and York, D. G. 1987, *Ap. J. (Letters)*, in press.
Yanny, B., Barden, S., Gallagher, J., and York, D. G. 1988, *Ap. J.*, submitted.
York, D. G. 1982, *Ann. Rev. A. A.*, **20**, 221.
York, D. G., Ratcliff, S., Blades, J. C., Cowie, L. L., Morton, D. C., and Wu, C. C. 1984, *Ap. J.*, **276**, 92.
York, D. G., Dopita, M., Green, R., and Bechtold, J. 1986, *Ap. J.*, **311**, 610.
York, D. G., Caulet, A., Rybskii, P., Gallagher, J., Blades, J. C., Morton, D. C., and Wamsteker, W. 1988, *Ap. J.*, submitted.
York, D. G., and Morton, D. C. 1988, *Ap. J.*, submitted.
York, D. G., Yanny, B., Williams, T. B., and Schommer, R. A. 1988, *Ap. J.*, submitted.

DISCUSSION

<u>Crotts</u>: I just thought I'd mention AO235+164 which shows a high-resolution profile for the Mg line corresponding to the redshift of the companion to the BL Lac, and it looks very much like a usual high redshifted MgII absorption line since there were 5 components seen at 22 km s^{-1} resolution spread over about 400 km s^{-1}.

<u>Wolfe</u>: Have you compared the MgII velocities with the 21 cm components? There are about 5 HI components also.

<u>Crotts</u>: I will be doing that soon.

<u>G. Burbidge</u>: What do you have to say about the other absorption line system in that object? Is that yet another accident?

<u>York</u>: You know, we have to do a much larger survey to make any sort of statistical statement.

<u>G. Burbidge</u>: The two objects are not enough?

<u>York</u>: Certainly.

<u>York</u>: I expected before starting the imaging program to look for the MgII interveners that either you would see a burst, or a diffuse glow. We looked very hard for a diffuse glow and could not find any, and we have very, very good numbers, 2 magnitudes times better than the numbers I've been quoting here. So, if this had anything to do with the spread-out galaxies Martin was talking about, you would have to postulate the bursts are short-lived, which is not surprising, and that they

leave behind a residual turbulence. In some cases, like the one you're talking about, you wouldn't have the burst that's necessarily observable. On the other hand, as Jacqueline pointed out, and as may be the case in 0453–423, it may be right on or near the the BL Lac that you have something. Now a QSO absorption line associated with a galaxy would be a rather large volume filled with gas with random bursts in it. Sometimes they may burst; sometimes they may not.

G. Burbidge: I'm just saying that this is an incredibly unlikely alignment of a BL Lac which is extremely rare with one of these objects you postulated, with yet another system which has got to produce the other absorption line system. You've got to have three in a row, where two of them are extremely rare even by quasar standards.

York: But if the BL Lac were at a cosmological distance, and there are two absorbers in front of it, there are others that have no absorbers in front of them, and it's not such an incredible accident if the galaxies in fact are spread out as far as has been suggested.

Bergeron: My comment concerns the velocity spread in low-redshift objects. In the QSO galaxy system 1327–306, which has very strong NaI, it has been studied at high resolution, and in fact we find at least two components. They are separated by 230 km s^{-1} on that line-of-sight. The galaxy is disturbed. You have a nearby galaxy well-defined with two systems maybe three, spanning at least 230 km s^{-1}. In one of the objects I showed (Table 3 in Bergeron 1988, this volume) I have also high-resolution studies on the MgII, and there are three components clearly present, also spanning something like 200 km s^{-1}. This is a mixed ionization system. As yet, those observed don't show a galaxy in which absorption systems span something like 500–1000 km s^{-1}.

Heckman: A comment on starburst identifications for quasar absorption line systems. I've been taking narrow-band images and spectra in a large sample of galaxies that are very luminous in the infrared. They're very commonly associated with large emission line nebulae, with velocities of 200 km s^{-1}. ... So, there are a number of cases where the nebula are nearly 100 kiloparsecs in size, so it's not totally out of the question that some counterpart should be identified as the quasar absorption line systems.

York: I should say that we've taken all the high signal noise gas rich dwarf spectra from IUE and measured absorption equivalent widths and they are 2–4 Å, just like quasar absorption lines.

Smith: I don't want to interrupt this discussion, but I have comments about two of the objects that you showed. The BL Lac object AO 0235+164, has a redshift for the BL Lac object of 0.94 Ross Cohen, Vesa Junkkarinen, Margaret Burbidge and myself. But from this point, it shows that the OII emission in that system is quite extended. It appears at a projected distance of something of the order of 20 kpc from the nebulosity.

York: This system is $z = 0.5$ or 0.8?

Smith: It is $z = 0.5$. The so-called galaxy, in the continuum, is stellar, to our ability to detect on high quality plates. The origin of the continuum is uncertain, but the OII emission is certainly highly extended. On another object, the double quasar, the *old* double quasar 1548, a recent paper, calls into question the reality of the absorption redshift that Margaret Burbidge and I originally recorded of the low redshift system in the spectrum of the high redshift quasar.

Shaver: Gordon Robertson and I looked at that a couple of years ago, and where you thought it was MgII, I think it turns out to be SiIV. There is a cluster of lines in that place, but it's not absorption at that redshift.

York: I should say that this has been a very hard area to get into, because after the claimed MgII absorption is erroneous—they don't exist! The point I want to make is that it's very hard to take absorption lines from the literature going back very far, which you can be sure are real in the sense that you want to spend 2 or 3 hours at the telescope. Secondly, you have to look at a very large area around the quasar. It's not fair to pick the nearest one and look at it and talk about it (assuming that any galaxy extensions are of a certain size) and not to look at objects over at least 30 arcsec. The third point is you've got to take a spectrum of every object. You can't assume that galaxies today are like galaxies in the past. You should be deriving that, not assuming it.

Disney: I wanted to return to the point that Stocke raised. I think that the selection effect against low surface luminosity dwarfs is very high ... and there are several surveys that have been undertaken which are turning up much larger numbers of dwarfs. They're predominately not in the general field. They are in clustered groups. And so the cross-section for intersection of the quasar is going to be large. I don't know what they look like.

York: Dwarfs will not necessarily produce 500 km s^{-1} absorption lines.

Tyson: I just want to comment that the fact that you find dwarfs in the vicinity of large galaxies might also be a selection effect, because if you consider the likelihood that a dwarf could be triggered to induce star formation in the presence of a large galaxy, then that's the most likely place you'll ever see it, which would not invalidate the fact that you have quiescent dwarfs somewhere else.

Felten: Mike Disney, when you said the surveys were turning up new dwarfs, you didnt' mean to imply that field results were negative, did you? Or are you simply saying that searches in clusters find them?

Disney: They're being found in clusters. The point is that we're not finding them outside quite large clusters.

Comment: But you have made a search. Did you think you would turn them up in the field?

Disney: Yes.

Wampler: I have two comments. One comment is that if you're going to make QSO absorption lines by starburst galaxies, you're going to have to do something about the reddening, because in a given object with a clear line-of-sight, really, you don't see the reddening, but many of these objects are very dusty. You could expect very often to see reddening. The other comment is that when you look at double quasars that are separated by a few arcmins, very often you find the same absorption line systems in both objects. So, you have to have a very high probability of hitting one of these small galaxies in a very short range of redshift difference.

York: Well, the universe is clustered. I've heard several people talk about this.

Boksenberg: Just a comment. The impression I seem to be getting, is that you expect all quasar absorption systems to be very complex, and they're not.

York: I showed that in a histogram.

Boksenberg: You have not made that point strong enough, because in most cases you don't have to get a very large velocity structure at all. In the extreme cases only, they're very complex.

York: It's so important to bring the literature together on this, and it's important for people to publish full results and instrumental profiles so that real things can be done. But I showed you an example of the four absorption line systems in Q1225+317. Two you would call simple. They would be unresolved. Two are complex.

Shull: Yes, I'm sitting here wondering how you can have a starburst and not see molecular hydrogen.

York: There may be quasars that we're not seeing, first of all. Those would be selected ones that are reddenned. The ones that are brightest, that can be well-studied, are certainly not going to be looking through dust clouds, so there are just all sorts of selection effects. Furthermore, and this has been pointed by several people this morning, it's very hard to make an argument that, say, [Fe/O] or [Fe/N] is not solar, though it's not clear that there are grains there. Now we have Ca in the low-z system which seems to be depleted, and we just need more data to find out. I mean, we don't know how grains form, we don't know exactly how they are destroyed, we don't know what their lifetimes are.

Savage: Yes, I just want to point out that you've got people viewing the same data and reaching somewhat different conclusions, and I would like to clarify what one might expect to see in terms of the velocity spread, if looking through the Milky Way, because you're representing it one way, and I would represent it somewhat differently. What I would say is the *minimum* velocity width would be approximately 150 km s^{-1}. That would be pole to pole. If you looked at any inclined angle which brought in galactic rotation effects, it's going to expand to the wider absorption depending on what that slant angle is. But I could easily imagine getting up to about 250 km s^{-1}, something like that, it would be very difficult to get it all the way up to 500 km s^{-1}. But 250 km s^{-1} would seem to me to be quite reasonable, even in the vicinity of the Sun.

York: Well, if you look at the Magellanic Clouds you only see gas with positive velocities. You emphasize that repeatedly.

Savage: Now, I showed you a line of sight towards North Galactic pole that show the velocity spread of 140 km s^{-1}. That's looking through half the Galaxy. That's seen in UV absorption lines. To see it at 21 cm you have to go way, way out on the wing, where the data becomes ambiguous.

York: And we have the statistics on the covering factor of the 21 cm line over the whole sky, and it's not 100%.

Savage: There are 5 or so objects at the North Galactic pole. They show this kind of spread, in UV absorption — let's not worry about 21 cm — and it is in the direction for which there is no galactic rotation. Now, if you look in directions where you have to worry about galactic rotation, you will get more spread, for example, suppose you look at an l of 90°, or an l of 110° or so. In those directions high velocity gas in the galaxy can be seen. You would get a spread, in those directions, looking again just through half the Galaxy of up to about 200 km s^{-1}.

York: As Weisheit pointed out long ago, it's very hard to use rotation to broaden QSO absorption lines, 60 km s^{-1} is all you can get when you're looking at it from a distance, unless you look right through the center which has zero probability.

Savage: Well, you're starting from 150 km s^{-1}, then add 60, 70, 80 km s^{-1}, you're getting up into a range —

York: I don't disagree, I don't disagree, but put two of them together, and then put five of them together, and ten?

Savage: Okay, but many of them, as Alec was saying, are around 200 - 250 km s^{-1} which is in my view what you would expect to get in the Milky Way situation.

THE EXTENT OF GALAXIES IN 21 CM

Renzo Sancisi
Kapteyn Astronomical Institute
Groningen University

Abstract. The density distribution of neutral hydrogen is known for a large number of galaxies. This is a brief review of the main properties of the large-scale HI density distribution in the outer parts of isolated spiral galaxies and in interacting systems which often contain large complexes of intergalactic gas. Also, attention is drawn to a category of isolated objects with peculiar gas distributions which may be the result of recent gas accretion or mergers. For the isolated galaxies the main aspects discussed here are the HI radial extent, the large-scale density distribution in the radial and vertical directions and the z-motions. Finally the discovery of high-velocity gas in the nearby spiral M101 is reported.

1. INTERACTING SYSTEMS

HI tails and bridges are known to exist around systems in tidal interaction. In addition to the well-known cases, such as the Leo Triplet, the M81 and the NGC4631 groups (cf. Sancisi 1981 and references therein), several new examples have been discovered in the last years. One of the most remarkable is that of the giant HI ring in the M96 group found by Schneider (1985) with the Arecibo radio telescope. HI tails and bridges have various shapes with sizes up to 100-200 kpc, masses up to 10^{9-10} $M_\odot$ and a broad range of densities. Sometimes they have optical counterparts and show recent star formation. It seems likely that these features are of tidal origin, although the hypothesis of primordial gas remnants, as proposed for the M96 ring, cannot be ruled out.

In addition to these cases of clearly interacting systems there are a few, puzzling examples of peculiar HI distributions which are found around systems apparently isolated and yet bear some resemblance with the distributions found in the tidal cases. They may be the result of ancient interactions followed by total disruption of one or more of the systems involved, *gas capture* by the present system and possibly final, complete *merger*. One of the most striking is the case of the S0 galaxy NGC1023, which is surrounded by a large complex of HI with an irregular and clumpy structure reminiscent of the tails and bridges found in interacting systems (see Figure 1, Sancisi et al. 1984). Since NGC1023 has no bright companion, it is thought that this gas has originated from an encounter of NGC1023 with one or more late-type hydrogen-rich systems, which would then have been tidally disrupted and would now be merging

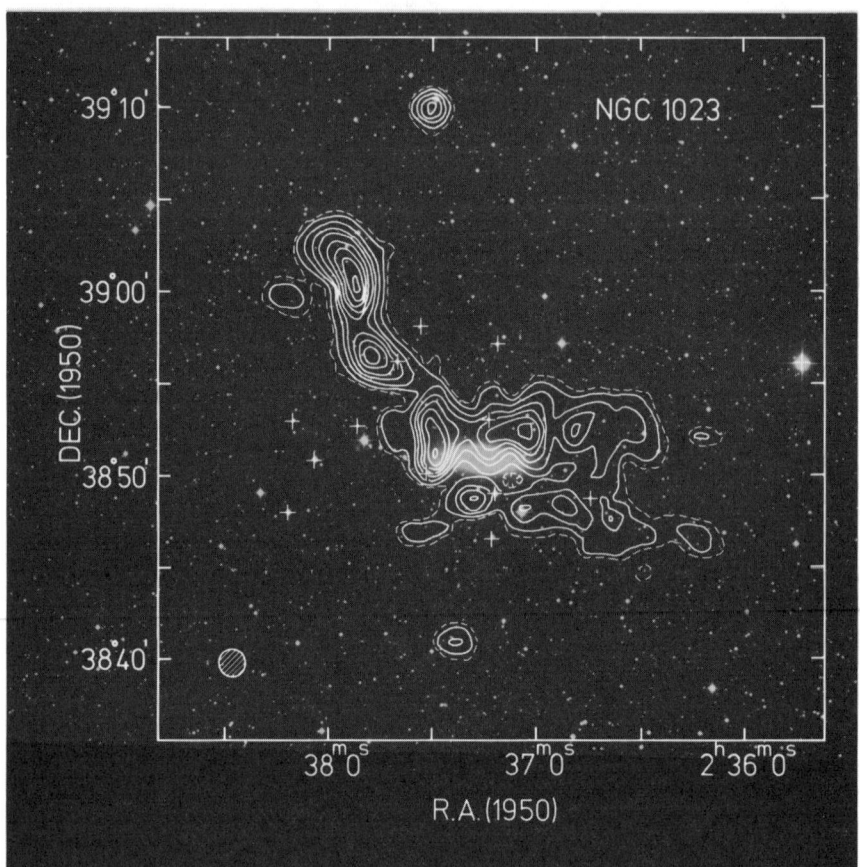

Figure 1. *Map showing the distribution of HI column density around the S0 galaxy NGC1023 (Sancisi et al. 1984). The contour values are 3 (dashed), 6, 12, 18, 24, 30, 36, 42, 48 × 10^{19} atom cm^{-2}. At an assumed distance of 10 Mpc, 1 arcmin = 2.9 kpc.*

with NGC1023.

Also the Seyfert galaxy Mkn348 may be such a case. This galaxy has one of the largest HI envelopes ever found (Heckman et al. 1982). Recent, more sensitive 21 cm observations with the Very Large Array (VLA) (Simkin et al. 1987) and the Westerbork Synthesis Radio Telescope (WSRT) (Mighell and Sancisi 1987) have revealed HI emission on one side of the galaxy extending out to 150 kpc (distance 63 Mpc). This is shown in Figure 2. The HI distribution is clearly very extended with respect to the optical image and very asymmetric. The outer HI may be, like in the case of NGC1023, a recent accretion from a gas-rich system which has been totally disrupted. In the present case, however, the origin of the gas might also be attributed to a past tidal interaction with a bright galaxy, NGC266, which is now seen at a projected distance of 23 arcmin = 415 kpc from Mkn348.

It should be noted that in the neighborhood of these puzzling 'isolated' systems, and sometimes also near the clearly interacting ones, there are dwarf, sometimes blue, elliptical companions which represent, perhaps, the remnant of the 'victim'. To this category of stripped systems may belong the small companion at the east edge of NGC1023 (Figure 1, see also Sancisi et al. 1984), the companion located 1.2 arcmin to the east of the nucleus of Mkn348 (Figure 2, also cf. Simkin et al. 1987), and the dwarf elliptical NGC4627 near NGC4631 (cf. Weliachew et al. 1978, Combes 1978).

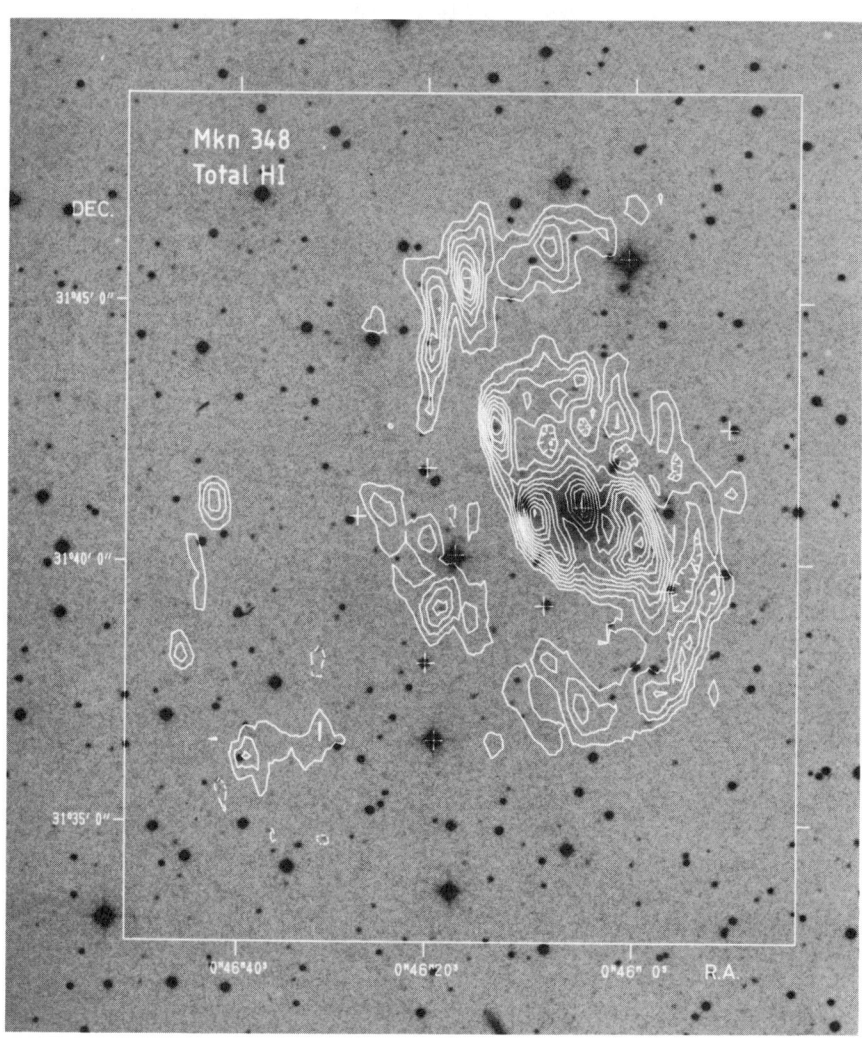

Figure 2. *Map of HI column density around Mkn348 obtained with the WSRT (Mighell and Sancisi 1987). The value of the lowest contour and the contour step are 6×10^{19} atom cm^{-2}. The beamsize is 25.1 arcsec $\times$ 47.7 arcsec. At assumed distance 63 Mpc ($H_o = 75$ km s^{-1} Mpc^{-1}) 1 arcmin = 18 kpc. The crosses mark the positions of field stars and of the centre of Mkn348. Coordinates are for 1950.0.*

2. ISOLATED SYSTEMS

We discuss here the radial and vertical distribution of HI in systems which do not show strong tidal interactions.

2.1. Radial Extent

The majority of spiral galaxies show HI emission out to 1-2 R_H, where R_H is the Holmberg radius at a level of about 0.1 to 1×10^{20} atom cm^{-2}. This is found in a large sample of galaxies studied recently at Arecibo by Briggs and Wolfe (1987) and also in a number of spirals mapped with the WSRT and VLA. Bosma (1981) found that the HI at a level of 1.82×10^{20} atoms cm^{-2} extends out to 1-2 R_H. An increase by almost a factor of 10 in the sensitivity of the observations of these last years does not seem to have led to a significant increase in the galaxy sizes. In a number of cases deep searches with the Arecibo radio telescope (Briggs 1982) seem to indicate that, as a rule, even at density levels as low as 1×10^{18} atoms cm^{-2} (for angular scales of 4 arcmin $\approx$ 10 kpc) galaxies do not extend beyond 2 R_H. At present a deep 21 cm survey of the spiral galaxy NGC3198 is being carried out with the VLA by van Gorkom et al. (1987) to test the extent of the HI layer down to levels of 1×10^{18} atoms cm^{-2} with a resolution of 1 arcmin (3 kpc). In some galaxies the HI layer seems to cut off and end at the edge of the optical disk, as in UGC2885 (Roelfsema and Allen 1985), or even inside the optical disk as in some cluster galaxies (van Gorkom et al. 1984; Warmels 1986).

There are, however, some well-known galaxies (Mkn348, M101, M83, NGC628, NGC2841, NGC2146, NGC4449, and others) with HI layers extending beyond 2 R_H (cf. Sancisi 1981, Huchtmeier and Richter 1982). A new, remarkable example of this kind is the dwarf irregular galaxy DDO154 studied recently (Krumm and Burstein 1984; Carignan and Freeman 1987). These systems seem, however, to be the exception. No clear correlation with morphological type, luminosity or other properties of the galaxy has been noted yet. It is not at all clear what the origin of the outlying gas is and why the extent is so much larger than in the large majority of cases, but our knowledge of the structure of the outer gas in these giant systems is now considerably improved. The surface densities are usually high, above 10^{19} atoms cm^{-2}. In some galaxies, like NGC628 and M101, a faint spiral structure is visible, extending from the inner to the outermost regions. In M101, well-known for its lopsided structure, the HI distribution becomes symmetrical again in the outer parts (Huchtmeier and Witzel 1979; van der Hulst and Sancisi 1987). There are also cases of irregular and asymmetric distributions as in Mkn348 and NGC2146, which seem to suggest a merger or a tidal origin as already suggested above.

2.2. Two-Component Structure

A general property seems to characterize the radial distribution of HI in spiral galaxies, both for the great majority with the 1-2 R_H extent as for the cases with the large envelopes. Near the edge of the optical disk, at R_{25} or R_H, the HI surface density drops off and a 'shoulder' shows up in the outer parts (see Sancisi 1983). Such distributions are found, for example, in NGC5055, shown in Figure 3, NGC3726, 5371, 2903 and 628 all of which are published by Wevers et al. (1986). The surface density

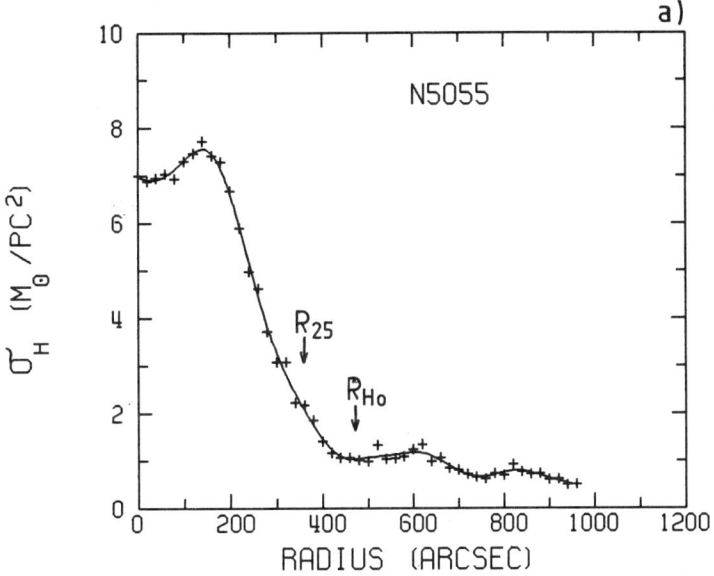

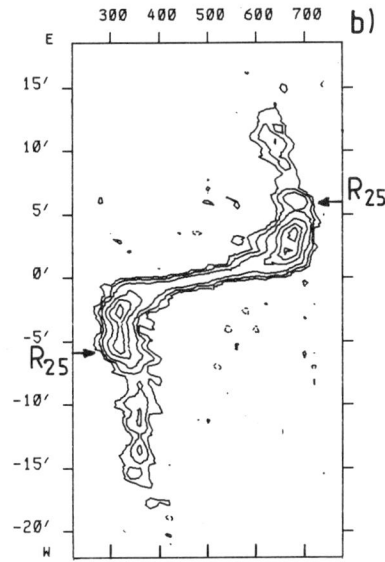

Figure 3. Diagrams showing the HI surface density drop off at the edge of the optical disk (near R_{25} or R_H) of the spiral galaxy NGC5055 and the faint 'shoulder' extending far out beyond the optical image (Wevers et al. 1986). (a) Radial distribution of HI surface density. (b) Position-velocity map along the major axis of NGC5055. The contour values are $-1.0, 1.0, 1.9, 3.6, 7.2, 10.8, 14.4$ and $18.0\,°K$. The radial velocities $(km\,s^{-1})$ are heliocentric.

in the region of the bright optical disk reaches large values above 5×10^{20} atoms cm^{-2}, whereas in the region of the shoulder it is usually below 1 or 2×10^{20} atoms cm^{-2}. This break in the HI distribution is sometimes visible in the radial density profile obtained after azimuthal averaging. In cases of asymmetric distributions the break may be smeared out by the averaging and become less noticeable. It is still recognizable, however, either in the HI maps before averaging or in channel maps or in position-velocity diagrams. At the location of this break in the HI density distribution there is usually a striking change also in other properties of the gas layer. A bending from the principal plane (e.g. NGC5907 and NGC4013) or a broadening of the HI disk (e.g. NGC891) set in. It is often possible by simply inspecting the density and kinematic structure of the gas in the radial and in the vertical directions to recognize such a characteristic radius and to predict where the bright optical disk ends. This may lead to the determination of a real 'physical size' for the stellar disk of a spiral galaxy.

The analysis of the HI observations of edge-on galaxies (NGC891 and NGC5907) leads to the suggestion that the gas in the shoulder may be part of an envelope surrounding a spiral galaxy also in the vertical direction. A two-component structure is suggested: (a) a bright, *thin layer* confined to the plane (thickness 300 pc) and extending radially out to the edge of the optical disk, similar perhaps to the 250 pc thick HI disk of our galaxy, and (b) a lower density, *thick envelope* extending vertically up to 5 kpc from the plane on either side, and stretching out in radius beyond the optical image. The two components can be recognized in the picture of Figure 4 which shows two HI channel maps of NGC891 obtained with the WSRT (Sancisi 1987). The dashed contour outlines the region of the broad component.

Figure 4 *(opposite page).* Two channel maps of neutral hydrogen (receding and approaching sides) superposed on a photograph of NGC891 (Sancisi 1987). The heliocentric systemic velocity and the radial velocities of the two channel maps are indicated. The contours show the distribution of beam-averaged brightness temperature. The dashed contour shows the extended emission at 1 mJy/beam area (= 2 × r.m.s. noise, 1 mJy/beam area = 1.2 °K, 25 arcsec × 37 arcsec beam). The other contour levels, 1.6, 3.2, 6.4, 9.6, 12.8, 16.0, 19.2, and 22.4 mJy/beam area, are from higher resolution data (1mJy/beam area = 3.1 °K, 13 arcsec × 19 arcsec beam, see hatched ellipse) and show the thin HI layer confined to the plane and coinciding with the dust lane. At the distance of 9.5 Mpc (H_o = 75) 1 arcmin = 2.8 kpc.

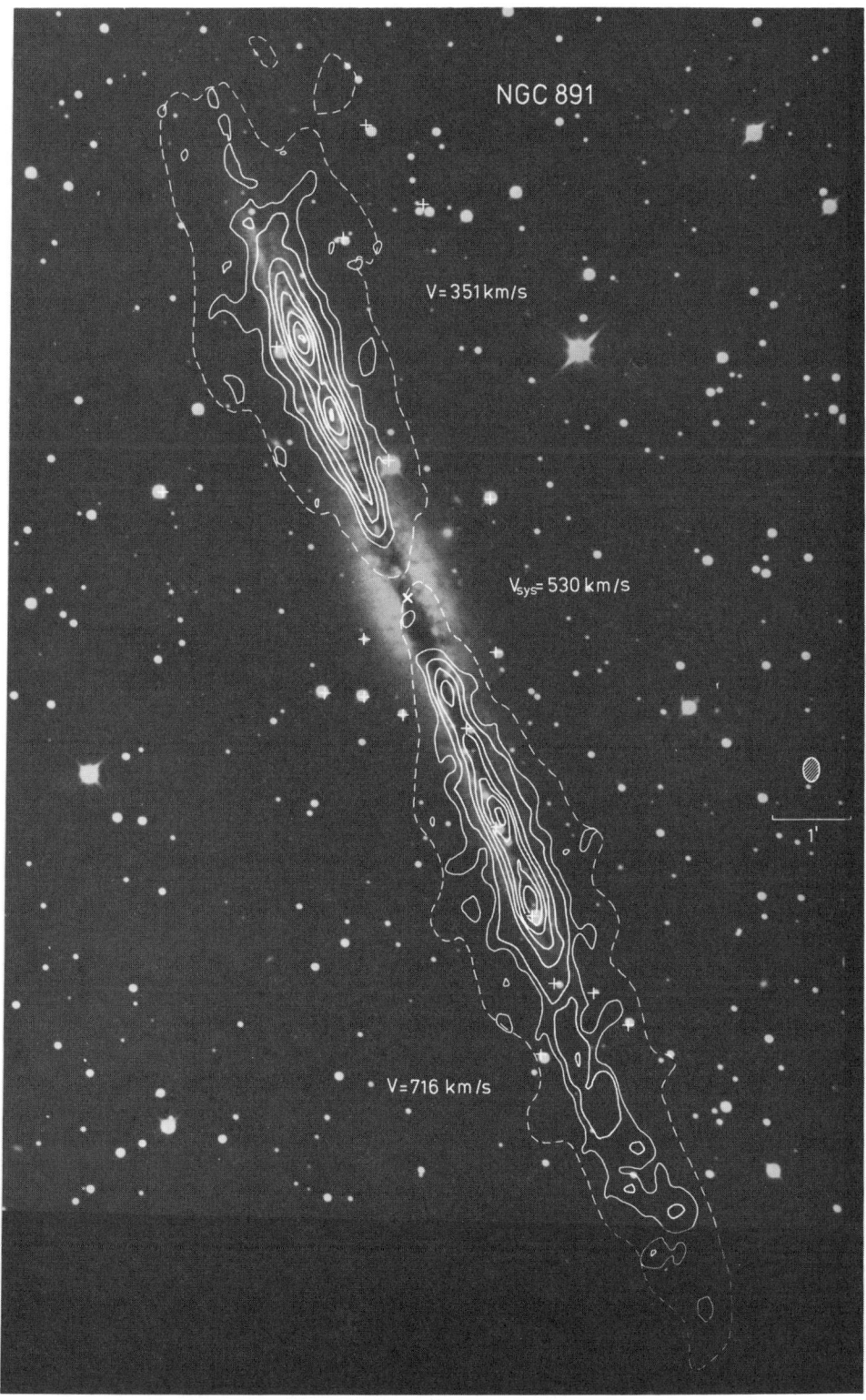

2.3. Dependence on Location in a Cluster

There is now clear evidence that in a galaxy not only the amount but also the extent of HI depend on its particular location in a cluster. The 21 cm line observations of the Virgo cluster of galaxies (van Gorkom et al. 1984; Warmels 1986) have shown that HI disks within 2 or 3 degrees of the cluster centre are HI deficient and much smaller than the optical disk (see the striking example of NGC4569 in Figure 12b p. 81 of Warmels 1986), whereas disks in the outer regions of the cluster are normal. Furthermore, galaxies close to the cluster centre show large asymmetries in their HI distribution. The gas depletion and the asymmetries are attributed to the effect of ram-pressure stripping by the hot intracluster medium.

2.4. HI Extent and Dark Halo

The outer HI layer or envelope is a tracer of the system kinematics and as such it provides crucial evidence on the presence of dark matter in the outer parts of spiral galaxies. It is therefore important to ask whether the gaseous layer ends at some radius or whether it simply becomes ionized. In the latter case it might still be observed by some other technique than the 21 cm HI line. One interesting property of the density distribution of the gas is that the ratio of the local mass surface density (from rotation curves) to the HI density remains approximately constant with radius in the outer parts (cf. Bosma 1981). This is in clear contrast with the well-known behaviour of the mass/luminosity ratio, which increases in the outer parts. This ratio of total mass density/ HI density needs to be investigated with more accurate measurements and a more careful study. If it is indeed constant with radius in the outer parts of a spiral, the question arises whether it is a coincidence or whether it implies that the distribution of gas and of dark matter are related and perhaps coextensive in a more or less flattened distribution.

2.5. The Vertical Distribution

The thin-thick component structure of the HI distribution in spirals has already been discussed above and the hypothesis has been made that the shoulder seen beyond the optical radius may be part of a low-density, flat system of gas enveloping the whole galaxy. This gas constitutes a large reservoir, often comparable in amount to that contained in the thin, inner HI disk, and may have an important effect on galaxy evolution. Its vertical distribution is characterized by large-scale deviations from the optical plane. Figure 5 shows an example of such a warped layer. Furthermore, HI layers of spiral galaxies seem to flare in their outer parts. The extent and shape of this effect, however, are not determined precisely. The evidence comes from the 21 cm line observations of our galaxy, M31 (Brinks and Burton 1984) and NGC891 (Sancisi and Allen 1979). A new 21 cm study of this edge-on galaxy with higher sensitivity and angular resolution is being carried out (Broeils et al. 1987) to investigate the 2-component structure, the outer thickening and the possibility of a line-of-sight warp.

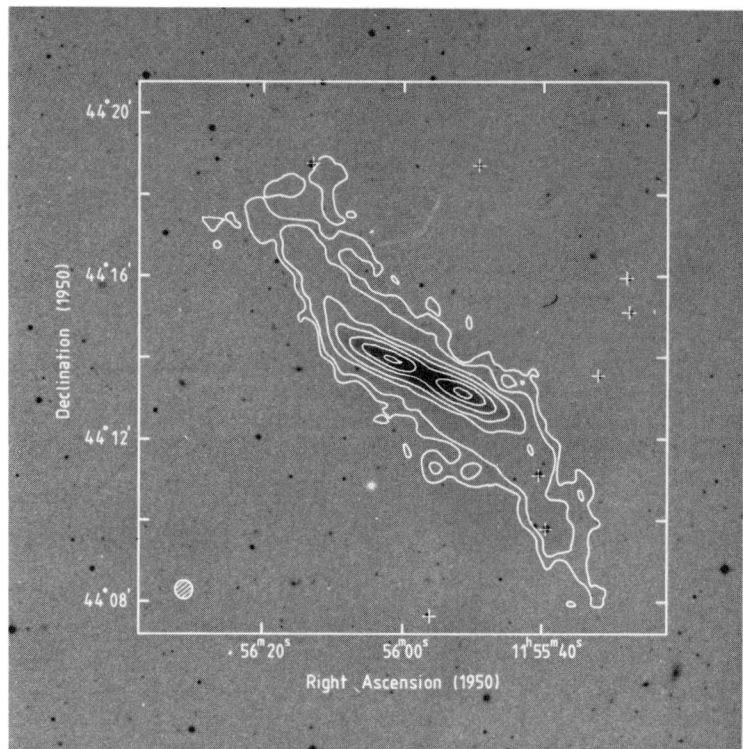

Figure 5. *Map of the total HI distribution showing the 'prodigious' warp of NGC4013 (Bottema et al. 1987). The beamsize is 30 arcsec × 30 arcsec. Contour levels are at 0.17, 1.1, 5.6, 11, 20, 29 and 38 × 10^{20} atom cm^{-2}.*

2.6. The Vertical Motion

In order to understand the origin of the gas at high z it is necessary to know its vertical motion and its connection to the HI in the plane. Observations of face-on galaxies such as those by van der Kruit and Shostak (1984), but more sensitive, may serve this purpose. It is possible now to carry out a deep search, in nearby spiral galaxies, for a population of gas clouds similar to the high velocity clouds (HVCs) of our galaxy. In fact, large gas complexes like the Magellanic Stream have already been detected in several nearby interacting systems (see above). A more 'normal' high-velocity gas component in isolated galaxies, presumably like at least some of the HVCs in our galaxy, may be too weak to be detected. Unfortunately it is not known how far away the HVCs are and consequently their sizes and masses are quite uncertain. They may well be several kpc away from the Sun, in which case a similar population of HI gas complexes in nearby spirals would be within detection range. A search for such objects has already been tried using existing Westerbork and VLA 21 cm observations of the nearby spirals M51, M83, and NGC628 with negative results (Wakker et al. 1987).

2.7. High-Velocity Gas in M101

A new, higher sensitivity 21 cm line study of M101 just carried out with the WSRT (van der Hulst and Sancisi 1987) has, however, revealed the presence of neutral hydrogen moving at high speed in the direction perpendicular to the disk. Two large complexes of 2×10^7 and of 2×10^8 $M_\odot$ moving with velocities of up to 150 km s^{-1} with respect to the rotating disk of M101 have been found in two different locations, in the direction of pronounced kinematical disturbances and holes in the spiral structure. They seem connected in velocity with the 'local' HI disk. The origin of these structures is not clear. It is unlikely that they are caused by supernova explosions in the disk of M101, but they could be the result of collisions of large extragalactic gas clouds with the disk of M101. These two high-velocity complexes, although quite massive, may be similar to the HVCs of our galaxy.

REFERENCES

Bosma, A. 1981, *A. J.*, **86**, 1825.
Bottema, R., Shostak, G. S. and van der Kruit, P. C. 1987, *Nature*, **328**, 401.
Briggs, F. H. 1982, in *The Comparative HI Content of Normal Galaxies, Proceedings of a Green Bank Workshop*, ed. M. Haynes and R. Giovanelli (NRAO Green Bank), p. 50.
Briggs, F. H. and Wolfe, A. 1987, private communication.
Brinks, E. and Burton, W. B. 1984, *Astr. Ap.*, **141**, 195.
Broeils, A., Sancisi, R. and van Albada, T. S. 1987, in preparation.
Carignan, C. and Freeman, K. C. 1987, private communication.
Combes, F. 1978, *Astr. Ap.*, **65**, 47.
van Gorkom, J. H., Balkowski, C. and Kotanyi, C. 1984, in *Clusters and Groups of Galaxies*, eds. F. Mardirossian et al. (Dordrecht: Reidel), p. 261.
van Gorkom, J. H., Cornwell, T., Sancisi, R. and van Albada, T. S. 1987, in preparation.
Heckman, T. M., Sancisi, R., Sullivan, III, W. T. and Balick, B. 1982, *M.N.R.A.S.*, **199**, 425.
Huchtmeier, W. K. and Richter, O.-G. 1982, *Astr. Ap.*, **109**, 331.
Huchtmeier, W. K. and Witzel, A. 1979, *Astr. Ap.*, **74**, 138.
van der Hulst, J. M. and Sancisi, R. 1987, preprint.
van der Kruit, P. C. and Shostak, G. S. 1984, *Astr. Ap.*, **134**, 258.
Krumm, N. and Burstein, D. 1984, *A. J.*, **89**, 1319.
Mighell, K. and Sancisi, R. 1987, in preparation.
Roelfsema, P. R. and Allen, R. J. 1985, *Astr. Ap.*, **146**, 213.
Sancisi, R. 1981, in *The Structure and Evolution of Normal Galaxies*, ed. S. M. Fall and D. Lynden-Bell (Cambridge University Press), p. 149.
Sancisi, R. 1983, in *IAU Symposium 100, Internal Kinematics and Dynamics of Galaxies*, ed. E. Athanassoula (Dordrecht: Reidel), p. 55.
Sancisi, R. 1987, in preparation.
Sancisi, R. and Allen, R. J. 1979, *Astr. Ap.*, **74**, 73.
Sancisi, R., van Woerden, H., Davies, R. D. and Hart, L. 1984, *M.N.R.A.S.*, **210**, 497.
Schneider, S. E. 1985, *Ap. J.*, **288**, L33.

Simkin, S. M., van Gorkom, J., Hibbard, J. and Su, H.-J. 1987, *Science*, **235**, 1367.

Wakker, B., Broeils, A., Tilanus, R. and Sancisi, R. 1987, in preparation.

Warmels, R. 1986. PhD. Thesis, Groningen University.

Weliachew, L., Sancisi, R. and Guelin, M. 1978, *Astr. Ap.*, **65**, 37.

Wevers, B. M. H. R., van der Kruit, P. C. and Allen, R. J. 1986, *Astr. Ap. Suppl. Ser.*, **66**, 505.

DISCUSSION

Rees: I have a question about the extent of these disks. You pointed out that it looked as though the mass of neutral hydrogen was increasing proportional to radius, in which case the column density would be going down as 1 over radius. If you bear in mind that this gas is being irradiated by the UV background, what is going to happen is that when the column density drops below some threshold value, it all becomes ionized. Therefore, even if the gas does continue all the way out, it will suddenly all become ionized and there will be a sharp cut off in the H I column density. Rough estimates suggest that may happen at the level you are seeing now. So would you therefore agree that even if you do see a sharp cut-off, this doesn't rule out the possibility that in fact gas does continue further out? This possibility was discussed by Bergeron and Felten about 15 years ago.

Sancisi: Yes, it may well be. I agree with you that the HI cutoff may already occur near the limit reached now, at about 2 Holmberg radii. This, if true, would also mean, of course, that we wouldn't be able to trace out the rotation curve in 21 cm. So we wouldn't be able to find out from HI rotation curves how far out the dark matter extends.

Rees: And there will be a sharp decrease in column density there?

Sancisi: Yes. So if you're right, we should see it from our VLA observations of NGC3198.

Shull: There is an ultraviolet image taken by the Goddard group of M101. If you look at it, is there something funny in that region?

Sancisi: I am not aware of any other peculiarity in the ultraviolet image of M101 in the regions of the just discovered high-velocity HI clouds, besides the presence of a bifurcation and a kink in the spiral arms, which are clearly visible in any optical photograph of M101 (see for the instance the picture in Arp's Atlas of peculiar galaxies).

P. Shapiro: Could you translate the null result that didn't make you famous in the beginning into some statement on what if you had a uniformly distributed component of matter in the voids which you would have been able to see.

Sancisi: That was the second reason why we didn't become famous. (Laughter) Very extended and uniformly distributed gas components, larger than 20 or 30 arcmin, would be resolved out in such interferometric measurements (with the WSRT or the VLA) and would not show up in our HI survey of the void with the WSRT. But we have been searching for discrete objects of galaxy structure and size. The limits we are able to set are not better than those already reached by single dish observations in the past, especially at Arecibo. The main advantage of our search, however, is that we have mapped a whole region of sky of about 40 arcmin × 40 arcmin over a velocity range of 8000 km s^{-1} so that we would be able to find in such a volume any structure of galaxy size (dwarf Magellanic to large spiral) that might escape when looking at specific directions. We have made tests for structures on scales of 0.5 to 3 arcmin (10 to 50 kpc) and reach 5σ upper limits of approximately 5 to 0.5×10^{20} atoms cm^{-2} and 2 to 50×10^8 $M_\odot$. We are still analyzing our data and hope to reach lower limits or detect something.

York: The large maps you show of extensions in interacting galaxies and so forth, do you actually measure full line profiles and then you integrate?

Sancisi: Yes.

York: Okay, so what's the typical full width of those lines?

Sancisi: They usually are narrow, 20–30 km s^{-1} full width at half power. But if you ask how broad are the HI line profiles from the galaxies themselves, the answer is that they may be rather broad, even for systems not seen edge-on, and sometimes as large as rotation. This result can be understood if galaxies have indeed a thick gas envelope as suggested in my talk. In such case the telescope beam would intersect a large portion of the galaxy in the line-of-sight and consequently see a broad range of velocities.

York: You mentioned yesterday in response to a question that the interacting galaxies at the moment don't fill a large part of the sky.

Sancisi: Yes. There are not many interacting systems.

York: So that when you take all these extensions and tidal interactions and so forth, what's a typical kind of average size for the extension around a galaxy in 21 cm, if you average over all the things you've observed?

Sancisi: Well, it may be of the order of 2 or 3 Holmberg radii, which means typically 100 kpc total size. Interacting systems may reach up to 200 kpc. But the point I have been trying to make is that even though interacting systems may seem to be too few at present to be able to explain the QSO lines, there may have been many more in the past, at redshifts of 0.5 to 2 or 3. I am sure Martin Rees could find a way to make them. The reason why we see so few of them now may be that they have merged and formed systems that look pretty much isolated or have some peculiar distribution like NGC 1023, Mkn 348 etc.

Davis: Murry Lewis has recently published in *Ap. J. Supplements* a careful study of face-on spirals studied at Arecibo. In a small fraction of those cases, perhaps 10–15%, he gets a profile which is indistinguishable from an edge-on galaxy in the hydrogen. There is no explanation or no understanding of what this is, but it is an unusual result.

Sancisi: What was the selection of the sample of face-on spirals based on?

Davis: Based on optical data.

Sancisi: The only thing that we can see here (in M101) capable of producing a broad profile is the high velocity gas. This gives a profile almost 200 km s^{-1} wide. Such a profile would be easily recognizable, however, because of its shape. It would show broad, low-intensity wings, instead of the usual two horns and sharp drop-off which characterize the global profiles of spiral galaxies seen at high inclination angles.

Lockman: This question is for Don Morton. You published a year or two ago some hydrogen 21 cm column densities for some of the components toward the Magellanic clouds. Do you remember what those column densities were?

Morton: I can't remember now, but probably 10^{19} cm^{-2}. Some of them went down to 10^{18} cm^{-2}.

Lockman: Okay, I was wondering about that in light of Martin Rees' question about why they're neutral at that column density.

Sargent: I mentioned to Renzo before his talk that in the case of the M81/M82 complex, where there's very extended gas around both systems, there is a QSO about a degree away from the center of M81 found by Arp in studies with a somewhat different design than the ones we've been considering here. As I mentioned at one of the symposia two years ago, this QSO shows H and K absorption of calcium at the redshift of M81, which is approaching us. It's at -120 km s^{-1} or so. It's much further out than in the picture Renzo showed us, I think. It's about 4 times the Holmberg radius of M81, and it shows directly that in these interacting systems you can get a very large cross-section for absorption.

Sancisi: It's a huge system.

York: I didn't quite understand what Wal said. There's a QSO a degree away and what does that show?

Sargent: Absorption in calcium as found by Boksenberg and myself.

Sancisi: At the velocity of the HI?

Sargent: Actually, it's slightly further out than the HI. It actually extends the HI rotation curve.

York: Do you know what the inferred HI column density is?

Sargent: The calcium column density is roughly 10^{-12} cm^{-2}. No, sorry 10^{12} (Laughter); even the IPCS can't do quite that well.

Disney: What are the limits in numbers per Mpc^3 on how many of these intergalactic clouds like Schneider *et al.* found there may be? Are they incredibly rare?

Sancisi: Clouds like the one found by Schneider *et al.* are probably rare, otherwise more of them would have already been discovered. But more modest HI tails and bridges are known to exist in at least 10 to 20 interacting systems. I think that the probability of finding intergalactic HI in such interacting systems, especially when they include spirals, is very high. I am not speaking here of loose groups, but just systems in close tidal interaction. If you are asking how many interacting systems there are, I don't know and I wouldn't dare to guess.

Dickey: (Summary) The brightness temperature of the 21 cm line in emission is proportional to the column density of atomic hydrogen if the line is optically thin, so emission studies are useful for comparison with optical and UV absorption lines which also depend on column density. However 21 cm absorption lines are also detected at high redshift; to compare the abundance of lines in this population with the local abundance requires some knowledge of the cross-section of a spiral galaxy in 21 cm absorption. We have little data on 21 cm absorption in nearby spirals (although it would not be difficult to get more, eg. Dickey and Brinks 1988, *Ap. J.*, in press), but at least in our own galaxy in the solar neighborhood we know very well the absorptive properties of the interstellar medium at 21 cm.

There are three ways to describe what we know about the statistics of 21 cm absorption in the solar neighborhood. The first is the mean absorption coefficient of the gas in the disk, $\langle \kappa \rangle$, which is defined by

$$\langle \kappa \rangle = \Delta V / L$$

where ΔV is the integral of $\tau(v)dv$ over the 21 cm line, and L is the line-of-sight length sampled by the observation. High latitude observations show (Kulkarni and Heiles 1987, *Galactic and Extragalactic Radio Astronomy*, 2nd edition, eds. K. Kellermae and G. Verschuur)

$$\langle \kappa \, h \, sin|b| \rangle = 0.75 \text{ km s}^{-1}$$

where h is the effective half width of the disk ($h = 130$ pc), so

$$\langle \kappa \rangle = 5 \text{ km s}^{-1}\text{kpc}^{-1}$$

in the solar neighborhood. So for a typical line-of-sight through a galaxy with inclination i we could expect absorption with equivalent width

$$\Delta V = \cos(i) \times 0.75 \text{ km s}^{-1}.$$

A more sophisticated way to describe the probability of hitting an absorbing cloud on an arbitrary line-of-sight is to analyse the abundance of clouds vs. optical depth in the solar neighborhood (Crovisier 1981, *Astron. Astr.*, **94**, 162; Payne, Salpeter and Terzian 1984, *Ap. J.*, **272**, 540). This abundance is described by the function $P(\tau)$, which is the probability per unit line-of-sight length of finding an absorption line with optical depth greater than or equal to τ. Over the range $0.25 < \tau < 2$ this function gives mean free path lambda

$$\lambda = P(\tau)^{-1} = 4\tau^{0.7} \text{ kpc}$$

in the solar neighborhood. Recent surveys by Garwood and Dickey (1988, *Ap. J.*, in press) suggest that the abundance of clouds decreases in the inner galaxy.

Finally, the abundance of absorbing HI can be compared with the total HI measured by emission surveys to study the mixture of warm and cool atomic phases. This is of astrophysical interest because it may imply the interstellar pressure and the balance between heating and cooling processes. Since the optical depth is proportional to column density divided by mean spin temperature (cf. Kulkarni 1984, Ph.D. Dissertation, Univ. of California at Berkeley; Dickey and Benson 1982, *A. J.*, 87, 278)

$$\tau \propto N/\langle T_{sp} \rangle$$

and for the solar neighborhood emission surveys show

$$\langle N \sin|b| \rangle = 2.7 \times 10^{20} \ cm^{-2}$$

we find that about half of the local atomic hydrogen is warm ($T_{sp} \gg 300 \ °K$ so $\tau \ll 1$) and half cool ($T_{sp} \sim 60 \ °K$). So the effective temperature of the HI in our galactic disk is

$$\langle T_{sp} \rangle = \langle N \sin|b| \rangle / \langle \Delta V \sin|b| \rangle = 200 \ °K.$$

A sample of six lines-of-sight through the disk of M31 show $\langle T_{sp} \rangle = 350 \ °K$. It would be very interesting to compare emission and absorption in a larger sample of galaxies covering a range of redshifts, but with the present meager sample it appears that roughly half of the total HI is cool enough to show up in absorption.

EVENING DISCUSSION

Shaver: This session should be a discussion of controversial issues. When the idea was first raised, I thought that there were two populations of issues: those that were controversial and those which were not. But as I look at these populations in more detail, I'm inclined to think that there is really one continuous population (Laughter) which exhibits a gradual transition in terms of controversiality. To start things off, Cyril Hazard has a few comments which were of a very general nature that he wanted to make.

Hazard: (Summary) Hazard noted that at the highest redshifts (e.g. $z \simeq 3.3$) about 15% of QSOs had BALs, and that that number may be consistent with, but is probably higher than, the fraction found at $z \simeq 2$ which is 10%. Hazard emphasized that there was a set of objects, apart from the more classically defined BAL QSOs, which show strong indications of having absorption intrinsic to the QSO. Hazard suggested that if all QSOs have BAL-type regions, which is often deduced from covering factor arguments, then high ionization intrinsic metal-line absorption (e.g. OVI and NV) might contaminate the Ly-α forest of absorption.

Hazard expressed concern that the sample from which Ly-α forest properties were derived may not be a representative sample as many of the spectra he has obtained do not appear to be consistent with the results which have been discussed so far at the Workshop. In particular, Hazard showed a spectrum of one of the most intrinsically luminous QSOs in the Universe with $z \simeq 3.5$ which showed an apparent surplus of absorption lines near the emission redshift, contrary to the 'inverse' or 'proximity' effect. Hazard also questioned the results that the line density generally increases with redshift in all objects and that apparent observed clumping of Ly-α forest lines could be easily explained by statistical fluctuations.

Hazard suggested that outflow from BAL-type regions in QSOs, or once active objects, may give rise to intervening type absorption in a background QSO spectrum, particularly in the Ly-α forest. Hazard showed one such proposed example.

Ostriker: I agree that its going to be very hard for the QSOs not to affect their environment. They should blow shells of up to Mpc in size.

Norman: The velocity of the shell is very, very high.

Hazard: From the data I've shown a shell of 200 kpc would be required. The velocity spread in the one that we actually see is 20,000 km s^{-1}.

Ostriker: Can you determine the grams cm^{-2} in that shell? Just to bring back the origins of explosive galaxy formation, this is exactly what Satouro Ikeuchi suggested. If you just take the kinetic energy in that shell and let it go on out, it will plow up more and more matter, and you'll get a dense shell which will be fragmented to galaxy-sized matter.

Turnshek: If you assume that the BAL region is about 50 pc from the central source and lasts 10^7 yrs, it will produce about 10^{59} ergs which is close to what you need for that to happen.

Hazard: We know these objects can eject material up to 50,000 km s^{-1}!

P. Shapiro: I thought in the earlier talk that there was some possibility that the material being ejected was being decelerated inside the QSO.

Turnshek: I think that's probably true in the PHL5200 type of BAL QSOs, but it may not be generally true.

Hazard: I'd like to show a composite picture of a 4 degree by 4 degree region at right ascension of 15 hours which we've been working on. There's about 15–20 QSOs per square degree found on the objective prism plate with magnitudes brighter than 19.5 and we may be missing some of the low redshift objects. First of all, what I wanted to show you is that there are very few QSOs which are not within a minute or two of arc of another QSO. Secondly, from a radio survey by Paul Coleman and Jim Condon only five objects are found to coincide with a radio source. However, many QSOs are very close to radio sources, so again one would expect in these particular cases perhaps that that QSO is along the line-of-sight to a once-active galaxy, at least there's an equal chance that there are galaxies between us and the QSO. So, it might be interesting to look and see what the absorption properties of some of these objects are. But there is a strong indication anyhow that you always have a galaxy or something very close to the QSO along the line-of-sight.

Ostriker: Are these QSOs closer together than they should be?

Hazard: No, but they're closer together than people once thought they might be. Also, there does appear to be some clumping. It is also relatively common to find close pairs which differ in redshift by 10,000 km s^{-1}, which leads one to believe that you might have physical associations of QSOs very close together in which the velocity differences are much higher than one would normally expect from a cluster.

Jauncey: I would suggest that the bulk of the objects we have seen within the last two days are radio QSOs. We are talking here about properties in the near vicinity of the QSO. A large number of the QSOs are from flat spectrum samples and in those QSOs you generally expect to be looking down the radio jet. These are the QSOs that show superluminal motion. So what you have is a set of radio selected QSOs, and your looking for example, particularly with the inverse effect, within a few degrees of down the radio jet. You might also be looking down a disturbed line with broad absorption line QSOs. So I might suggest that the degree to which you look down a particular line-of-sight might be an answer, or at least part of the answer, to some of the issues Cyril has raised.

Hazard: I think there is a lot in that. One thing I didn't say anything about is that there is a problem in much of the discussion concerning comparison of radio samples and optical samples. People do not realize, I don't think, that whenever one talks about an optically selected QSO, it covers a very wide range of objects. The objects we are finding, and I'm not saying we are finding the whole population, cover a tremendous range. For example, the width of Ly-α emission can range from 600 km s^{-1} to 20,000 km s^{-1}! That is just one sort of total range of objects. In a lot of these objects we really don't know about the ionizing process which produces the lines. I think that the range of properties of the radio selected objects tends to be much less than those of the optically selected samples. I think if you want to make a comparison between radio samples and QSOs in optical samples, one should be careful that one really takes an optical sample which is likely to have similar properties. I don't know if you can do that, but I certainly think it is unfair to make comparisons with the total population. I think you are right. I think there is a lot to be said about the fact that people are using radio selected QSOs and its not giving us the whole story. I think the situation is much more complicated than people think. I'm not saying that anything is wrong with the analysis up till now, but I think you might be finding a simple solution to the problem, perhaps because of the objects selected out. But if one looks at the totality of the observations which are available, there may be a much more complicated story involved in the whole process.

Sargent: I'd like to make a comment on what Cyril has said. I think it is very important to science generally, and particularly in Astronomy, to discover things in the framework of a process which is inhibited by the method by which time is allocated on telescopes. This is going to be particularly true, in my opinion, with the Space Telescope for which one to some extent has to know the answers to your questions before you are allowed to get the time. Therefore, you have to produce a reason for doing things. Rather than just observe random objects in the sky, I elected to observe objects which Cyril discovered, which is pretty much the same thing. It's certainly true that the variety of QSO spectra that has been produced by Cyril in his objective prism work has been quite remarkable. Cyril deserves a lot of credit for that. Nevertheless, I don't agree with many of the conclusions that he has expressed this evening. One of them is the suggestion that there is a wide variation in the Ly-α forest from one object to another. This was actually tested in a paper by Stiedel and I where we actually took eight relatively bright QSOs, most of which were discovered by Cyril and one or more of his allies. We actually measured the decrement that occurs on the blue side of Ly-α emission relative to the red side due to the accumulation of all the narrow lines. We find that, with the exception of one broad absorption line object which was obvious upon examining the spectra, that this decrement didn't differ statistically from any one object to another. And so while it is true, if you look at low resolution spectra, you'll find what appears to be a wide variation in the behavior of the Ly-α forest, I believe that is often an illusion, and I still retain confidence that the Ly-α forest is fairly uniform from one object to another at a similar redshift.

Rees: There is no reason to believe that all the objects identified on objective prism plates do have active nuclei. They could just be young galaxies that have emission lines with broad wings. Many people have said that they can't just be young galaxies because the line widths are large, but you can certainly imagine that radiation driven or supernovae driven winds could give you broad lines in young galaxies. I'd like to know if there is any argument against the hypothesis that some fraction of the objects identified with objective prisms are just young galaxies.

Sargent: Of course, the nature of the background source shouldn't influence the nature of the intervening material, except if its very close. The only statement that I can make is that the Ly-α forest looks fairly uniform. As to the nature of the background object, I have little to say about it.

Ostriker: Martin, could it not be that in some sense most are not really quasars, but young galaxies with only 10% of them having nuclear activity, because in general they don't have radio emission, and in general they are not highly polarized.

Gondalekar: If you look at galaxies which are undergoing star formation today, they don't show these very broad emission lines.

Rees: There would have to be some mechanism that doesn't apply now whereby you can get broad lines due to winds driven by, for example, radiation pressure in the lines or supernovae. I think this is not inconceivable.

Ostriker: The presently active galaxies are very inactive now compared to what galaxies must have been in order to make the metals that we now see in galaxies.

Sargent: Not all the objects have broad lines. Cyril has found a class of objects, which I can confirm and many others like Don Morton can confirm, which have quite narrow lines with redshifts above 2.5. These might very well be a different class of objects.

Heckman: Can you explain the x-ray emission if the average quasar is actually just a young galaxy?

Sargent: For most of these objects at high redshifts nothing is known about the x-ray region.

Ostriker: If I compare the spectrum of 3C273 with that of the Crab you can't tell me which is which.

Norman: Remember, in Arp 220 you have 100 $M_\odot$ yr^{-1} going into star formation and a very high supernovae rate, but its optically undistinguished. This does not look like a quasar—that is the point. It does not produce lines of 10,000 km s^{-1}!

Tytler: If we were looking at young galaxies I would be very surprised if they were optically thin in the Lyman continuum.

Sargent: I guess that I would like to make a comment about one of the spectra that Cyril showed in connection with what Tytler was saying this afternoon. I have often suspected that the Ly-α forest knows that Ly-β absorption is there, and of course if it does, then the clouds have to be small. I think it would be an interesting statistical study to look at a lot of unbiased quasars and count the number of absorption lines, not only in the Ly-α emission lines but also in the Ly-β emission lines. In one spectrum Cyril showed this evening it seemed to me that the Ly-β emission line sticks out quite clearly, and then there are absorption lines on either side. That might imply, if God is very unkind, that the absorbing objects are actually very small, smaller than the emission line region.

Hazard: The object in question appears to have a narrow emission line region component, about 600 km s^{-1}.

Rees: In a young galaxy there will be a lot of gas out at several kpc, and of course some of the continua is going to be scattered by that gas. It would be interesting therefore to know whether the narrow core of those emission lines did in fact come from a more extended region which could be the whole body of the galaxy.

Hazard: I'd like to make the point that in the narrow line objects the high ionization lines like NV are extremely weak indeed. So the gas in the narrow line objects may be ionized by a different input spectrum. Also, I'd like to say that sometimes in broad absorption line QSOs I have my doubts about how much of what we see is NV emission. Maybe its just the broad wing of Ly-α emission.

Boksenberg: Let me first say that when you see NV emission its absolutely there when you see it. It is clearly not due to anything else. But also the impression I have is that you see narrow lines in much greater contrast than you see the broad lines, so I don't think that there is anything inconsistent about narrow lines sticking up like they do even though, like Martin says, they come from an extended region.

E. M. Burbidge: When you see NV emission, it is very definitely there.

Turnshek: In broad absorption line QSOs, where we know the material is ejected, we see that these objects have different emission line properties than the non-broad absorption line QSOs. It would be interesting to see if narrow absorption line properties depend on emission line properties. In other words, if we get a sample of non-BAL QSOs with strong NV emission, it would be interesting to see how the Ly-α forest or narrow metal line system properties compared in those two samples. If there are differences, then we know that there is some ejection activity going on there. I've never seen a paper address that issue.

Tytler: This is serious because the objects that aren't BAL quasars also may have signs of ejected material.

Turnshek: Yes, all I am saying is that the emission line properties differ in the broad absorption line quasars versus the non-broad absorption line quasars and we could take advantage of this.

Tytler: That's right, but even if all quasars have BAL regions, that doesn't mean to say that there is going to be any ejected material along our line-of-sight.

Turnshek: The point is, what we need to do is determine dN/dz for non-BAL quasars that have strong NV emission and dN/dz for those that don't, okay? If dN/dz is the same in both cases, then its pretty conclusive that we are looking at intervening material. For example, Craig Foltz and collaborators are doing that as a function of the radio properties.

Shaver: I think we better move on, we have the rest of the universe to cover. Don York had some comments on another general issue which brings us into the subject of the study of the Ly-α forest in general, and that is the measurement problem.

York: Once you find an absorption line system that has all the heavy-elements in it, the number of lines above some limit, even in a nearby star, is enormous: OI and NI have 20 lines each below Ly-α and I know of at least one case where people simply didn't look back at the UV spectra and completely miss some systems. A line which is called Ly-α may be a metal-line from a known system. There are an enormous number of Ly-α lines, one line every 5 Å or so, and when you get to high redshift, if you have too many heavy element systems, there are bound to be lines in the Ly-α forest which are metal-lines.

Boksenberg: Look, metal lines are looked for, and you don't see them. Of course you see the rare cases, but really, the systems you're talking about from Copernicus have relatively large column density and you also see metal-lines from those ones longward of Ly-α. These other systems have been searched for. This is a valid question, but it isn't part of the problem.

York: My point is that maybe it is a small thing, but there are a lot of other small things. It is just a very, very complicated process.

Wolfe: It sounds like this effect should be a function of equivalent width too.

York: The other thing is that the Ly-α column density is, particularly when you get below 5×10^{19} cm^{-2}, very poorly determined, even in gas near the Sun where there is simple component structure. There are a few cases, like Zeta Oph, where Don Morton and I have published completely different column densities: I use Ly-β and he used Ly-α, and there are things that just mess-up a line, like Gaussian components out in the wing of a damped line and so-forth. So a factor of 2 error in what should be very precise data is not at all unusual. As you go down to column densities less than 10^{20} cm^{-2}, like in Zeta Pup, things just get very bad.

Shull: It has to do with the number of components that make up a line. I'm very worried that going from equivalent width to column density is very uncertain because b-values are 30 or 40 km s^{-1} and that's too high to be thermal, otherwise you collisionally ionize the hydrogen. Therefore, it must be made up of 5, 10 or 20 components. It is crucial to know whether those components are distributed exponentially, Gaussion-like or whatever. I did a simple analytic calculation just comparing the column density you would get if it were exponential and the ratio goes as e^{x^2}/e^x. If x is large, x being the ratio of equivalent width to thermal-width, say 4 or 5, then the difference between e^{x^2} and e^x is large.

Boksenberg: It is, of course, true that you can get serious errors. In isolated systems that's more true than say when you have a pattern of systems with different species. In some of the data I've looked at, you can certainly get a range in column densities of over 2 orders of magnitude from the equivalent widths. Assuming complex profiles, you can hide very massive systems in the middle of a complex system. But if you take a pattern which involves several species, then the self-consistency of those ties you down a bit more, and in the end you find you can't be that much wrong if you take everything into account. In isolated cases, I'm sure you could find serious errors, but in general I don't believe you can be that far off. However, for the Ly-α forest that it not true if you can't fit to the series lines.

Ostriker: Suppose there were systematic errors in the sense that when you get larger equivalent width lines it is just because you get more and more blending and they are not really larger equivalent widths. Perhaps that would make the power law that many people have found purely a statistical effect, because remember you are taking the exponential of something to get this because you think you're on the Doppler part of the curve of growth, not the damping part. This problem would not affect the damped systems.

Wolfe: What you are saying is that in that case you would be underestimating the column densities.

Ostriker: Well, it depends on how you do it.

Carswell: How many components one has that cannot be resolved at the moment is a worry. But, in some sense you can try and fit simultaneous Ly-α, Ly-β and Ly-γ and in some cases you may have 5 or 6 decent Lyman lines which are clean and not blended. You do get reasonable consistency between the Lyman series. In those cases the single component analysis gives b-values of 30–40 km s^{-1}. So if you tried to make these things up with a number of quite narrow components, you are asking for clustering on a very small scale which is actually quite sharply peaked.

Ostriker: Couldn't it just be chance coincidence on a Poisson basis? This is consistent with finding fewer and fewer lines at large equivalent width.

Carswell: It depends on the circumstances. If we have temperatures of 10^4 °K you are going to have a Doppler parameter of 13 km s^{-1}. So depending on how I am going to do it, you can't stack 2, 3 or 4 components in there and get what you observe. You should also see a range of Doppler parameters in that case which is larger than what we do see.

Ostriker: Your data shows a large range in line widths.

Carswell: I know, but its not large enough. What I'm saying is that in the case you've described I believe you should see some narrow single components and you don't.

Jenkins: You can analyze ensembles of lines, if they are not badly saturated, i.e. they're only moderately saturated. Then the distribution function of b and N albeit large is not bimodal. You can use a normal curve of growth technique and come out with the right answer for N. The number you get for b is of course completely meaningless because the b that you get is just proportional to the number of components times the b of each component, but the end value has a relatively small error of the order of 10 or 20 percent, provided you don't have a really bizarre distribution in the N or the b values.

Wolfe: There are counter examples to that. Just look at the optical lines where you measure b-values of 30 km s^{-1} and you get a low column density. Then you go and measure the 21 cm and you find a tremendous amount of gas buried in 2 km s^{-1} components. I'm not saying they're all like that, but you run into a number of cases where the naive application of the curve of growth leads you to really under estimate the column densities.

Jenkins: That's a special case where you have a bimodal distribution with dense clouds having b-values of 2 km s^{-1} emersed in a medium that has b-values of 15 km s^{-1}. All bets are off in that case.

Bergeron: For the metal systems we know they can have low b-values. I suspect that at low enough equivalent width detection limits, 5 – 10 % of the Ly-α forest will be weak low b-value CIV lines. Therefore, I think statistically these low b-value lines should be present in the Ly-α forest. I'm surprised that analysis of Ly-α forest lines doesn't show some low b-values.

Shaver: Let's move on to a discussion of the distribution of column densities, which is central to what David Tytler was talking about today. What do people think about this single slope describing the distribution of column densities?

Ostriker: There is one thing that is surprising: you normally expect that there are fewer massive stars than there are low mass stars, and that there are still fewer super massive stars. It is similar with galaxies. Everything we find is just distributed like power laws. It's hard to find anything that isn't. Sometimes there are breaks and wiggles, but in your case, you don't find any, but you have terrible resolution. My questions is: is it surprising that there are fewer more massive clouds and fewer still, even more massive ones? Now, the theoretical expectation would have been that when they become opaque to ultraviolet radiation, at that point the physical conditions in them will change and you can have star formation. So at neutral column densities near 10^{17} cm^{-2} it's not surprising that at that point you start seeing metal lines. Because at that point you can have cool globules in them or whatever, which you may or may not see in line-of-sight, and at that point you can also have some star formation—little bits that contaminate them. We see intergalactic HII regions.

Tytler: But in that description it's exactly one population. You can bump the column density up a little bit, to make a cloud neutral, to make stars, and I've done practically nothing to it. The way you just described it, the only discriminant is the neutral column density so a slight change can make metals form.

Ostriker: Right. But I did say isn't that what you expected?

Tytler: Yes. But that doesn't fit into the basic picture that intergalactic clouds form primordially and are not connected to galaxies.

Ostriker: What you say is true only if galaxies were sort of made by the hand of God separately, rather than made out of gas clouds. If you assume that galaxies are made out of intergalactic matter, the way stars are made out of interstellar matter, then at some point they form. And presumably star formation occurs when they become opaque to ultraviolet radiation.

Tytler: There are systems that are transparent in the Lyman continuum which do show metals. Probably at least most of all metal line systems are in that category. We have trouble finding them as the metal lines tend not to be as strong.

Ostriker: What it shows is that there is a characteristic break at that point. It's more common on one side and less common on the other. And that's just what you'd expect. But if it were homogeneous clouds, then it would be impossible. With inhomogenous systems then it just means that one piece of it is a little thicker and one piece is a little thinner, and that you had some star formation in the thicker one.

Tytler: Yes, but I think that is the remarkably important point. It means these systems are making metals, rather than the metals having been made somewhere else, at some other time, then forming after.

Wolfe: But isn't it surprising that at the lower column densities in the standard picture most of the gas is ionized, so that the HI is a small fraction of the total column density and yet you see the same power law from the point where the HI is a tiny fraction of the column density until where it is practically all HI?

P. Shapiro: I thought there was an original distinction between the two populations of clouds based on the clustering that occurs in one and doesn't occur in the other. Are you willing to say that that is no longer true?

Tytler: You're asking the wrong man.

Sargent: I think that David himself said today that the clustering is different in the two categories of objects.

P. Shapiro: So then why are we struggling to make them a homogeneous population?

Tytler: We don't understand the clustering.

P. Shapiro: But they're not homogeneous. You can't call them homogeneous if they're not homogeneous.

Wolfe: But they follow the same power of distribution of column density.

P. Shapiro: But they're different.

Ostriker: Again, it would be just what you'd expect. You see more and more massive objects as you go down to a larger and larger column depth—then there would be more clustering for the massive objects. That is what happens in every conceivable theory of galaxy formation—that more mass will cluster more. And we know it's true on all counts.

York: Well, can I ask what the evidence is for clustering in the metal line systems? Isn't that just very small scale stuff?

Sargent: Yes, but its very different between the two.

P. Shapiro: Can you argue that the larger column density metal line systems should be more clustered than the lower column density systems?

Ostriker: Yes.

Wolfe: Has a two point correlation of function ever been done for Ly-α in the metal line systems. I remember in your paper that was done for CIV.

Sargent: No. It would be very difficult due to the saturation and lack of observations.

Shaver: From an observational point of view, how happy are the other workers with this single power law?

Boksenberg: Well, it isn't a single power, is it? The metal sytems have a slope which is different from the non-metal sytems, but they meet. Isn't that right David?

Tytler: Yes, but they seem to be predictive. What I'm saying is that from the frequency of Ly-α only clouds. I can tell you how frequently metal line systems will occur.

Boksenberg: Yes, but at a given column density, not equivalent width, there are two populations, one of which has certainly got metals, and its' high ionization, and the other has no metals, and you don't know what the ionization is. But they're two populations, and they come together.

Sargent: It seems to me that Boksenberg's point is a critical objection to Tytler's. That is, we know that the relatively high column density systems are fairly ionized, most of them. We don't know what the ones below 10^{16} or 10^{15} cm^{-2} are, but Tytler claims that they're neutral, in order to preserve the continuity of the distribution. But it seems to me that that's rather artificial. Since you know that the high end of the distribution is not neutral, why then claim that the low end has to be?

Tytler: I'm trying to make a distinction between an ionization of say 10^5 or 10^6 and 10^2.

Ostriker: Why not just make the ionization fraction change continuously for the column density?

Tytler: Well, that's what seems to me to be what the data are telling us.

Ostriker: Why is it surprising. If you just do the simplest naive thing: make them spherical clouds with a mass like globular clusters, which is just about the minimum mass you can expect which happens to be a Jean's mass at decoupling and then as you go on, you get up at a certain point to the mass of low mass galaxies, and at that point they start having metals.

Tytler: Yes, but we don't yet know if the Ly-α only systems have a higher or lower mass than the metal line systems, even if you do have the ionization.

Ostriker: If you just take a standard ionization for them all ...

Tytler: ... then they have about the same total column densities, so you haven't convinced me.

Ostriker: You have this typical break and above this column density, they're mainly metal line systems and below this column density, they are mainly Ly-α only systems. It is approximately that point where they become predominantly neutral and you can start having metals.

Tytler: Yes, that's what I'm happy with, but what I'm unhappy with is the jump from having the HI column density translate directly to the mass distribution. It is not clear to me that a typical Ly-α only cloud will have either a higher or lower mass than a metal-line cloud. We don't really know that.

Ostriker: Well, you just take the temperature and the column density and you get a size and you cube it and you get a mass.

Wolfe: But it depends on the ionization fraction.

Boksenberg: But I'd be surprised if the low column density systems are the ones that are neutral compared to the high column densities—it's the other way around.

Green: Essentially, there is a significant difference in the results which comes from dealing with the multiplicity of components. When you try to get a column density from a Lyman limit system alone, it can differ from the case where you actually know how many components there are in one of these optically thick Lyman limit systems. It really seems to make a significant difference as to how you go about turning equivalent width sums and Lyman limit column densities into Lyman limit optical depths in the columns. This is what Jill Bechtold is finding.

Shaver: So it's not completely agreed then by all workers in the field that its a single power law.

Tytler: Maybe not even by myself. This is the first time that's ever been attempted. If you work out how much of my observing time went to this it's zero. I took no data for that graph. I'm just suggesting that we ought to think about it.

Ostriker: What's the heavy whether about. We don't expect anything to be a perfect power law.

Shull: I want to bring up a point on the survivability of these things. If you take Martin Rees point of view that they are gravitationally bound, and throw away a confining intergalactic medium, then this 10^{17} cm^{-2} column is important, because below that in column density they are transparent to ionizing radiation. They are probably 10,000 or 20,000 or 30,000 °K. Above that, if they have metals, they can cool. So if you photoionize them you over pressure them, and they can dissipate. Now if it is an intercloud medium as Jerry thinks, which pressure confines, then you won't have that difference, so maybe this is evidence in favor of gravitational binding. It makes them more prone to evaporations if the core potentials are not too deep.

Rees: I must say I don't understand that argument (Laughter) because the temperature doesn't get above 20,000 or 30,000 °K.

Shull: That's my point. If there are metals it would be below that. Therefore, the potential wells aren't too deep. If you heat it up to 20,000 or 30,000 °K then it would expand on a sound crossing time providing there is no intercloud medium confining them.

Rees: I see. So you're saying the high column density ones are in even shallower potential wells.

Shull: They have shallower potential wells, and none of this intercloud medium to pressure confine them. Then you just do away with them by photoionization. Do you see anything wrong with that?

Ostriker: Is this for all clouds, or for some fraction of them?

Shull: Ones with greater than 10^{17} cm^{-2} column densities which have metals.

Ostriker: I think there is no objection with it.

Shull: I thought you were still for a confining medium.

Ostriker: No, the work that we did was for the lower column density clouds, and then it is absolutely necessary. We also wrote a paper—talked about—including gravity from background matter and it affects the massive clouds. It doesn't affect lower mass clouds at all. But its just like the Jean's mass. For low mass it is the thermal energy dominating and it doesn't matter what gravity says, but as you approach the massive galaxies, once again, you expect that the effect of gravity will be important and that will be important for the massive clouds. That's why I said, the simplest thing is to just say that you are approaching dwarf galaxies or intergalactic HII regions. I mean, is there gravity in those, I presume so? They make metals and there are stars in them. What is the column density in those by the way?

Sargent: Somebody recently showed me an HI map of one of those objects in which I think there was 10^9 $M_\odot$ of hydrogen, over an enormous area, and then the emission object was a tiny little thing in the middle there.

Ostriker: Now that would be a good example of what I would like the metal line systems to be.

Norman: Why not extend it all the way to very low column density? It's just a prejudice you have to pressure confine them.

Ostriker: There is a physical problem, but let's go on with high masses. I think it works out well and is quite nice.

Sancisi: But look, those extragalactic HII regions are rotating very very fast, so they must have a lot of mass. I don't know about the gas mass, but certainly the total mass must be a lot. People who say that it is just velocity dispersion there, they are wrong. You see there quite clearly a systematic rotation. I'm convinced they have a lot of mass.

Wolfe: Are they self gravitating?

Ostriker: Yes they must be. My guess is if you would see a quasar behind them, it would be like the metal line systems. Some star formation would contaminate them.

Felten: (Summary) There was a heated discussion this afternoon involving John Stocke, Artie Wolfe, Jacqueline Bergeron, Neil Tyson, and me. I asked for several minutes this evening to make clear that there is really more agreement than disagreement here. The question is this: of the cross-sectional absorbing area presented by galaxies for QSO absorption, what fraction of it belongs to high-luminosity galaxies, and what fraction to low-luminosity galaxies? John Stocke assumed that the effective absorber radius for a galaxy of luminosity L is $\tilde{\propto} L^{0.4}$. Then we have to calculate percentage points of the integral $\int L^{0.8}\phi(M)dM$. If the luminosity function $\phi(M)$ is of Schechter form, this may be done from a table of the incomplete gamma function. For the integral $\int L\phi(M)dM$, I happen to know some results, viz.: about 45% of the integral lies more than one magnitude fainter than M^*, and about 25% lies more than two magnitudes fainter than M^*. For Stocke's integral, these numbers would be somewhat larger, perhaps 50% and 30%. Stocke will have to agree to these numbers, because it's his integral! Bergeron must have made rather similar assumptions, because the percentage points she quoted this afternoon were very close to these. Therefore I think she too will agree to this. This afternoon, Stocke showed only a differential curve, while Bergeron only quoted integral percentage points. This is why the agreement wasn't immediately apparent.

However, an important point is hidden here. The numbers I quote above are for a Schechter parameter $\alpha = -1.25$, corresponding to a faint-end slope of 0.1. All the discussers were in fact assuming this α. But α isn't well known, particularly for field galaxies. In the Virgo cluster, Sandage finds slope 0.2, corresponding to $\alpha = -1.5$. Holmberg found the same result in small groups. If α is < -1.25, the percentages above increase! If the faint-end slope were to reach 0.32, implying $\alpha = -1.8$, the integral $\int L^{0.8}\phi(M)dM$ would diverge at the faint end; i.e. essentially 100% of the area would be due to small (and even to dwarf) galaxies.

I want to emphasize that 30% is already a number of order unity! Since so much contribution is coming from small and dwarf galaxies, we must allow the possibility that if the faint end of $\phi(M)$ is steeper than we think, or if it becomes steeper in the past, or if small galaxies contain more gas in the past, they could dominate the integral, and make it much larger than we think.

There's at least one factor that could make things go in the opposite direction. Sandage believes that there are no dwarf spirals and few dwarf irregulars, implying that dwarfs generally would have very little gas. Neil Tyson, however, maintains in his poster paper that there could be a lot of gas-rich dwarfs. Mike Disney, in his poster paper, claims that there may be a lot more dwarfs, and low-surface-brightness galaxies generally, than are listed in present catalogs.

Bergeron: I have a question for Allan Sandage. The question to Allen is the following. In the building of a luminosity function are the dwarf galaxies which are included in the luminosity function randomly distributed or do the galaxies which are very close to bright ones, say within 50 or 80 kpc of a bright one included in the luminosity function. I think it is a crucial point.

Sandage: There is no correlation between the dwarf ellipticals and the bright ellipticals. So the dwarf ellipticals are distributed like a King function in the Virgo cluster and the bright ellipticals are distributed like a King function. The dwarf ellipticals are not satellites of the bright ellipticals.

Bergeron: This is very important to what I was discussing this afternoon. For the detection of galaxies I've made, it will be very difficult to claim then that the real absorber is a satellite of a bright galaxy which I've detected. One approach would be to forget about the known luminosity from imaging of galaxies causing low redshift QSO absorption lines. The morphology could be obtained with Space telescope.

Sandage: I think it's very important to emphasize what Jim said. The luminosity function that we get in the Virgo cluster that has this slope of 0.2 is dominated entirely by the dwarf ellipticals, and I think they have nothing whatsoever to do with what we've been talking about the last two days. Now Neil Tyson and John Scalo have a preprint out which says that our luminosity function for the Magellanic cloud types, which in the Virgo cluster we said have a maximum at a certain magnitude and do not increase in numbers towards fainter magnitudes, they say that's wrong. If they are right, then the Magellanic Cloud types may go like that, but there is no evidence for that except for the argument, if I understand Tyson right, of our incompleteness. The second point is the one that Disney raised. Are the Magellanic Cloud types in the general field? He says no. We have a catalog of potential dwarfs in the general field that look like the regular luminosity function. The redshifts are being measured at Arecibo to see if they are really dwarfs or not. So I think the question of whether or not the dwarfs in the general field are connected with giants is still open.

Felten: I'd like to ask Mike Disney: do you have any ideas in your own mind of what the slope should be for dwarfs, say in the field and in clusters separately? Is it 0.1, 0.2 or more?

Disney: I think I'll know better in six months, but I'd like to comment that the number 0.8, the relationship between the luminosity and the radius, is also a very difficult and very important number here. It's a very difficult number to get out. It takes a long time, and I think we will know the answer to that in about a year's time down to some level. But I think if you look at the raw data, 99% of the galaxies that go into the Schecter function are all around L^* or about a magnitude on either side of that. To extrapolate is very uncertain and that's why I think John Stocke's point is well taken. I think probably the best survey that we have at the moment is an incredibly ancient one. That's the one of Holmberg. I think Allan's survey as well, but I haven't put the numbers in. But we don't really know.

Shaver: George Miley had something he wanted to say relevant to absorption systems near the emission redshift.

Miley: I might have an actual picture of an absorption line cloud, a relic of an absorption line cloud at low redshift, and even maybe some evidence for Cyril's 200 kpc shell. Now knowing that these systems are associated with strong steep spectrum radio sources, I think it is legitimate to ask: what do we know about radio sources at low redshifts which might be relevant to the absorption line question. The first thing is that there is a real difference between the optical morphologies of very high powered radio galaxies and the weaker ones. The high powered ones are really peculiar. They've got tails and bridges and they really suggest mergers. The other difference between the low and the high powered radio galaxies is their infrared luminosities. IRAS has shown that the high powered radio galaxies are much, much stronger elliptical infrared sources, again evidence that they are not normal elliptical galaxies as the old forklore says. The next thing is a new result. I think it may be very relevant because we are seeing more and more low redshift radio galaxies which have extra-nuclear emission line regions. It looks as if the higher redshift objects are definitely optically aligned along the radio axis. I'll show you some evidence for that which is hot off the press. Essentially, Ken Chambers drew this histogram today (shows figure). These are objects in a survey of ultra-steep spectrum radio sources. The ultra-steep spectrum is a way of picking out the high luminosity radio galaxies. I think it is very clear that the optical position angles are definitely aligned along the radio axes. We don't know yet if it is gas or stars but it is evident that there is something peculiar about these powerful radio galaxies and this is changing with redshift, it's evolving. I am not saying that this is a characteristic of ultra-steep spectrum radio sources. I am saying that it is a characteristic of more distant radio galaxies, because the ultra-steep spectrum is a way of pinpointing the very luminous ones. Independently, McCarthy, van Breugel, et al. have done the same thing for the known steep spectrum selected sample of Spinrad and they get the same results. So two completely independent groups have done this. I think this is very important.

We saw yesterday that the bending increases as a function of redshift. Now what bends radio sources? This is a low redshift example of a bent radio source (shows figure). The radio source is 3C277.3. You see a radio jet hitting this Hα cloud and bending and decollinating. Now those are exactly the sort of types that we are seeing at high redshift. See the Barthel poster paper upstairs for work that Barthel and I have done. So, I think it is a very reasonable argument that this is an example of one of these clouds at high redshift, but it is a relic. If you do the sums and you ask what the mass of the cloud would have to be to bend this jet in 3C277.3, you find that a mass of about 10^6 $M_\odot$ or more is needed by very simple arguments. We find from the Hβ emission that we are talking about a much more massive cloud. So I'm just trying to convince you that this is a reasonable scenario for what's happening at high redshifts. If you extrapolate this to high redshifts you get a typical mass needed to bend the radio sources of about 10^8 $M_\odot$ and I think that it is perfectly reasonable. What we see in 3C277.3 at low redshifts is an example of the sort of thing that is responsible for the systems close to the emission line redshift.

Sandage: Is alignment along the major or minor axis?

Miley: The major axis.

Disney: I was going to ask Martin Rees and perhaps Renzo to comment on two kinds of different scenarios. Martin wants galaxy formation to have taken place recently and one would therefore expect that there was still a lot of hydrogen around which hasn't formed stars. Renzo knows how much hydrogen there is floating around in space which may or may not be associated with optical images. My impression was that if the scenario that Martin Rees wants is true, then we should be seeing quite a lot of messy hydrogen clouds around.

Davis: For every 'on' spectrum taken at Arecibo there is an 'off' spectrum taken which is quite capable of detecting any sort of random hydrogen that's around. There is only one such cloud (the Schneider cloud) detected in 7 or 8 years.

Disney: Does that worry you Martin, that there is very little hydrogen around now which has not formed stars?

Rees: Not very much, because I think it may be ionized.

Shaver: There is one subject which is well represented in the posters and something that Wal Sargent briefly touched on. Gordon Robertson is going to say a few things about the pair Tololo 1037/1038 (see poster papers for details).

Robertson: (Summary) The question is whether the absorption is physically common to the two QSOs. That is, are we look at one very large coherent structure? Some of the absorption looks very much like BAL absorption. There is strong NV.

Turnshek: Was there much Ly-α broad absorption?

Robertson: None at all. If we believe that the object is a BAL QSO then we certainly expect that the absorbing cloud is close to the QSO and not a large cloud which could be in common with the other QSO. Some absorption looks like one of the associated complex systems that Craig Foltz told us about and its not too far from the emission redshift. If we perhaps agree that that is what those systems are, then it's not so terribly exceptional that we have this high line density. Certainly we have many more lines than the Young, Sargent and Boksenberg average value, but then they excluded the BALs by design and, as Craig told us, they didn't get the associated complex systems. I think perhaps the next point to turn to, with regards to the question of whether the sytems are in common, is the question of what level of coincidence is required to see four systems at about the same redshift in the two QSOs. Here we need to consider the question of possible line-locking. You may notice that in one QSO particularly, systems are very nearly equally spaced and this may be an example of line-locking. If it is, then you have a much reduced level of coincidence required. Whether that is the case or not, one doesn't know, but it is a possibility which we really ought to think about.

Tytler: Some of us don't understand what type of line-locking you are talking about.

Robertson: Well, the velocity difference is very close to the separation between Ly-α and one member of the NV doublet. That means that Ly-α absorption line in one of these systems will fall on one of the NV absorption lines of one of the other systems, but it's not even a perfect match.

Shaver: As Wal Sargent said, the probabilities depend on how things are calculated and range from 10^{-5} to 20% or so. I estimated something like a 2%. I'd like to ask Dave Turnshek, would you classify at least those two systems in each QSO as BAL systems, assuming you saw the absorption in the absence of the rest of the systems.

Turnshek: I suppose I would. However, with regards to the intervening/ejected question in QSOs in general I have always been concerned about whether or not there is evidence for a continuous progression of morphological absorption types. If one were to collect hundreds of spectra, and then order them in terms of amount of absorption, it would be interesting to see if there were a clear way to distinguish between ejected and intervening types. It is not clear that there is. I don't think you could answer your question for sure until you really could see the distribution of absorption types and exactly where one stopped and the other started. There just aren't enough spectra to make a good point empirically one way or the other. Certainly at least one of the Tololo pair looks like a BAL QSO to me, but I suppose it depends on how many objects it took to discover that quasar.

Robertson: Isn't there a critical difference in that BALs generally don't break-up into recognizable doublets.

Turnshek: That's not true. BALs often break-up into doublet components. You can recognize those.

Foltz: Let me show the spectrum of Q1303+308 which shows recognizable doublets and has SiIV line-locking. Dave showed this during his talk. (See Turnshek's paper on BAL QSOs in this volume for spectra of Q1303+308.)

Turnshek: Do the two Tololo quasars have strong NV emission? That is a characteristic of BAL QSOs.

Shaver: Yes, they do.

Turnshek: That might be a clue suggesting they are BAL QSOs.

Shaver: In a sense there are two points. One is whether both quasars by chance happen to have BAL systems, in which case it is rather implausible to talk in terms of massive superclusters. Also, it seems slightly interesting, or should I say perversed, that if this superclustering was suppose to be commonplace on angular scales of 20 arcmin at these redshifts, that it is not so frequent amongst quasar pairs with separations of 4 or 5 arcmin. We have several such pairs now and we find very few cases of common absorption. So if there are superclusters like this, they must be relatively infrequent, but I understand there are other searches going on now for similar cases elsewhere in the sky. Well, we have pretty well run out of time. We've still got a whole list of controversial issues. It is clear that every issue that we have touched on has been controversial, but is there any extremely pressing controversial issue that has to be addressed in the next 1 minute and 46 sec? Thank you.

21 CENTIMETER ABSORPTION LINES

F. H. Briggs
Department of Physics and Astronomy
University of Pittsburgh

1. INTRODUCTION

Considerable effort, dating from the discovery of absorption lines in the spectra of QSOs, has gone into redshifted HI 21 cm observations. Theoretical discussions (Bahcall and Ekers 1969) were shortly followed by radio observations of the strongest absorption systems that had been identified optically (Shuter and Gower 1969; Miley and Heiles 1970). Since then, a number of observations have been performed, involving a large number of participants. In some sense, the results from all this work might seem disappointing, since the number of QSO absorption line systems detected in the 21 cm line is small (see Table 1). There are 9 detections of redshifted 21 cm absorption in the radio spectra of QSOs. Two of these occur at such low redshifts that a galaxy close to the line-of-sight can be observed to have the same redshift as the absorber. Two upper limits have been included in the table for systems that have been central to discussion during this Workshop, but there are many more objects for which radio surveys have failed to detect 21 cm absorption. Statistics describing the success rate are difficult to assemble, since many of the non-detections have gone unpublished. The published work describing non-detections shows a wide range of sensitivities and observational strategies (cf. Galt 1977; Perrenod and Chaisson 1979; Peterson and Foltz 1980), making a systematic analysis nearly impossible.

There is a sharp contrast between the short list of 9 systems detected in the 21 cm line and the line lists compiled by observers of optical and ultraviolet light, where hundreds, perhaps thousands, of lines have been observed. Why are the 21 cm lines so hard to find? Have the results been worth the considerable effort? The answers will be that the systems capable of 21 cm absorption are by their nature rare; in order to properly appraise the statistics of their detection, a wide variety of approaches and much observing time has been required; and the effort has produced unique and occasionally surprising information that could not have been obtained in any other way.

TABLE 1. The 21 cm Absorbers

QSO	z_{em}	z_{abs}	N(HI)* 10^{21} cm^{-2}	FWHM† km s^{-1}	Method of Selection	Ref.
PKS 1229-021	1.038	0.395	0.9	12	MgII	1, 2
3C196	0.871	0.437	1.2	2 × 17	21 cm	3
AO 0235+164	0.94	0.524	9	~ 5 × 8	MgII	4, 5
3C286	0.849	0.692	0.9	9	21 cm	6, 7
MC 1331+170	2.081	1.776	0.8	20	Ly-α	8
PKS 1157+014	1.986	1.944	6	42	Ly-α	9, 10
PKS 0458-020	2.286	2.038	6	2 × 8	Ly-α	11
3C232	0.513	0.0047	0.12	3.6	nearby galaxy	12, 13, 14
PKS 2020-370	1.050	0.02	0.15	< 11	nearby galaxy	15
			non-detections!			
PKS 0215+015	1.715	1.345	$\tau < 0.04$		MgII, Ly-α (IUE)	16
PKS 0528-250	2.765	2.811	$\tau < 0.1$		Ly-α	17, 18, 19
+ many, many more						

Notes:

* For z_{abs} less than 1.6, a value of 300 °K for T_s has been assumed in order to compute N(HI).
† Multiple components are designated by, for example, 2×17, to indicate 2 components which are each 17 km s^{-1} wide.

References:
(1) Brown and Spencer (1979), (2) Briggs and Wolfe (1983), (3) Brown and Mitchell (1983), (4) Roberts et al. (1976), (5) Wolfe et al. (1978), (6) Brown and Roberts (1973), (7) Davis and May (1978), (8) Wolfe and Davis (1979), (9) Wolfe, Briggs and Jauncey (1981), (10) Briggs, Turnshek and Wolfe (1984) (11) Wolfe et al. (1985), (12) Hashick and Burke (1975), (13) Wolfe (1979), (14) Rubin, Thonnard and Ford (1982), (15) Carilli and van Gorkom (1987), (16) Briggs and Wolfe, (unpublished), (17) Jauncey et al. (1978), (18) Brown (unpublished), (19) Briggs, Wolfe and Schaeffer (unpublished).

2. THE NATURE OF THE 21 CM LINE

The principal cause of the rarity of 21 cm absorption line systems lies in the weakness of the transition. A straightforward comparison of the strengths of the 21 cm line and the electric dipole transition for Ly-α (cf. Spitzer 1978) shows that the optical depths along a line-of-sight are in the ratio

$$\frac{\tau_{Ly-\alpha}}{\tau_{21}} = 2.4 \times 10^8 \left(\frac{T_s}{100}\right) \quad (1)$$

Here, T_s is the spin temperature measured in Kelvin. While column densities of only N(HI) ~ 10^{13} to 10^{14} cm^{-2} can cause a cloud to be optically thick to Ly-α, it is clear that only the highest column density systems will be capable of producing detectable 21 cm absorption. This fact is reflected in the measured HI column densities listed in Table 1 for the 21 cm absorbers. Absorption systems of such high column density constitute a very small fraction of the systems that have been discussed in this

Workshop.

We are accustomed to finding these large column densities of HI at the present epoch where lines-of-sight pass through the planes of spiral galaxies. Therefore, there is a predisposition toward thinking that a high z system with large N(HI) probably represents an interception by a gas-rich galaxy.

The spin temperature T_s is a compact way to specify the relative populations, n_0 and n_1, of the ground state hyperfine levels

$$\frac{n_1}{n_0} = \frac{g_1}{g_0} e^{-\frac{h\nu_{21}}{kT_s}} \qquad (2)$$

where g_i represents the degeneracy of the ith level, h is Planck's constant, k is Boltzmann's constant, and ν_{21} = 1420.4057 MHz. In every astrophysical situation that is well observed and understood, the population balance for these levels is determined by collisions in the gas. Therefore, T_s takes on the value of the gas kinetic temperature, and one can think of performing experiments to infer the temperature of a high redshift cloud simply by measuring its optical depths at Ly-α and 21 cm. A great deal is known about nearby neutral gas, and in this Workshop Dickey has presented evidence that T_s = 300 °K is a typical value measured for lines-of-sight through our Galaxy. [For galactic gas, T_s is determined from a comparison of the line profiles measured in emission and absorption at 21 cm (cf. Dickey, Salpeter and Terzian 1979), unlike QSO absorption line studies in which a measurement of Ly-α is crucial.]

One can imagine situations in which the radiation field competes with collisions in determining the level populations. The presence of a very strong radiation field, such as might be expected in the environment close to a quasar, has been examined by Wolfe, Davis and Briggs (1982) in an effort to account for the behavior of the complex 21 cm absorption line system in the spectrum of AO0235+164. This unique object will be discussed later in this article. Another case where the influence of radiation in determining T_s has been closely examined and which is also relevant to QSO absorption studies is the "intergalactic" HI cloud in Leo (Schneider 1985; Schneider et al. 1983). Upon its discovery, some astronomers visualized it as a large diffuse cloud of such low density that the microwave background radiation would determine T_s. Subsequent radio studies at higher resolution have shown that the cloud is more compact than originally thought and that the cloud is actually part of a still larger ensemble of clouds which are bound to an eccentric orbit around two early type galaxies (M105 and NGC3384). In this sense, the Leo cloud is not really what astronomers have in mind when they daydream about intergalactic clouds floating in space, isolated from galaxies. However, the Leo cloud may be nearer to our ideal picture than anything else we have to study since:

1. it lies in a long period orbit that may have contained this gas since the formation of the cluster; and
2. it shows no optical emission to within quite sensitive limits (Skrutskie, Shure and Beckwith 1984) implying that there are no stars present which have been polluting the primordial material from which the cloud originally formed.

None of the clouds that have been called 'intergalactic' are sufficiently isolated from galaxies to be considered members of a separate population of extragalactic object. In all cases, they are bound to a group or cluster of galaxies and may represent debris left from either galaxy formation or galaxy interactions that are common (see Sancisi 1988). Several surveys have searched substantial volumes of intergalactic space for isolated clouds (Haynes and Roberts 1979; Lo and Sargent 1979; Shostak 1977). Furthermore, a significant volume has been searched inadvertently during the calibration process for HI observations of nearby galaxies (cf. Tully and Fisher 1981). While numerous HI features have been discovered in fairly isolated volumes of space, they must be classified as galaxies since deep photographs inevitably reveal optical counterparts indicating the presence of stellar emission. It is probably safe to say that HI clouds are less than one tenth as numerous as galaxies at any given detection level for HI mass, and that they can not be a major component of the universe at the present. If they were abundant in the past, very efficient mechanisms must have been at work to sweep them up, evaporate them, or cause them to collapse to galaxies.

3. THE REDSHIFTED 21 CM LINE ABSORBERS

Beyond sharing the common trait of very high HI column density, the specific characters of the nine systems in Table 1 follow closely from the method by which they were selected. For the high redshift systems, three methods of selection have been used: radio surveys at 21 cm, the presence of narrow metal absorption lines from low ionization transitions such as the MgII doublet at 2792,2803 Å, and the identification of a damped Ly-α absorption line system (Ly-α + narrow metal lines). The two very low redshift systems were discovered through sensitive spectroscopy of radio QSOs chosen because lines-of-sight to them pass close to nearby foreground galaxies.

3.1. Radio Surveys at 21 cm

Two of the absorption systems in Table 1, those associated with 3C196 and 3C286, were discovered during a large survey at Green Bank that Brown described during this Workshop. The sample was constructed of the brightest radio QSOs of sufficiently high redshift. Neither of these 21 cm absorption systems would have been selected for study based on either of the other techniques. Both absorption systems lie at too low redshift for Ly-α to be shifted to the optical window, and neither system shows strong metal lines. In fact, 3C286 has only very weak MgII absorption (Spinrad and McKee 1979) at the 21 cm redshift, and 3C196 shows no absorption at all in the usual UV and optical transitions. The implication for 3C196, which is an extended radio source, is that the absorbing cloud covers a substantial fraction of the extended source but does not cover the QSO nucleus.

The 21 cm absorption profiles commonly have components of narrow velocity spread (Table 1). The weakness of the MgII lines observed in the optical spectrum of 3C286 would be predicted for a cloud with the velocity spread indicated by the 21 cm observations. The lines are highly saturated but are very narrow, creating a very small equivalent width.

Pure radio searches for 21 cm lines are difficult for several reasons. Since the probability for intercepting an absorber of sufficient column density is low, a large "redshift path" involving many QSOs must be surveyed in order for the cumulative probability to reach a significant level. These surveys are complicated by the nature of radio receivers and spectrometers which cover relatively narrow bands (typically 10 to 20 MHz per integration). Therefore, the radio spectrometer must be repeatedly stepped in frequency, a process which may require re-tuning a receiving system or even a receiver change. A further complication arises because the $z_{abs} = 0$ to 4 range places the observed frequencies in radio bands that are allocated for military, commercial and miscellaneous civilian use. Thus, a random choice of frequency is likely to place the frequency of observation in conflict with such radio interference as airport and military radar, TV, citizens band, time services, paging beepers, and harmonics of FM stations. Under some circumstances, the interference can be so strong as to make any observation impossible. The interference environment becomes worse every year.

The program at Green Bank conducted by Brown and colleagues involved stepping through the radio spectra from 1420 MHz down to 750 MHz for nearly 100 bright radio QSOs (Brown, unpublished). There were two detections, 3C196 and 3C286. Following the usual arguments for absorption probability, one can infer that this frequency of detection could be by created by a population of disk galaxies with the comoving number density presently observed for spirals if their cross sections are about twice the cross section defined by disks with the present Holmberg radius, R_{H_o}. Derivation of this statistic assumed $R^* = 23$ kpc and the normalization $f = 1.8 \times 10^{-6}$ Mpc^{-3} for the luminosity function proposed by Schecter (1976) and revised by Felten (1977) [$H_o = 50$ km s^{-1} Mpc^{-1}; $q_o = 1/2$]. The radio observations imply fairly high column densities N(HI) $\sim 10^{21}$ cm^{-2} for these two systems. Nearby large spiral galaxies have typically fallen off to N(HI) $\sim 5 \times 10^{19}$ cm^{-2} (measured perpendicular to the disk) at 1.4 R_{H_o} (Briggs and Wolfe, in preparation).

A survey at higher redshift was carried out using the most sensitive receiving system at Arecibo (Briggs, Wolfe and Silverglate, unpublished). Every radio bright QSO with redshift greater than 2.3 that lies within the declination range accessible to the Arecibo telescope was surveyed over the redshift range $z = 2.22$ to 2.37, which corresponds to the 20 MHz band of sensitive response of the 430 MHz system (which is actually centered at 431 MHz). There were 21 QSOs in the sample. No 21 cm absorption lines were detected. The sensitivity of the survey varies across the band and as a function of brightness of the background radio sources. About 3/4 of the total redshift path was surveyed at a sensitivity to optical depth

$$\tau_{3\sigma} < 0.03(4 \text{ km s}^{-1}/\Delta V)^{1/2} \text{ for } \Delta V < 100 \text{ km s}^{-1}. \tag{3}$$

This limit translates into a limit on column density

$$N(HI) < 2.2 \times 10^{19}(T_s/100)(\Delta V/4 \text{ km s}^{-1})^{1/2} \text{ cm}^{-2}. \tag{4}$$

For the redshift path surveyed, one would expect galaxies to be able to provide one or more interceptions with probability of 0.6 to 0.4 for $q_o = 0$ to 1/2 respectively if the interception cross section were determined by a disk of radius 3 times the presently observed Holmberg radii. This limit becomes interesting in light of the discovery discussed by Wolfe et al. (1986) that the detectability of damped Ly-α indicates that at $z = 2$ to 2.5, the effective cross section of spiral galaxies corresponds to 3 R_{H_o} for N(HI) $> 2 \times 10^{20}$ cm^{-2}.

3.2. Systems selected by MgII Absorption

Two of the systems in the list of 21 cm detections (Table 1) were discovered by tuning a radio receiver to the redshift determined from a prior optical observation of the MgII doublet. The presence of the low ionizations states of common elements is a logical characteristic on which to key when searching for high column densities of neutral gas since these ions (CII, OI, MgII, FeII, SiII, etc) are abundant in neutral gas along typical lines-of-sight through the disk of our own Galaxy. A better indicator of the total column density of neutral gas can be determined from observations of Ly-α so identifications based on these metal ions are clearly most relevant to absorption systems with $z_{abs} < 1.6$. The optical spectrum for the $z_{abs} = 0.524$ system in AO0235+164 (Table 1) has been discussed in detail by Wolfe and Wills (1977). Several papers discuss radio observations of this object (Wolfe et al. 1978; Johnston et al. 1979). A discussion of the absorption line-variability observed in the $z_{abs} = 0.524$ system of AO0235+164 can be found in §4.2.

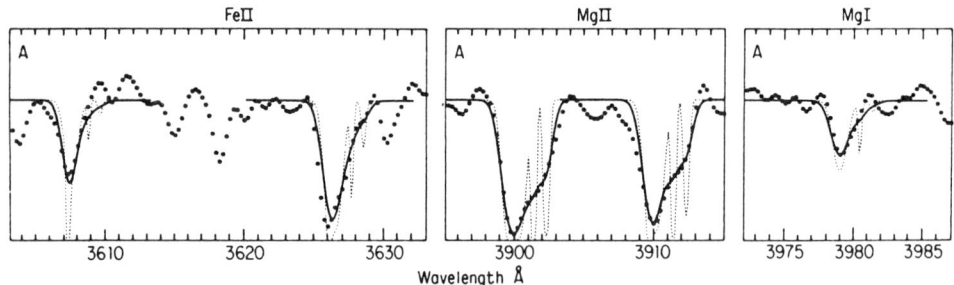

Figure 1. *Observed and synthetic profiles for the FeII, MgII and MgI lines at z = 0.395 in the spectrum of PKS1229–021. Filled circles indicate the observed data points. Solid lines indicate the synthetic profiles after convolution with the instrumental profile. Light dotted lines indicate the model line shape prior to smoothing by the instrumental resolution. The model in discussed by Briggs et al. (1985).*

The absorption system at $z_{abs} = 0.395$ in PKS1229–021 was one of the first QSO absorption systems discovered (Kinman and Burbidge 1967). It has been closely examined in several studies (Peterson and Strittmatter 1978; Brown and Spencer 1979; Briggs et al. 1985). The most detailed analysis of the optical spectrum is contained in the paper by Briggs et al., which reports 1 Å resolution MMT observations of the MgI, MgII, and two FeII lines. The findings serve to illustrate the relation between the high column density cloud of HI that is responsible for the 21 cm line and the gas that creates the strong, distinctive MgII absorption by which the system was selected for study at 21 cm. Figure 1 focuses in on three small regions of the spectrum in order to examine structure within the optical absorption lines. The 21 cm line width, 12 km s^{-1} in this case (Table 1), is much narrower in velocity spread than the optical profile which extends over ~ 200 km s^{-1}. Furthermore, the profile is distinctly asymmetric, with the blackest part of the MgII lines centered on the 21 cm redshift and a wing extending to the red that is of lower optical depth. In the weaker transitions of FeII and MgI, the red wing is much less pronounced. Although the signal to noise ratio is not very good, profile fitting reveals that a component with the velocity width of the 21 cm profile would not be capable of creating a nearly black MgII line at 1 Å spectral resolution. Additional absorbing gas must be present with a velocity spread that encompasses the very thick component that causes the 21 cm line. Of less certainty is the deduction that the red wing of the profile might be composed of several clouds. The fit to the line is improved somewhat by the assumption of two clouds, but clearly the data would permit a wide range of possibilites. The background QSO is bright enough to make this object a good candidate for high resolution spectroscopy which might go a long way in illuminating the multi-component nature of this absorber.

Briggs et al. assumed that, as a working hypothesis for the nature of the $z_{abs} = 0.395$ system in PKS1229−021, our line-of-sight may pass fairly close to the center of a spiral galaxy that is inclined more than $\sim 60°$ to the plane of the sky. The closest approach of the line-of-sight to the galaxy's nucleus would occur above the plane but in a region where corotating halo clouds would be capable of creating the red wing of the MgII profile. The line-of-sight would intersect the plane of the galaxy fairly far from the center at a point (close to the projection of the galaxy's rotation axis onto the sky) where the galaxy's rotation is carrying the gas in a direction that is nearly transverse to our line-of-sight. Thus, the 21 cm redshift would be more closely representative of the true redshift of the galaxy. The close passage of the line-of-sight to the brightest part of the galaxy would make it very difficult to distinguish the galaxy's image from the bright background QSO, which is consistent with the fact that the sensitive searches for galaxy images near this QSO have yielded negative results (Weymann et al. 1978; Bergeron 1988).

The hypothesis designed for the absorber in PKS1229−021 can also be applied on a broader scale to account for the detection rate of 21 cm lines in a sample of systems selected on the basis of low ions such as MgII. There have been a number of studies of this kind of system performed with a wide range of sensitivities. The most sensitive is the survey of Briggs and Wolfe (1983) which reports Arecibo results for the study of 18 absorption system with 3σ upper limits to optical depth at the level of a few %. No new 21 cm detections were added to Table 1 by this study implying that only about one in ten of the clouds capable of creating MgII absorption is capable of creating 21 cm absorption. This fraction is of interest when compared to the cross section for MgII absorbers, σ_{MgII}, normalized to Holmberg-sized galaxy disks. This cross section has been shown by several observers (Weymann et al. 1979; Bergeron 1988) to be $\sigma_{MgII} \sim 10\sigma_{H_o}$ where σ_{H_o} represents the effective cross sections of disk galaxies at the current epoch. Clearly then, the frequency of occurrence of 21 cm absorption that is associated with MgII is consistent with interception by disks of approximately Holmberg size. The column densities that are estimated (by assumption of $T_s \sim 300$ °K) for systems selected in this way are also consistent with the column densities expected for lines-of-sight through spiral galaxies.

Thus, the number of 21 cm absorbers in a sample of MgII systems appears to be well understood. On the other hand, there are approximately 10 times as many MgII absorbers without 21 cm absorption, and the origin of these systems, which are indistinguishable in 1 Å resolution observations from the 21 cm absorbers, also needs to be identified. The working hypothesis discussed above for PKS1229−021 can be extended in the manner suggested by Briggs and Wolfe (1983): strong MgII systems appear strong because the line-of-sight passes through a gas of wide velocity spread (possibly turbulent; possibly spread by differential rotation) and not necessarily because there is a high column density of neutral gas. Only occasionally (1 in 10) does the line-of-sight chance to hit a high column density of low turbulence (a cloud in a galactic disk) that is embedded in the more extensive medium (a galactic halo). A comparison of the optical and radio data for the 21 cm absorbers shows that, in nearly all cases, the gas within the 21 cm profile alone cannot account for the strength of the optical lines. Therefore, an additional 'diffuse' component is required. The notable exception is the system in 3C286 in which both the radio and optical results are consistent with gas of very low velocity spread, implying that the line-of-sight passes through a cloud situated in a place where halo gas is not extensive.

3.3. Systems selected by Ly-α Absorption

The presence of a strong, damped Ly-α absorption line has proven to be the most successful predictor of a detectable 21 cm line. Although all information concerning the velocity structure is obscured by the opacity in the damping wings of the Ly-α profile, the determination of the total N(HI) column density is unambiguous. All of the systems for which Ly-α indicates N(HI) $> 10^{21}$ cm^{-2} have detectable 21 cm lines when they are examined with a sensitivity to optical depth at the 0.01 level. For N(HI) below that threshold the detectability will depend on how the gas is clumped in velocity, since a given column density of gas is easiest to detect when it is confined to a narrow line. Apparently, for lines-of-sight with N(HI) $> 10^{21}$ cm^{-2} the probability is high that at least one high column density cloud of narrow velocity dispersion will be encountered.

The QSO PKS0528–250 has a well-studied optical absorption system at $z_{abs} \sim$ 2.810 that has N(HI) $\sim 3 \times 10^{21}$ cm^{-2} (Morton et al. 1980; Smith, Jura and Margon 1979; Jauncey et al. 1978) and has a tentative detection of molecular hydrogen (Foltz, Chaffee and Black 1987). This system appears to be an ideal candidate for a 21 cm absorption line. Nevertheless, several radio observations (Jauncey et al. 1978; Brown, unpublished; Briggs, Wolfe and Schaeffer, unpublished) have produced non-detections. Since the background radio source is weak (~ 0.35 Jy), and the source is too far south in declination for observation by either the Arecibo or 300 Foot telescopes, the observational upper limit to the 21 cm optical depth in this system is not very strict.

The existence of both Ly-α and 21 cm data for three systems in Table 1 allows the computation of T_s, provided the assumption can be made that a negligible amount of HI gas lies outside the 21 cm profile. Neutral gas spread out over more than 100 km s^{-1} would be difficult to detect in the 21 cm line observations, especially if it is hot, but it would still contribute to the N(HI) that is measured in Ly-α. Thus the temperatures $T_s \sim 1000$ °K (MC1331+170), 1500 °K (PKS1157+014), and 600 °K (PKS0458–020) represent upper limts to the T_s of the gas contained within the 21 cm profiles.

3.4. QSOs near Galaxies

The selection of QSOs with celestial coordinates that place them close to visible, low redshift galaxies can pick out absorption systems that may be distinctly different from the high redshift systems for which the absorber cannot be seen. While optical observations of such systems have produced a number of CaII and NaI lines at the redshift of the nearby galaxy with impact parameters of a few Holmberg radii (Blades 1988), there are few published statistics on the detection rate. The 21 cm line in PKS2020-370 was discovered by tuning the radio spectrometer (Carilli and van Gorkum 1987) to the absorption redshift determined optically (Boksenberg et al. 1980). The system in 3C232, on the other hand, was the only successful detection in a purely radio survey of a number of radio sources that lie close to foreground galaxies (Hashick and Burke 1975) and was only later observed to have CaII absorption (Boksenberg and Sargent 1978).

In one respect, the low redshift systems do resemble the usual population of high redshift absorption lines. The line width of the 21 cm profile is much narrower than the velocity spread inferred from the optical measurements, implying that there is a cool, neutral cloud immersed in a more turbulent medium.

The HI column densities of the low redshift systems are in sharp contrast with the other systems in Table 1. The column densities of the low z systems are very low and would not have been detectable in most of the high z QSO spectra. The proximity to a foreground galaxy and the availability of very sensitive receiving systems at these frequencies have led to the detection of these two weak systems. It is probably not fair to expect the low z systems to serve as models for the high z ones when they are typically down by a factor of 10 in column density.

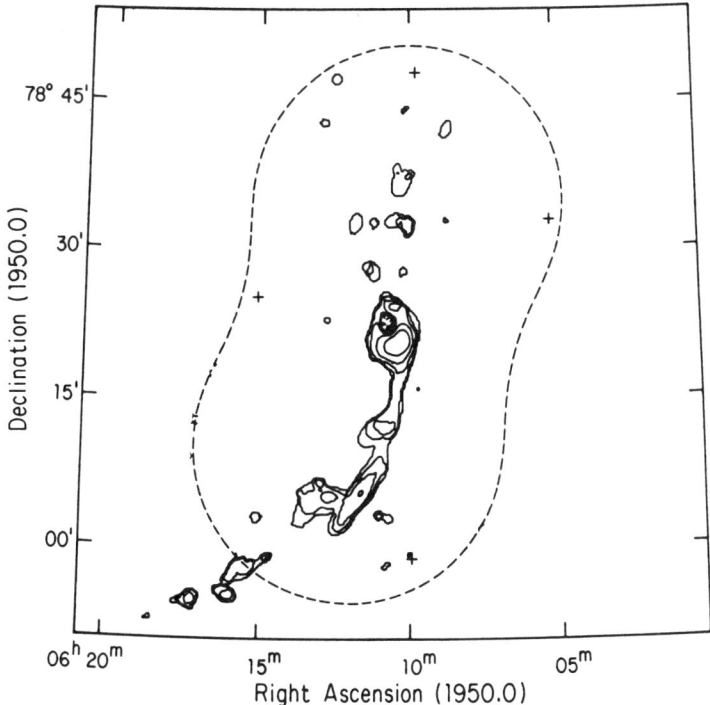

Figure 2. *Contours of neutral hydrogen column density for NGC2146. The column densities are integrated over a 300 km s^{-1} velocity range centered on 900 km s^{-1}. Contours are plotted at 3, 6, 12, 24, 44 × 10^{19} cm^{-2}. Resolution of the map is 2 arcminutes. The galaxy is located at the center of the map and has blue diameters of 5.9 arcmin × 4.9 arcmin. The dashed curve represents the half-power line for the combination of two overlapping VLA primary beams.*

The discussion by Sancisi (1988) may be relevant here to explain the presence of neutral gas outside a galaxy's Holmberg radius. The tides produced during galaxy collisions are very effective at dispersing gas that has been part of a galaxy into 'intergalactic' regions. Interactions can thus account for metal enriched gas at considerable radii from galaxies. Another example that appears to illustrate this process well is the galaxy NGC2146, which had been labelled as the owner of an 'extended HI envelope' (Fisher and Tully 1976). Higher resolution VLA maps in the 21 cm line (Figure 2) now show extended gas lying in long trails that extend to $\sim 10\ R_{H_o}$. The distribution is quite asymmetric and not at all like a 'spherical halo' or an 'extended HI disk.' This galaxy is an extreme example of the kinds of effects that are observed during collisions between spiral galaxies. Despite the large extent of the filament, the presence of the gas probably increases the cross section by only a factor of ten in this case, since large expanses of empty space remain around the galaxy where the gas abundance has not been enhanced at all by the collision.

4. UNIQUE OBSERVATIONS IN THE 21 CM LINE

The ability of radio spectrometers to provide high spectral resolution for the high column density absorbing clouds is evident from the narrow line widths in Table 1. Measurements of HI spin temperatures, which are probably indicators of the kinetic temperatures of the absorbers, were mentioned above. The following discussions demonstrate two other examples of unique information that can be obtained by radio techniques.

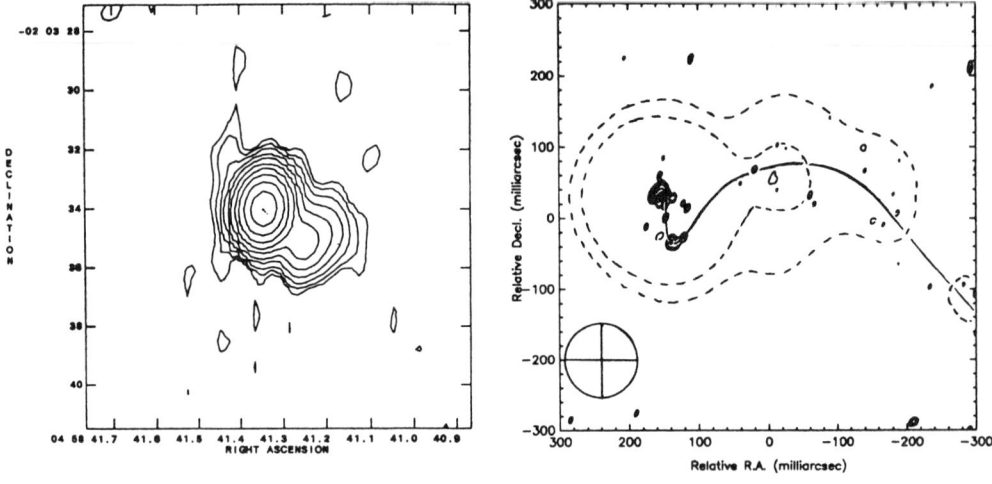

Figure 3. *Radio Structure of PKS 0458–020 (from Briggs et al. 1987). Left frame: VLA A-Array map with 1 arcsec resolution at 1590 MHz. Right frame: Detailed 608 MHz VLBI Network map of the inner part of the source with 10 milli-arcsec resolution. The compact part of the source is a complex, sinuous jet which emanates from an unresolved core. Two low-lying contours from a map made with a 0.1 arcsec beam have been superimposed to verify that the scattered dots at the center of the frame are just the peaks of a more diffuse, extended component. The two beams are indicated at the lower left corner. A line has been drawn to aid the eye in following the wavy path of the extended emission.*

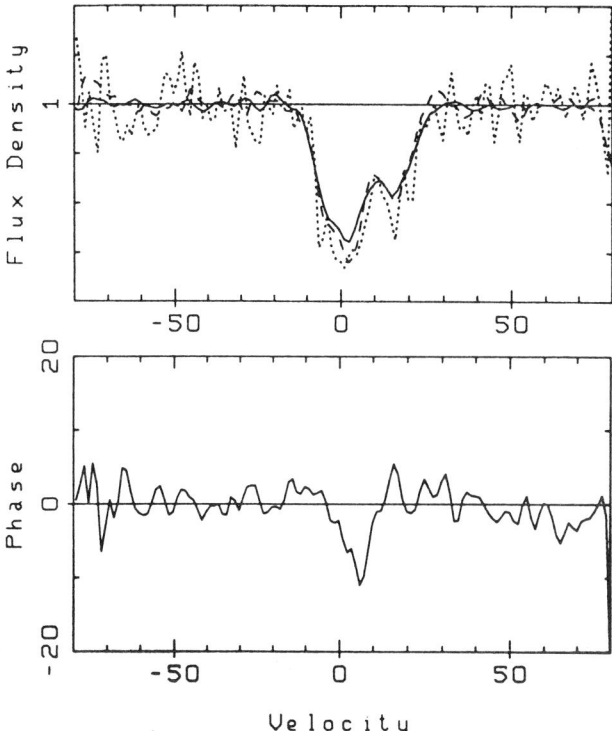

Figure 4. *Single dish and VLBI spectra of the PKS0458–020 HI 21 cm absorption line at $z = 2.04$ as a function of velocity in km s^{-1} (Briggs et al. 1987). Top Frame: Normalized Flux Density: (solid) Arecibo single dish (Wolfe et al. 1985), (dashed) Arecibo single dish at time of VLBI, and (dotted) cross-correlation spectrum between Arecibo and the Green Bank 140 arcmin. In each case the amplitude has been normalized to unity continuum strength. Bottom Frame: Phase spectrum in degrees for the Arecibo to Green Bank interferometer baseline.*

4.1. Spatial Extent of the Absorber in PKS0458–020

Very Long Baseline Interferometry (VLBI) techniques allow radio astronomers to examine the absorbing clouds with angular resolution as fine as a few milli-arcseconds. In order for the technique to be effective, the background radio source must have structure on appropriate scales. Observations reported by Wolfe et al. (1976) of 3C286 and by Brown et al. (1987) of 3C196 have produced interesting results because both these sources have extended structure. Wolfe et al. found that the absorber at $z = 0.692$ in 3C286 extends ~ 1 kpc across the line-of-sight, while Brown et al. deduced limits on the transverse size of the absorber at $z = 0.424$ of greater than 2.25 kpc but less than 14 kpc.

The $z = 2.04$ absorber in the spectrum of PKS0458–020 is particularly interesting since the background source exhibits structure on a range of scales from 10 milliarcsec to 2 arcsec and the absorber has the highest redshift of any 21 cm absorber observed to date. Figure 3 summarizes the structure of the radio source determined from maps made at 1420 MHz with the VLA and at 608 MHz with the European and US VLBI

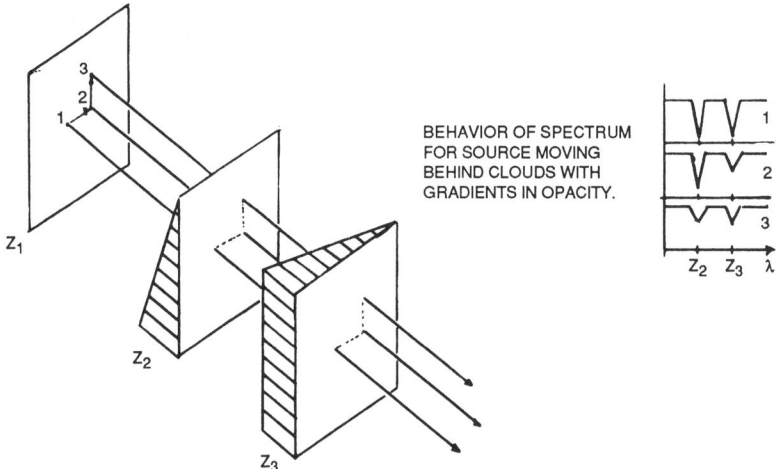

Figure 5. *Simplified schematic describing a moving source (at z_1) behind two absorbing clouds (at z_2 and z_3) that have opacity gradients transverse to the line-of-sight.*

Networks. The 21 cm absorption line is observed at 467 MHz, and a preliminary VLBI experiment has been conducted using telescopes at Arecibo and Green Bank. Figure 4 makes a comparison between the spectrum observed at Arecibo in 'single dish' mode and the spectrum measured with the VLB interferometer. The interferometer is sensitive only to the absorption by the clouds in front of the most compact structure (< 0.1 arcsec) while the single dish spectrum indicates the net effect of the extended absorber in front of an extended background source.

There is a striking similarity both in optical thickness and velocity structure between the absorption spectrum measured at Arecibo and the spectrum sensed by the interferometer. The most naive interpretation is that the absorber extends continuously and uniformly over the entire radio source ($\sim$ 10 kpc in the absorber's reference frame). A simple error analysis shows that only 0.3 Jy of the 2.4 Jy radio source could be peaking around the edge of such a cloud when it is modeled as a uniform 'slab.' On the other hand, a more realistic picture of the absorber would be expected to show considerable variability on scales of 0.1 to 1 kpc. The phase shift detected at the center of the absorption line in the VLBI spectrum is strong evidence that there is variability in optical depth on $\sim$ 0.1 kpc scales. Clearly this system is a superb candidate for a more detailed mapping program.

4.2. The Variable Absorption Lines in AO0235+164

The radio source AO0235+164 has attracted considerable observational effort over the past decade due to its violent variability at all wavelengths and the extreme compactness of its radio source. Cohen *et al.* (1987) review the history of the observations and present new data which define the redshift of the background object to be $z_{em} = 0.9399$ and confirm that the strong absorption system at $z_{abs} = 0.524$ is associated with a highly extended and very luminous line emitting region which is

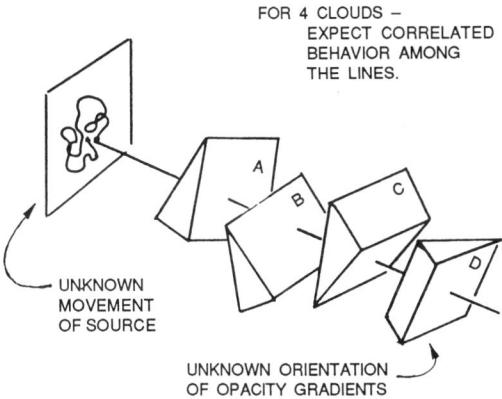

Figure 6. *For the absorbers in the spectrum of A00235+164, neither the motions of the source nor the orientation or strength of the opacity gradients are known in advance. However, changes in the features should be observed to be correlated after the source centroid has wandered in either a random or a systemtic way.*

centered about 2 arcsec to the south and slightly west of radio source. A very strong, complex 21 cm absorption profile is associated with the $z_{abs} = 0.524$ system (Roberts et al. 1976). The degree of compactness of the background radio source makes VLBI probes of the spatial extent of the absorber nearly impossible (Johnston et al. 1979), but the spatial extent of the absorbing object can be inferred from the extent of the optical emission to be ~ 10 kpc.

Wolfe, Davis and Briggs (1982) found that the relative strengths of the four principal features observed in the 21 cm profile vary on time scales of ~ 6 months. A detailed model was explored in which the radio source was assumed to be in close proximity to the absorber. In this picture, the large variability of radio flux can drive the populations of the hyperfine levels and thereby affect the opacity of the four features with gains that depend on each cloud's density and distance from the radio source. The model was tightly constrained by the published observations, and subsequent monitoring of the line variability and source strength have produced feature variations that do not agree with predictions based on radio flux variations (Wolfe, private communication).

Briggs (1983) formulated a test of a class of models for the line variability. The test is statistical in nature and independant of physical mechanisms. The general character of the variability is that the individual features vary in depth and velocity about their mean values but the general appearance of the spectrum remains the

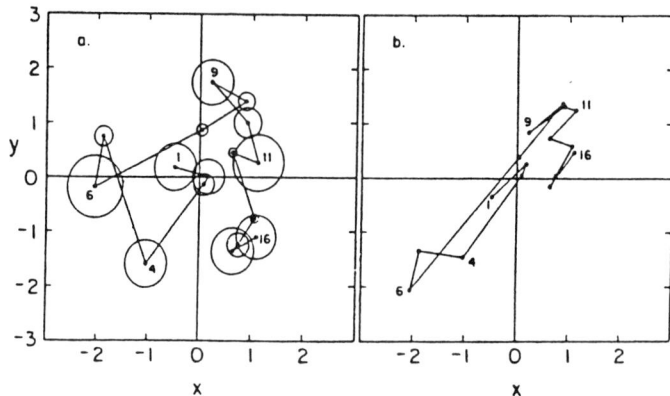

Figure 7. *Pseudomotions of the source centroid for AO 0235+164 (Briggs 1983). Two examples of possible paths are shown. Both assume unit variance in x and y, but the left frame assumes no correlation between the x and y components, while the right side assumes a strong correlation. The ellipses in the left frame represent uncertainties determined from the goodness of fit at each epoch.*

same. There have been no monotonic trends over the five year period that the profile was monitored. If the variability occurs because the background source is moving about behind a screen of four absorbers that have gradients in their opacities, then the source positions must be constrained to stay in the same general area. Physically, this situation might arise if the background source ejects radiating blobs which have a component of motion perpendicular to our line-of-sight. Their motion would shift the centroid of radio emission until the blob's radio brightness dies out and the centroid shifts back toward the nucleus of the radio source. A class of models which invokes a variable bending of the average light path in the absorber naturally fits into such a scheme.

A schematic for a simplified, two absorber system is illustrated in Figure 5. Changes in the absorption spectrum can be predicted if the source's centroid moves between epochs 1, 2 and 3. Conversely, a program to monitor the line depths could be used to infer the position of the source centroid relative to the absorbing foreground clouds as a function of time.

A more realistic picture of the AO0235+164 is depicted in Figure 6. At the outset, we know only that the line depths change, but we know nothing of the orientation and steepness of opacity gradients. Nor do we know the movement of the source centroid, although, in principal, this could be observed with an appropriate VLBI experiment that would be synchronized with the observations to monitor the absorption profile. The statistical test arises from the fact that certain patterns should recur in the measurements of feature depths as the source centroid retraces the same region behind the absorbers, which are assumed to retain constant properties over time. The data do pass the statistical test. Furthermore, Briggs (1983) shows that the feature variability can be inverted in order to infer 'pseudomotions.' The pseudomotions of the source centroid for the data set analyzed by Wolfe, Davis and Briggs (1982) are shown in Figure 7. The motions are arbitrarily scaled and oriented since these parameters cannot be specified by single-dish data alone. The physical mechanism causing the variability remains speculative.

CONCLUSIONS

The sites of redshifted 21 cm absorption lines are probably gas rich galaxies. The requirement of high column densities, coupled with measurements of spatial extent, spin temperature and finely-scaled velocity structure, mesh very well with our understanding of the spiral galaxies that are observed nearby. Observations of this class of absorption system provide in formation that could not be obtained in any other way.

REFERENCES

Bahcall, J. N., and Ekers, R. D. 1969 *Ap. J.*, **157**, 1055.
Bergeron, J. 1988, *This volume*.
Blades, J. C. 1988, *This volume*.
Boksenberg, A., and Sargent, W. L. W. 1978, *Ap. J.*, **220**, 42.
Boksenberg, A., Danziger, I. J., Fosbury, R., and Goss, W. 1980, *Ap. J. (Letters)*, **242**, L145.
Briggs, F. H., and Wolfe, A. M. 1983, *Ap. J.*, **268**, 76.
Briggs, F. H. 1983, *Ap. J.*, **274**, 86.
Briggs, F. H., Turnshek, D. A., and Wolfe, A. M. 1984, *Ap. J.*, **287**, 549.
Briggs, F. H., Turnshek, D. A., Schaeffer, J., and Wolfe, A. M. 1985, *Ap. J.*, **293**, 387.
Briggs, F. H., Wolfe, A. M., Liszt, H., Davis, M. M., and Turner, K. L. 1987, *Ap. J.*, submitted.
Brown, R. L., and Roberts, M. S. 1973, *Ap. J. (Letters)*, **184**, L7.
Brown, R. L., and Spencer, R. E. 1979, *Ap. J. (Letters)*, **230**, L1.
Brown, R. L., and Mitchell, K. J. 1983, *Ap. J.*, **264**, 87.
Brown, R. L., Broderick, J. J., Johnston, K. J., Benson, J. M., Mitchell, K. J., and Waltman, W. B. 1987, *Ap. J.*, in press.
Carilli, C. L., and van Gorkom, J. H. 1987, *Ap. J.*, **319**, 683.
Cohen, R. D., Smith, H. E., Junkkarinen, V. T., and Burbidge, E M. 1987, *Ap. J.*, in press.
Davis, M. M., and May, L. S. 1978, *Ap. J.*, **219**, 1.
Dickey, J. M., Salpeter, E. E., and Terzian, Y. 1979, *Ap. J.*, **228**, 465.
Felten, J. E. 1977, *A. J.*, **82**, 861.
Foltz, C. B., Chaffee, F. H., Jr. and Blades, J. H. 1987, *QSO Absorption Lines: Probing The Universe, A Collection of Poster Papers*, eds. J. C. Blades, C. A. Norman, D. A. Turnshek (ST ScI Publication), p. 136.
Galt, J. A. 1977, *Ap. J. (Letters)*, **214**, L9.
Hashick, A. D., and Burke, B. F. 1975, *Ap. J. (Letters)*, **200**, L137.
Haynes, M. P., and Roberts, M. S. 1979, *Ap. J.*, **227**, 767.
Jauncey, D. L., Wright, A. E., Peterson, B. A., Condon, J. J. 1978, *Ap. J. (Letters)*, **221**, L109.

Johnston, K. J., Broderick, J. J., Condon, J. J., Wolfe, A. M., Weiler, K., Genzel, R., Witzel, A., and Booth, R. 1979, *Ap. J.*, **234**, 466.

Kinman, T. D., and Burbidge, E. M. 1967, *Ap. J. (Letters)*, **147**, L59.

Lo, K. Y., and Sargent, W. L. W. 1979, *Ap. J.*, **227**, 756.

Miley, G. K., and Heiles, C. 1970 *Ap. J. (Letters)*, **160**, L83.

Morton, D. C., Chen Jian-sheng, Wright, A. E., Peterson, B. A., and Jauncey, D. L. 1980, *M.N.R.A.S.*, **193**, 399.

Peterson, B. M., and Foltz, C. B. 1980, *Ap. J.*, **242**, 879.

Peterson, B. M. and Strittmatter, P. A. 1978, *Ap. J.*, **226**, 21.

Perrenod, S. C., and Chaisson, E. J. 1979 *Ap. J.*, **232**, 49.

Roberts, M. S., Brown, R. L., Brundage, W. D., Rots, A. H., Haynes, M. P., and Wolfe, A. M. 1976, *A. J.*, **81**, 293.

Rubin, V. C., Thonnard, N., and Ford, W. K. 1982, *A. J.*, **87**, 477.

Sancisi, R. 1988, *This volume*.

Schecter, P. 1976, *Ap. J.*, **203**, 297.

Schneider, S. E., Helou, G., Salpeter, E. E., and Terzian, Y. 1983, *Ap. J. (Letters)*, **273**, L1.

Schneider, S. E. 1985, *Ap. J. (Letters)*, **288**, L33.

Shuter, W., and Gower, J. 1969, *Nature*, **223**, 1046.

Shostak, S. 1977, *Astr. Ap.*, **54**, 919.

Skrutskie, M. F., Shure, M. A., and Beckwith, S. 1985, *Ap. J. (Letters)*, **282**, L65.

Smith, H. E., Jura, M., and Margon, B. 1979, *Ap. J.*, **228**, 369.

Spinrad, H., and McKee, C. F. 1979, *Ap. J.*, **232**, 54.

Spitzer, L. 1978, Physical Processes in the Interstellar Medium, (New York: John Wiley & Sons).

Tully, R. B., and Fisher, J. R. 1976, *Astr. Ap.*, **53**, 397.

_____. 1981, *Ap. J. (Letters)*, **243**, L3.

Weymann, R. J., Boroson, T. A., Peterson, B. M., and Butcher, H. R. 1978, *Ap. J.*, **226**, 603.

Weymann, R. J., Williams, R. E., Peterson, B. M., and Turnshek, D. A. 1979, *Ap. J.*, **234**, 33.

Wolfe, A. M., Broderick, J. J., Condon, J. J., and Johnston, K. J. 1976, *Ap. J.*, **222**, 752.

Wolfe, A. M., and Wills, B. J. 1977, *Ap. J.*, **218**, 39.

Wolfe, A. M., Broderick, J. J., Condon, J. J., and Johnston, K. J. 1978, *Ap. J. (Letters)*, **208**, L47.

Wolfe, A. M. 1979, in Active Galactic Nuclei, ed. C. Hazard and S. Mitton (Cambridge: Cambridge University Press), p. 159.

Wolfe, A. M., and Davis, M. M. 1979, *A. J.*, **84**, 699.

Wolfe, A. M., Briggs, F. H., and Jauncey, D. L. 1981, *Ap. J.*, **248**, 460.

Wolfe, A. M., Davis, M. M., and Briggs, F. H. 1982, *Ap. J.*, **259**, 495.

Wolfe, A. M., Briggs, F. H., Turnshek, D. A., Davis, M. M., Smith, H. E., and Cohen, R. D. 1985, *Ap. J. (Letters)*, **294**, L67.

Wolfe, A. M., Turnshek, D. A., Smith, H. E., and Cohen, R. D. 1986, *Ap. J. Suppl.*, **249**, 304.

DISCUSSION

[Editors note: At this point in the Workshop, Robert L. Brown (NRAO) was encouraged to describe the program that he and several colleagues conducted to survey the radio spectra of about 100 quasars from 1420 to 750 MHz. His presentation involved the use of many viewgraphs and is therefore not reproduced here verbatim. In the survey the 21 cm absorption line was discovered in two members of the sample (see 3C196 and 3C286 in Table 1 of the preceding paper by Briggs). One radio source, 3C196, was sufficiently extended that an interferometer experiment could place constraints on the size of the absorber. The results of the VLBI investigation are are reported in Brown, Broderick, Johnston, Benson, Mitchell and Waltman (1987, *Ap.J.*, in press).]

Wolfe: I just want to emphasize some results concerning the work that Frank just discussed on VLBI measurements of 0458-020. This comment also possibly applies to Bob Brown's object. The point is that we really can get more information out of the structure of this extended object, because, independent of the detail, that admittedly we don't know very well, it does extend transverse to the line-of-sight more than 10-15 kpc, and the velocity dispersion is effectively 10 km s^{-1}. Therefore it's very unlikely to be a sphere, because if we assumed that the object is also extended along the line-of-sight 10-15 kpc, it turns out that in order for the object to hydrostatically support itself against its own self gravity, it would have to have a velocity dispersion on the order of 100 km s^{-1}, and that's not seen. So, independent of the details of the model, we really are looking at something that is a disk-like structure. I haven't done this calculation for Bob Brown's object because I think the column density is a factor of 10 less, but I also think it's going to be difficult to make that object a sphere.

Brown: That's right, and that's on a view graph they wouldn't let me show. I've done exactly the same thing. It's actually quite interesting. You can put limits on the Hubble constant if you assume it's in hydrostatic equilibrium.

Wolfe: You know that the free-fall time is something like 10^8 years, so it would have to be in hydrostatic equilibrium. I think we really must be looking at big disk-like structures.

York: I still don't have the answer to the broader question of how many have been looked at and how many have been found. What are the statistics?

Brown: In our particular case, we looked at 120 objects and found two 21 cm absorption systems. I should add at least one more sentence to that in deference to veracity. We chose objects based on how strong they were and how small they were, but we didn't select necessarily those whose emission redshifts extended to $z = 1$, thinking that we'll search even for absorption at redshifts considerably higher than the emission redshifts. In normal cosmology, these things can't exist. Some various examples are: in 3C48 we searched out to $z = 1$; in BL Lac we searched out to $z = 1$. So those objects are among the 120, as are another half dozen or so low redshift objects.

G. Burbidge: What obstruction does the absorber in AO0235+164 have to have to let the absorber move steadily in one direction? Can you do it that way?

Briggs: I suppose you could do that, but then, if you did that you would turn up getting something that just went right off the page. The four absorbers would have thick and thin regions as required to produce the observed variability rather than simple linear gradients that are consistent with a source that just appears to move around in a restricted region. My model is very restricted in its applicability. I think it worked okay as far as it went. Of course I probably ought to go try to fit the more recent data in the same scheme I've avoided doing, because I didn't want to put it out of business. So, I guess my argument then in addressing the questions I started out with which were that I think it is worth the effort of coming up with just a few things. We have learned some things. If you really want to find these, it's best to start out with systems that you understand pretty well and that have lots of Hydrogen in front of them, really 10^{21} per square cm. The other point was that you really can come up with some unique observations with radio wavelengths.

York: This concerns the question about line variability in AO0235+164 that I started to ask a while ago. I think Geoff was getting at the same thing, but I didn't quite get the answer. If you put a very clumpy interstellar medium at a redshift of 0.5 which would be indicated by the OII emission, then how do you know that you don't have intrinsic source variations, never mind moving across the path?

Briggs: Are you talking about changing the properties of the absorber?

York: Yes, how did you rule that out?

Briggs: Well, you can't, but you would have to coordinate the properties of a fairly large cloud, or several very large clouds that extend at least tens of pcs so that they change on a time scale of months. Art's original notion was sort of like that. The question is: how can you control them? Well, you might control them by bathing them in variable 21 cm continuum. That would be one way to do it. However, it's hard to coordinate the change of opacity of a big cloud.

York: What do you mean by coordinate?

Wolfe: The clouds have to be more than 10 pcs, and they would have to change something on the order of 10 pcs in a month. How are you going to do that?

Norman: You could have internal star formation.

Wolfe: But, you would have to affect the entire cloud in a month.

Norman: Sort of like a Christmas tree model.

Wolfe: Alright, then you'd have to have something very odd.

Jenkins: I think the important thing is comparing the variability with the equilibrium time for establishing the temperature, which depends on collisions. Therefore, you have to assume a density. Once you know the density, and you know the collision frequency, you have to have that time scale shorter than the variability time scale. Have you worked that out?

Wolfe: Yes, that's possible. I don't think that's the problem. When we first did this work, we did that. That's the spin temperature relaxation time. Nonetheless, you still have to do it over 30 light years and I don't see how to do that.

Jenkins: Suppose the source is illuminating the cloud, and it pumps it up to a higher temperature. How long does that take to relax?

Wolfe: Oh, that's no problem, but that's the model that Frank ruled out. That relaxation time is the radiative relaxation time.

Jenkins: That's 10^{13} seconds.

Wolfe: No. It's one over $B_{21}J_\nu$. It's not one over the spontaneous emission coefficient. It's driven by stimulated emission. It's like a maser, but the population's are not inverted. However, I think Frank showed that that doesn't work.

York: Is there any independent observation of this thing moving around? Is the only evidence this variable interstellar line?

Briggs: This is a very compact radio source, and if you look very closely at it on the longest VLBI baseline, you can just start to see little jet-like protruberances, but nothing like the long jet in 0458-020. However, its compactness also means that it's likely to have the radio equivalent of an optical seeing effect where it could appear to wiggle around a little bit. That's kind of what they're talking about in Barney Ricket's refractive scintillation models.

Jauncey: But that's still easy to measure because 0235+164 can be measured astrometrically with respect to other quasars nearby.

Briggs: That's what I've been arguing should be done.

Jauncey: The other thing that clearly should be done is to look for changes on the VLBI baseline. The changes were not on the VLBI line, they were just in the single dish observations.

Briggs: Oh, there have been changes on the VLBI line.

Jauncey: The VLBI line observations are really what's going to tell you what's going on, because you've got the spatial resolution. Here we're arguing about what's happening due to lack of spatial information.

Briggs: The trouble is that the longest baselines we have to observe VLBI in the line don't really resolve the radio source enough. There are just hints that we're just starting to get a little bit of spatial information. So that means that with VLBI you essentially see the same spectrum that you see with a single dish.

Black: Under some circumstances the Ly-α pumping could affect the spin temperature. Is that a possibility here? The line variability might be correlated with the UV variability of the source rather than radio variations.

Wolfe: I don't think so because the Ly-α optical depth of this cloud is going to be 10^8 or 10^9, and once the Ly-α photons start scattering around they lose their memory. So, I don't think that can work. However, I agree with you that Ly-α pumping could dominate the spin temperature.

Smith: I have a comment about the optical absorber in 3C196: Ross Cohen and I have been doing a project similar to that of Jacqueline Bergeron and have imaged and obtained spectroscopy of things in the field of 3C196. There are a number of galaxies of the appropriate magnitude. The brightest of them unfortunately doesn't have the same redshift, but there are other possible candidates.

Jauncey: Bob, do you know which of the radio lobes of 3C196 that you see the absorption in?

Brown: No.

Jauncey: What frequency was it?

Brown: 988 MHz.

Jauncey: That must be close to one of the Merlin frequencies.

Brown: It is.

Jauncey: That really will tell you what's going on rather than this sort of calculation, because it will give you the whole structure, and you can identify what's happening.

Brown: That right.

E. M. Burbidge: I guess you've tried hard to get the AO0235+164 0.85 redshift absorber in 21 cm, but the optical lines are so much weaker than in the 0.524 absorber. Is there any hope of getting that?

Briggs: Well, I think that we did work on that pretty hard when we were selecting on the basis of Mg II absorption, and we did not see anything. So, my suspicion is that if you were to observe that system with Space Telescope, you would not find a very strong Ly-α absorption line and that it's not a very high column density of neutral hydrogen. That would account for it. In fact, considering those other two measurements on the bottom of the Table that I presented, you can pretty easily explain why you don't see anything. 0215+015 has a measured Ly-α absorption line from IUE, and there just isn't very much neutral hydrogen there. The neutral hydrogen column density is less than 10^{20} cm^{-2}. For this other one, PKS0528−250, we know that there is a tremendous amount of neutral hydrogen in front of it, but the object turned out to be a very weak radio source — the spectrum just plummets when you get down to 367 MHz which is where the 21 cm absorption line would be. We would just love to have an Arecibo type telescope to point in that direction.

Davis: Fat chance for that. (Laughter) However, I think there is some chance of doing better work at Arecibo on a somewhat long time scale. Almost all of the work that you've seen here was done at Arecibo operated in a very non-optimum fashion, typically with a single polarization and a short line feed, because these feeds have to be tuned-up for each new redshift. We have to manufacture these feeds relatively inexpensively. The upgrade that we're proposing to the National Science Foundation would allow us to cover the entire range of 300 MHz up to 8,000 MHz, so in particular, redshifts up to $z = 2.5$ would be covered. This would be a fully optimized system, about a 20° K system temperature in both polarizations and about 12° K per Jy. Another extremely important aspect is that this would be enclosed in an aluminum ray dome, so that all of these beepers do not give us problems. So Bob Brown's kind of experiment, which I think should continue to be done, would be far, far better if we could get this upgrade project done. Then we will certainly look for the 0.8 redshift system in AO0235+164 at that time.

Sancisi: This was done several years ago, and now you have improved sensitivity. Have you tried again to see if there is more velocity structure or weaker absorption lines?

Brown: You could do better now, but not so significantly better, not a factor of 2 better. So the answer is no.

York: Do you know about the paper by Blades, Cowie and Songalia on the Magellenic clouds? Well, Chris could speak to this better than I, but there are regions several kpc in size where you see very similar absorption line structure in CaII. It would be very interesting to look at that to see if the size scales are anything close to what you'd need.

Blades: Yes. I think McKee on the basis of 21 cm observations interpreted this in terms of sheets, and we seem to see this sort of structure.

York: These are multiple sheets or components that cover a region of several kpc.

Blades: Within those several kpc, there also is certainly velocity structure over 30 - 40 km s^{-1}.

York: The point is that hydrostatic equilibrium may not be the relevant assumption. It may be a non-stable region, that's all.

Wolfe: It depends on what the free-fall time is.

Norman: It's not a free-fall time. It's probably an expanding shell, a very narrow super shell.

DAMPED Ly-α ABSORPTION SYSTEMS

Arthur M. Wolfe
Mt. Wilson and Las Campanas Observatory
and the University of Pittsburgh

Abstract. QSO absorption systems with damped Ly-α lines belong to a distinct population of absorbers with properties that bring to mind the HI disks of spiral galaxies. A survey for damped Ly-α systems is described. It is shown that they occur more frequently along the line-of-sight than do normal spiral galaxies. The damped systems contribute a mass density per unit comoving volume that (a) dominates the baryon content of the universe at redshifts $z = 2-3$, and (b) is comparable to the mass per unit comoving volume contributed by luminous matter in present-day spiral galaxies. It is suggested that the damped absorbers are large HI disks which are the progenitors of normal spiral galaxies.

1. INTRODUCTION

QSO absorption systems with 21 cm lines have attracted attention because they exhibit physical conditions found in the HI disks of spiral galaxies (see Briggs 1988). The resemblance to HI disks is not suprising. The large HI column densities, $N(HI) \geq 10^{21}$ cm^{-2}, low velocity dispersions, $\sigma \approx 10$ km s^{-1}, and low spin temperatures, $T_S \leq 300$ °K, typical of HI disks lead to 21 cm optical depths, $\tau_{21} \approx 1$. Moreover, the fact that the signal-to-noise ratio of a resolved 21 cm feature with velocity width, Δv, is proportional to $N(HI)/(T_S \times \sqrt{\Delta v})$, indicates that HI layers which are cold and quiescent are the easiest to detect. Consequently, 21 cm observations provide a limited field of view that focuses on absorbers with the essential properties of HI disks. In order to decide whether or not these really are the disks of high-redshift galaxies, we need a broader perspective, one that reveals how the 21 cm systems fit into the wider context of the universe at large redshifts.

For these reasons we began a survey for a large, unbiased sample of absorbers with 21 cm properties. The survey was carried out at optical rather than radio wavelengths mainly because of the limited frequency response of the most sensitive line feeds at Arecibo. At present Arecibo is the only radio antenna capable of detecting a statistically meaningful sample of HI layers with $\tau_{21} \geq 0.01$, the threshold set by the detection of some very weak 21 cm lines at large redshifts (see Briggs 1988). The problem is that the phase-correcting line feeds available at Arecibo provide a 21 cm redshift path given by $\Delta z_{21} = 0.1 - 0.2$. On the other hand, $\Delta z_{opt} \approx 1$ for prominent resonance lines detected at optical wavelengths. Thus the optical surveys should be more efficient, provided a suitable signature of the 21 cm systems is found.

The ideal signature is an absorption feature whose strength is sensitive to N(HI), but insensitive to Δv; i.e., a feature that stands out despite the low velocity dispersion expected in HI disks. The advantage of metal line doublets such as MgII is that they are easy to recognize redward of Ly-α emission. The disadvantage with this, and other doublets arising in abundant metals, is that the lines have low equivalent widths, W(MgII) $\approx$ 0.3 Å, because they are saturated transitions that form in gas with low velocity dispersion. Morever, it is probable that most MgII lines do not form in HI gas, but rather arise in ionized gas which need not be related to material in HI disks (Wolfe 1986a). On the other hand, Ly-α lines that form in HI disks will be very strong. The combination of large N(HI) and low velocity dispersion assures that Ly-α will be broadened by radiation damping. In that case W, the rest-frame equivalent width, is uniquely determined by N(HI) according to $W(Ly-\alpha) = 7.3 \times [N(HI)/10^{20}cm^{-2}]$ Å. For the N(HI) that typify the 21 cm systems, $W(Ly-\alpha) \geq 10$ Å. Thus while the absorber is hardly noticeable in the metal lines, it stands out amongst the 'confusion noise' of the Ly-α forest. The appearance of damped Ly-α, 30 times stronger than associated low-ion transitions, is a unique spectroscopic imprint not found in any other type of absorption-line cloud.

2. THE LICK SURVEY

The search began with a low-resolution ($\Delta\lambda = 10$ Å) survey for absorption features in the Ly-α forest that were candidates for 'Ly-α disks.' A total of 68 QSOs with emission redshifts, $z_{em} \geq 2.25$, were observed with the Lick 3-m telescope. The results, reported in Wolfe et al. (1986) and Wolfe (1986b), are that 47 out of 476 statistically significant absorption features qualify as candidate 'Ly-α disks', with $W(Ly-\alpha) \geq 5$ Å. Using accurate follow-up spectroscopy described below we have determined the nature of 31 of the candidates. We find that 14 of the Ly-α features are not damped. What appeared as a single strong feature at low resolution, breaks up into a series of multiple narrow velocity components at high resolution. The velocity intervals spanned by the components range from 1000 to 3000 km s^{-1}. In addition the equivalent width distribution of the multi-component systems peaks near the 5 Å threshold. The remaining 17 candidates are in fact damped Ly-α. Accurate spectroscopy shows that these Ly-α profiles do not break up at higher resolution. Rather, the observed features are well fit with Voigt damping profiles. Moreover in each case, damped Ly-α is associated with narrow low-ion absorption lines at the same redshift. In contrast to the multi-component systems, the damped absorbers dominate the equivalent-width distribution above $W(Ly-\alpha) \approx 10$ Å.

The detection of so many damped Ly-α systems is suprising. Voigt profile fits to the Ly-α features show that 15 out of the 17 damped systems have $N(HI) \geq 2 \times 10^{20}$ cm^{-2}, the HI isophote out to which spiral galaxies have been extensively studied (Bosma 1981). By contrast, no more than 3 disks are predicted for (a) the redshift path ($\Delta z_\alpha = 55$) of the survey, and (b) the cross-sectional radius corresponding to this N(HI). The implication of this discrepancy is that the HI radius $R(HI) \geq 3.5 \times R_{Ho}$ (R_{Ho} is the Holmberg radius), if there is a one to one correspondence between the damped systems and present-day spiral galaxies. Another possibility is that the high detection rate is caused by an inflated population of small objects, such as dwarf galaxies, at large redshifts.

That a particular class of redshift system occurs along the line-of-sight more frequently than predicted for intervening galaxies is nothing new. Every known class of QSO absorption system shares this property. The CIV systems (Young, Sargent and Boksenberg 1982), the MgII systems (Lanzetta, Turnshek and Wolfe 1987), and the Ly-α forest systems (Sargent et al. 1980) all occur more frequently per unit redshift interval than the damped systems. Thus it is entirely possible that the damped systems lie on the tail of the N(HI) distribution characterizing any one of these populations. But the evidence indicates otherwise. For example, the frequency distribution of Ly-α equivalent widths above 2 Å has been derived from the Lick data (cf Wolfe 1986b), and indicates a change in slope above W(Ly-α) $\approx$ 10 Å. Below this equivalent width, the distribution is indistinguishable from the exponential Ly-α-metal distribution proposed by Sargent et al. (1980) to describe Ly-α lines associated with CIV and perhaps MgII absorption systems. Above 10 Å, where the damped systems dominate, the distribution is noticeably flatter, indicating a different parent population. The evidence for a population shift is further supported by the kinematic structure of the absorbers. The sequence in Ly-α-metal equivalent widths is actually a sequence in velocity spread, with the weak features consisting of one or two narrow components distributed over a small velocity spread, and the stronger systems made up of many components distributed over much larger velocity spreads. It is also likely that the total N(HI) in these systems is correlated with velocity spread, owing to the correlation of number of components with the total width of the blended feature. On the other hand, the sequence of the damped Ly-α equivalent widths is independent of velocity spread, and soley reflects a sequence in N(HI). In this case N(HI) is uncorrelated with velocity spread. Therefore the evidence suggests that the multi-component systems originate by a different mechanism than the damped systems.

3. PROPERTIES DIRECTLY DEDUCED FROM ACCURATE SPECTROSCOPY

We now turn to properties of the individual absorbers. Accurate spectra have been obtained at the MMT and CTIO 4-m in order to (a) test the damping hypothesis for the Ly-α profile of each 'disk candidate', (b) confirm the redshift with the wavelengths of narrow metal lines, and (c) derive physical parameters for the absorbing gas. A full description of the observations and analysis is given in Turnshek et al. (1988) and Wolfe et al. (1988). The essential results are summarized in what follows.

All of the spectra blueward of Ly-α emission have been obtained with the MMT at a resolution of $\Delta \lambda = 1 - 2$ Å. Figure 1 is a sample of spectra in which redshifts determined from the metal lines have been used to shift the wavelength scale to the restframe of each damped Ly-α absorber. In QSOs with two damped systems, a restframe spectrum is shown for both redshifts. In every case Ly-α takes on the characteristic shape of a profile with damping wings. Quantitative tests show that most of the damped fits to the observed profiles are quite good. The wide range in W(Ly-α) illustrated in Figure 1 indicates that the damped systems exhibit wide variations in N(HI). Figure 1 also shows that in cases where the metal lines are predicted to occur redward of Ly-α emission, CII λ1334 is detected. When low-ion metal lines such as CII are predicted to occur blueward of Ly-α emission, confusion with the Ly-α forest lines is inevitable. To determine redshifts in those cases, we obtained the FeII $\lambda\lambda$ 2367–2600 multiplets and the MgII doublet with red spectra acquired with the CTIO 4-m.

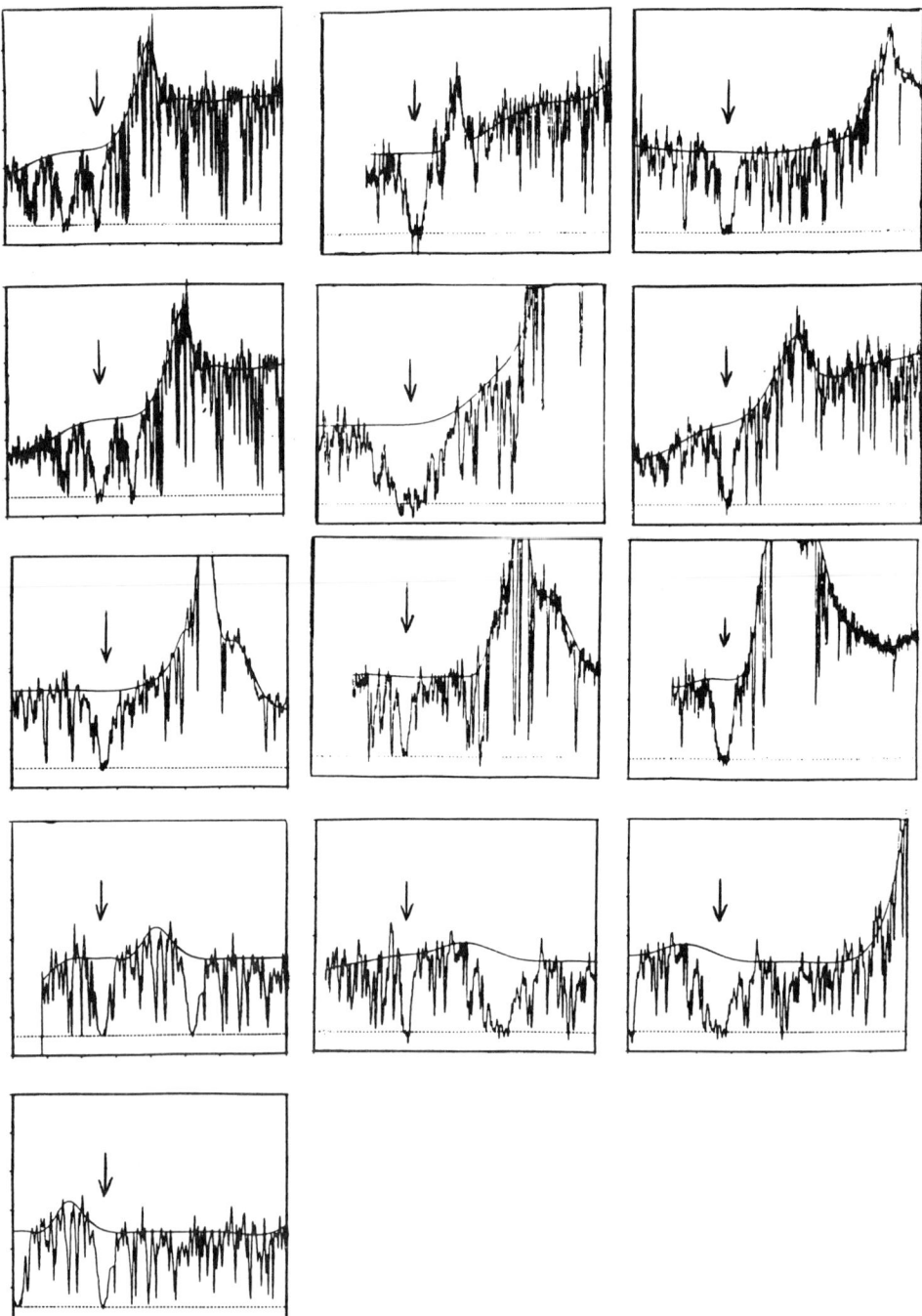

Figure 1. *Restframe Ly-α forest spectra of some sample QSOs with damped Ly-α absorption plotted on the same scale. Damped Ly-α is labeled with an arrow.*

Comparison between the metal-line and Ly-α profiles on a velocity scale is particularly valuable, as it provides a sensitive test of the damping hypothesis, even in cases where the signal-to-noise ratio of the data is not optimum. In every case the metal-line profiles resemble the example in Figure 2, in which a narrow, unresolved ($\Delta v \approx 150$ km s^{-1}) metal line is located at the center of a symmetric Ly-α feature which appears to have $\Delta v \geq 3000$ km s^{-1}. The broadening mechanisms for the two transitions are obviously different. The CII line is broadened by Doppler motions, while quantum mechanical effects dominate the Ly-α profile. This is in contrast to CIV-selected systems with wide Ly-α lines. In those cases the metal lines break up into multiple narrow components (a) which span velocity intervals not much smaller than that covered by Ly-α, and (b) which are not symetrically displaced with respect to the center of Ly-α (Bechtold, Green and York 1987; Bechtold 1987). In this instance both Ly-α and the metal–lines are Doppler broadened, with Ly-α assuming a larger width due to an increase in saturation.

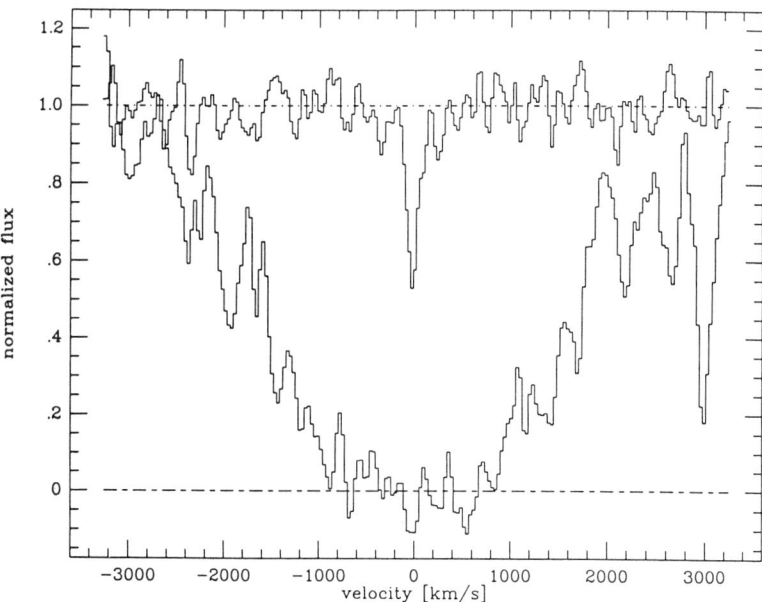

Figure 2.   Comparison between $z = 2.8$ damped Ly-α and CII profiles for Q1337+113.

The N(HI) inferred from the fits of Voigt profiles are better determined than in most absorption systems, because the fits are independent of velocity dispersion. The errors are dominated by inaccuracies in setting the continuum, and are unlikely to exceed 50 %. Above 2×10^{20} cm^{-2}, where the distribution of N(HI) is complete, we find that the mean column density, $\langle N(HI) \rangle = 10^{21}$ cm^{-2}. While this is typical for the HI disks of spirals at $R \leq R_{Ho}$, it greatly exceeds the upper limit of 10^{18} cm^{-2} set at $R \geq 2.5 \times R_{Ho}$, for a complete sample of spiral galaxies (Briggs et al. 1980). If the spirals have the same comoving number density as the damped systems, the latter are greater than the former in HI content as well as size.

Because the damped Ly-α sytems are selected primarily on the basis of their HI properties, the resulting distribution of metal-line properties is unbiased with respect to metal-line selection criteria. Comparison between the detected W(CIV) and W(CII), for example, should provide a fair sample of the actual metal-line properties of the damped systems. We find that for the vast majority of damped systems, W(CII) ≥ W(CIV). This distinguishes the damped systems from most CIV-selected clouds in which W(CIV) ≥ W(CII) (Wolfe 1983; Danly, Blades and Norman 1987). Rather the damped systems resemble the MgII–selected clouds in this respect (Lanzetta, Turnshek and Wolfe 1987). This is an interesting result in view of the recent identification of MgII absorption systems with the halos of galaxies at $z \leq 0.8$ (Bergeron 1988). The large W(CII) that are found for some damped systems may then indicate the presence of a turbulent halo associated with the quiescent disk.

In principle the damped systems are ideal for abundance determinations. Unlike most QSO absorption systems the ionization state of the abundant elements is known; i.e, $H^0/H \approx 1$, $C^+/C \approx 1$, etc. However, a spectral resolution of $\Delta v \approx 10$ km s^{-1} is required to seperate the quiescent disk component from the turbulent halo component. Failure to do this leads one to underestimate the ionic column densities deduced from lines that are invariably saturated. A detailed description of the dilemma is given in Briggs et al. (1983). At the same time we (Lanzetta, Turnshek and Wolfe) are attempting to search for heavy elements locked up in grains. A program of absolute spectrophotometry was initiated in order to compare reddening in QSOs located behind damped systems with a control sample that shows no sign of damped Ly-α absorption. The absence of observed reddening leads to a preliminary constraint on the dust-to-gas ratio given by $(D/G)_{damped} \leq 0.5 \times (D/G)_{Galaxy}$ (see also Pei and Fall 1987). This limit is consistent with the strict upper limits placed on the presence of H_2 in one of the damped systems in our sample (Black, Chaffee and Foltz 1987).

4. OTHER PHYSICAL PROPERTIES

We continue our discussion of the physical properties of individual damped systems. Some very interesting results have been derived with techniques other than optical spectroscopy, and these are considered first.

Observations of the polarized radio continua of a sample of 116 QSOs reveals the presence of Faraday rotation in $\approx$ 20 % of the objects. Furthermore the incidence of Faraday rotation is linked statistically with the occurrence of metal-line absorption (Kronberg and Perry 1982; Welter, Perry and Kronberg 1984). The implication is that the foreground gas is a magneto-ionic medium with a significant rotation measure; i.e., a **B** field with a significant component along the line-of-sight $B_\parallel$ is associated with partially ionized gas. In many cases the detected gas is intrinsic to the QSO. Since one expects one or more intervening CIV clouds per line-of-sight (Young, Sargent and Boksenberg 1982), it is safe to assume that less than 20 % of these clouds exhibit Faraday rotation. On the other hand 4 out of the 5 QSOs in the sample with 21 cm absorption exhibit statistically significant Faraday rotation. Although we are dealing with small number statistics, the implication is that **B** fields are a generic feature of the damped Ly-α population. Table 1 shows the results. The Kronberg and Perry (1984) formalism has been used to calculate $x_e B_\parallel$ at the absorption redshift, z, from RM, the rotation measure corrected to exclude the Galactic contribution: $x_e B_\parallel$ is the fractional ionization of hydrogen times the density weighted $B_\parallel$. Since x_e is unlikely to exceed 0.1 in gas with a Lyman-limit optical depth of $\approx 10^4$, we conclude that

$B_\| \approx 10$ μGauss in gas with $z \leq 2$. The existence of significant **B** fields in a galactic disk with $z \approx 2$ may be difficult to reconcile with the dynamo theory for the formation of magnetic fields. In the dynamo theory, an infinitesimal seed field is amplified to μGauss strength after a few rotation periods of the Galactic disk (Parker 1979). The problem is that the rotation period of a giant disk at $z \approx 2$ is comparable to its age, and as a result there may not be enough time for significant amplification.

Table 1

QSO	z	RM[rad m^{-2}]	N(HI)/[10^{20}cm^{-2}]	$x_e B_\|$ [μGauss]
0458-02	2.040	100	50	0.7
3C 196	0.437	145	$(0.86$–$3.5) \times (T_S/100$ °K$)$	14.0–0.33
1229-02	0.395	-56	$3 \times (T_S/100$ °K$)$	1.3
3C 286	0.692	≤ 3	$3 \times (T_S/100$ °K$)$	≤ 0.11
1331+17	1.776	-23	15	0.45

Radio investigations have also been valuable in providing detailed information on the kinematics and spatial structure of damped systems with 21 cm lines. Since these topics are discussed by Briggs (1988) in this volume, they will not be repeated here. It is important to stress, however, that the VLBI investigation of the $z = 2.040$ 21 cm line in PKS0458-02 reveals (a) that the absorber extends more than $(8h^{-1})$ kpc transverse to the line-of-sight ($h = H_0/100$ km s^{-1} Mpc^{-1}), and (b) that the self-gravity and low velocity dispersion of the HI imply a hydrostatic scale-height of less than 500 pc. That the transverse dimension greatly exceeds the line-of-sight dimension is direct evidence for a disk-like structure with a radius of curvature large compared to the radii of dwarf galaxies.

We now consider the density parameter. The global mass density of the damped systems is obtained by averaging $\langle N(HI) \rangle$ over the spacetime volume occupied by these absorbers. A mean molecular weight of $\mu = 1.4$ is adopted to convert from number density to mass density. This mass density is then divided by the critical density corresponding to $\langle z \rangle$, the average absorption redshift, and the result is redshifted to the present to obtain the density parameter, Ω_{damp}. The results, which depend on q_0 and h, are summarized in Table 2. Ω_{damp} is relatively well determined, because the ratio of total H to detected H^0 is of order unity. By contrast the density parameter of the Ly-α forest clouds is highly uncertain, because H/H^0 is unknown. In most cases Ω_{damp} dominates the mass contributed by model Ly-α forest clouds (see Tytler 1988). Ω_{damp} takes on added significance when it is compared with $\Omega_{lum-disk}$, the density parameter due to luminous matter (stars) in the disks of spiral galaxies. Combining the luminosity density of galaxies, $L_B = 3 \times 10^8 h L_\odot$Mpc^{-3} (Gunn 1982) with a $(M/L)_{lum-disk} \approx 2$ we find that $\Omega_{lum-disk} \approx 2 \times 10^{-3} h^{-1}$. The rough agreement between the mass per unit comoving volume of the damped Ly-α absorbers at high redshifts and the stellar disks of spiral galaxies at the present epoch might be a coincidence. More likely it indicates that we have detected the same baryons at different stages of galactic evolution. As a result it is possible that we have detected galactic disks soon after the collapse of the protogalaxy, but before most of the gas was converted into stars.

Table 2

q_0	Ω_{damp}
0.05	$1.5\times10^{-3}h^{-1}$
0.50	$2.5\times10^{-3}h^{-1}$

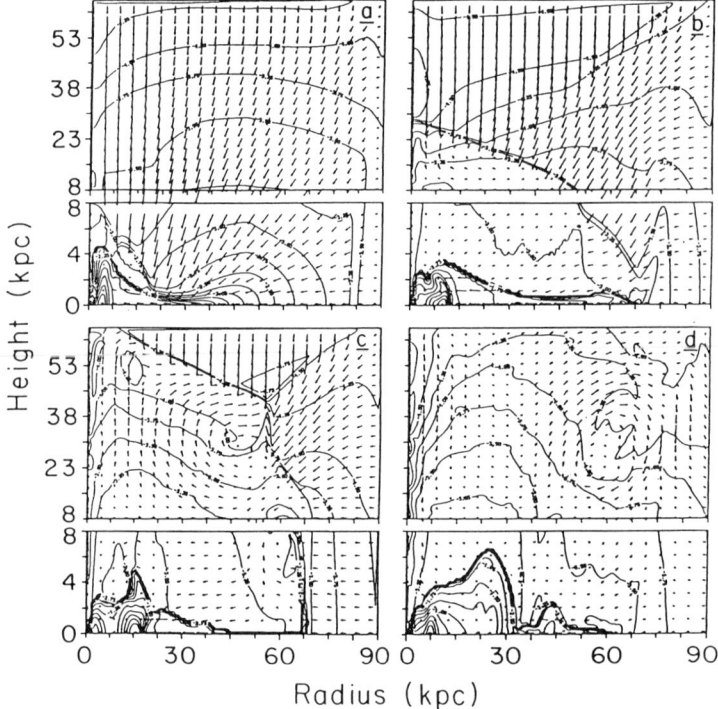

Figure 3. *Logarithmic density contours (cgs units) and velocity fields in the r–z plane for a model with $r_c = 3$ kpc. Elapsed times are (a) 0.30, (b) 0.49, (c) 0.70, (d) 1.03 Gyr.*

5. IMPLICATIONS FOR GALAXY FORMATION

If the damped Ly-α systems are the progenitors of galactic disks, then our findings restrict scenarios suggested for disk formation. For example if the disk-like structure indicated in the case of PKS0458-02 is a generic feature of the damped systems, then their detection out to $z = 2.8$ indicates that galactic disks have formed by $t \approx (0.9 - 2.1)\ h^{-1}$ Gyr. This pushes the epoch of disk formation back to the time when globular clusters first formed, and consequently argues against a significant age gap (Mihalas and Binney 1981) between the disk and the halo. Furthermore, the early formation of disks restricts models in which disks form slowly, on a Hubble time scale (Gunn 1982b).

If, in addition, the radii of the high-z disks are large compared to R_{H_0}, then a natural sequence of events is one in which the protogalaxy first collapses to a giant disk which subsequently contracts in the plane. Schiano, Wolfe and Chang (1988) recently ran a series of hydrodynamical calculations to check the plausibility of this scheme. The protogalaxy was simulated by axially symmetric spheroids of gas and dark matter in which the ratio of masses, $M_{dm}/M_{gas} \approx 10$. The gas and the dark matter obtain their spin from tidal torques induced by neighboring galaxies (Peebles 1969; Fall and Efstathiou 1982). Initially both spheroids are hydrostatically supported by the random motions of isothermal density distributions which fall off like r^{-2} beyond the core radius, r_c. The gas collapses because it cools on a time-scale short compared to the free-fall time. Figure 3 shows the density contours in a meridional slice of the spheroid at early stages of collapse. The important point is that collapse to a large disk is possible by $z \approx 3$ in most cosmological models.

In the second phase of the calculation we examined the contraction of the centrifugally supported disk that formed out of the initial collapse shown in Figure 3. We considered a mechanism which is a dynamical consequence of the infall of mass shed by halo stars (Ostriker and Thuan 1975). When the infall increases the mass of the disk by a factor, f, the radius of the disk must contract by the same factor: this occurs because the radius, R, of the disk times M(R), the mass within R, is an adiabatic invariant. Our numerical calculations showed that the disk did contract by a factor f in about 1 Gyr, although the contraction was accompanied by axisymmetric instabilities. The rapid shrinkage of the disk implies that the detection rate of damped Ly-α systems at $z \leq 1.5$ should be much less than the rate reported in our survey, even after cosmological effects are taken into account.

This work was supported by NSF grant AST 84144-14.

REFERENCES

Bechtold, J. 1987, private communication.
Bechtold, J., Green, R. F., and York, D. G. 1987, *Ap. J.*, **312**, 50.
Bergeron, J. 1988, *This volume*.
Black, J. H., Chaffee, F. H., and Foltz, C. B. 1987, *Ap. J.*, **317**, 442.
Bosma, A. 1981, *A. J.*, **86**, 1825.
Briggs, F. H., Wolfe, A. M., Krumm, N., and Salpeter, E. E. 1980, *Ap. J.*, **238**, 510.
Briggs, F. H. 1988, *This volume*.
Briggs, F. H., Turnshek, D. A., Schaeffer, J. and Wolfe, A. M. 1985, *Ap. J.*, **293**, 387.
Danly, L., Blades, C., and Norman, C. 1987 *QSO Absorption Lines: Probing the Universe, A Collection of Poster Papers*, eds. J. C. Blades, C. A. Norman, D. A. Turnshek, (ST ScI Publication), p. 8.
Fall, S. M. and Efstathiou, G. 1980, *M.N.R.A.S.*, **193**, 89.
Gunn, J. E. 1982a *Astrophysical Cosmology*, eds. H. A. Bruck, Coyne, G. V., and M. S. Longair (Specola Vaticanum: Vatican City), p. 557.
Gunn, J. E. 1982b *ibid*, p. 233.
Kronberg, P. P. and Perry, J. J. 1982, *Ap. J.*, **263**, 518.
Lanzetta, K. M., Turnshek, D. A., and Wolfe, A. M. 1987, *Ap. J.*, **322**, 739.
Mihalas, D. and Binney, J. 1981 *Galactic Astronomy* (San Francisco: Freeman), p. 448.

Ostriker, J. P. and Thuan, T. X. 1975, *Ap. J.*, **202**, 353.
Parker, E. N. 1979 *Cosmical Magnetic Fields* (Oxford: Clarendon Press), p. 616.
Peebles, P. J. E. 1969, *Ap. J.*, **155**, 393.
Pei, Y. C. and Fall, S. M. 1987, in *QSO Absorption Lines: Probing the Universe, A Collection of Poster Papers*, eds. J. C. Blades, C. A. Norman, D. A. Turnshek, (ST ScI Publication), p. 158.
Sargent, W. L. W., Young, P. J., Boksenberg, A., and Tytler, D. 1980, *Ap. J. Suppl.*, **42**, 41.
Schiano, A. V. R., Wolfe, A. M., and Chang, C. A. 1988, *Ap. J.*, submitted.
Turnshek, D. A., Wolfe, A. M., Lanzetta, K. M., Briggs, F. H., Cohen, R., Foltz, C., Smith, H. E., and Wilkes, B. 1988, in preparation.
Tytler, D. 1988, *This volume*.
Welter, G. L., Perry, J. J., and Kronberg, P. P. 1984, *Ap. J.*, **279**, 19.
Wolfe, A. M. 1983, *Ap. J. (Letters)*, **268**, L1.
Wolfe, A. M. 1986a *Gaseous Halos of Galaxies*, eds. J. N. Bregman and F. J. Lockman (NRAO: Charlottesville), p. 259.
Wolfe, A. M. 1986b, *Phil. Trans. Roy. Soc. Lond. A*, **320**, 433.
Wolfe, A. M., Turnshek, D. A., Smith, H. E., and Cohen, R. D. 1986, *Ap. J. Suppl.*, **61**, 249.
Wolfe, A. M. et al.*1988*, in preparation.
Young, P. J., Sargent, W. L. W., and Boksenberg, A. *1982*, *Ap. J. Suppl.*, **48**, *455*.

DISCUSSION

Tyson: How sensitive is the radial collapse of your disk to the amount of the mass that is infalling? I'm interested, because you said 10^{11} $M_\odot$ and that sounds quite big.

Wolfe: It's not that sensitive to it. However, that's not a lot, actually, because you need 10^{11} $M_\odot$ to make a big disk galaxy. Remember, most of this gas eventually winds up in stars.

Tyson: I thought you have the disk, and now you want to collapse the disk some more, and so you're getting more infall.

Wolfe: Oh, I didn't realize you meant secondary infall. The mass increases by a factor of 2, and I can't believe we know the mass of the disk to better than a factor of two.

Ostriker: Artie, I'm only giving the complaints. I agree with almost everything. A couple of complaints. One concerns the dust. When you make comparisons with the Galaxy it's a real strawman. No one would have ever expected that the dust to gas ratio is normal. It is much more likely to be like the SMC. Given the state of the galaxy at that time, according to the rest of your talk, you would have expected the dust to gas ratios to be something like that in the SMC which is completely consistent with everything. That is what one should be looking for. It's a factor of ten below.

dust to gas ratios to be something like that in the SMC which is completely consistent with everything. That is what one should be looking for. It's a factor of ten below.

Wolfe: No, you're right. We are much less sensitive to that sort of thing.

Ostriker: The second point is that on modeling the formation of a galaxy, (I don't want to be one of those cold cranks who says, "Look back at my data of 1937."), but (Laughter) the one prediction in the paper I did with Cowie (1981, Ap. J. (Letters), 243, L127) that we did on explosions was that galaxies should be formed in two-dimensional surfaces. That was going to be a test, because that was not the prediction of other theories, whereas a clear prediction of this theory — and I should say of the Zeldovich theory as well — was that they have in common two features, and that is that (1) they automatically give you two-dimensional surfaces, so you should see things forming that way, and (2) they automatically give you the Faber-Jackson result. Well, that was known, so it's not a prediction, but it's almost a prediction. It could have been predicted and was predicted by both that you'd find galaxies forming in sheets, and then the sheets collapse. So therefore, when it's found, it seems odd that you then go back in your modeling to something that absolutely doesn't follow naturally from the hierarchical picture.

Wolfe: I guess I should ask Mike Fall about that. Where do those initial conditions come from? I thought they were reasonable in some sense.

Norman: They were reasonable.

Fall: I think you will get flattened structures in almost any picture of galaxy formation. As Jerry Ostriker emphasizes, blast waves and pancakes are, almost by definition, flattened. In the hierarchical picture, protogalaxies start out fairly round but with some rotation induced by tidal torques. The rotation just before the collapse of the protogalaxies (as measured by the spin parameter λ) is almost independent of the perturbation spectrum at recombination (e.g., $\delta_k \propto k^n$ with different indices n or the 'cold dark matter' spectrum). The gaseous component can then collapse almost indefinitely along the axis of rotation but is stopped in the orthogonal direction by an angular momentum 'barrior'; the result is a centrifugally-supported disk. What is far less certain is how much time a proto-galaxy spends in the round and flat configurations, which depends on how bumpy the gas is, how fast stars form and other theoretically unknown factors. We may be able to use the observations by Art Wolfe and his collaborators to place some important constraints on the relevant time scales.

Rees: I'd like to repeat what I think other people are now saying, which is that it's not obvious that what you are seeing is a rotationally supported structure. So I think even in your models, the system collapses to the plane by the Lin, Mestel and Shu effects at a larger radius than what you need for rotational support. If one thinks that in general these things won't be axially symmetric anyway, what one would expect is that they will collapse to some sort of sheet, not necessarily a planar sheet. The endpoint will then be either a disk galaxy or maybe even an elliptical. But I'm sure all one can really say is that you are seeing a planar structure in which some star formation has already occured. If you think of it that way, then the large extent does not become surprising, whereas I think, still, if you believe in tidal torques and if you believe that the rotational velocity that you get from tidal torques is only a few percent of the gravitational support, then it's still rather hard to get collapse of a very large disk early enough to explain your data.

Wolfe: Well, can I just make a point about that. The answer to that question depends on the core radius of the density distribution that I showed. The one that I showed during my talk has a small core radius. If you made it bigger, you'd get a bigger structure.

Rees: But is there any evidence against your saying it's a structure which is flattened to some sort of caustic surface?

Wolfe: No. You're correct. I don't want to push that point. I don't think we can say that what we're looking at is a centrifically supported structure. All that I'm saying is that that's what comes out of these models fairly rapidly, although I guess that if it's not centrifically supported, it's going to oscillate on a time scale of less than a billion years.

Rees: My remark applies in the context of the cold dark matter model, not in explosions, etc. If you just consider a general halo which is going to be experiencing mergers, then you would expect that you will have this sort-of sheet-like structure.

Wolfe: Let me say one more thing. I do think that the 21cm line does indicate that what we're looking at is big. It's a flattened structure which is large compared to the size of the beam.

Ostriker: How big is it?

Wolfe: Unless you want to take a very contrived face-on absolutely flat structure (which I think is crazy), I think what we're looking at is something that's at least 50 kiloparsecs. I can't believe that it's any less than that.

Norman: In principle you can really test this, because with a covering factor on the order of 20 % on the sky, almost all of the large extended radio quasars will have one of these things along the jet itself. So you can actually do dynamics. You will have the disk with the extended jets coming across the object, and eventually one will be able to do the dynamics.

Wolfe: That will be lovely.

Disney: The other explanation of course is that maybe the disk hasn't contracted, and maybe there is a population of very large galaxies of low surface brightness which have a pretty high column density in hydrogen.

Wolfe: You mean that are still around locally.

Disney: Yes. Those would be very difficult to see. If you work out the selection effects against finding them, they're enormously high. It seems there's no reason whatever that you couldn't have such a population. There's a very interesting paper that's just come out from Bothun *et al.* (preprint) where they found just such an object.

Wolfe: I heard Bothun talk about that. It's fascinating. The question is, how often do those occur?

Disney: They could be very common, and we wouldn't know about it.

Wolfe: I think we'll know the answer to that very soon from their survey. I did discuss this with him, and he felt it would be hard to make them common enough to explain the damped Ly-α systems, but I agree that we should wait until the results of the survey are in.

Heckman: It seems like what you're saying is that a typical galaxy with a redshift of 2 or 3 looks like this, but this is also the epoch where the highest density of QSOs occur. Have you considered what one of these things would look like if the nucleus of the galaxy turned-on as a QSO which photoionized its disk?

Wolfe: Yes, I have. It depends critically on if there is dust in the disk.

Heckman: There doesn't seem to be dust, does there?

Wolfe: Not at high redshifts, but I thought you meant at current epochs.

Heckman: No, I mean looking at a typical QSO say at a redshift of 2.5. What would the QSO nucleus do to the galaxy?

Wolfe: It would set up an R-type ionization front which would rapidly ionize the entire disk.

Heckman: What would be detectable? What could you actually see?

Wolfe: I don't know the answer to that.

Rees: I think that if it were a planar disk, then the covering factor would be small, but certainly one constraint on the type of model I discussed in my talk is that if you have gas falling in with a large covering factor, and you switched on a QSO in the middle, then quite a large fraction of the QSO light would be used up and reprocessed. So indeed, the fact that there is not much more conspicuous fuzz around all high redshift QSOs puts a constraint on these infall models with large radii.

Wolfe: Do you think it puts an upper limit on the covering factor?

Rees: Yes.

Felten: I'm interested in the question of the space density and the effective cross-section of these things, and so I have several questions about this effective path length in redshift of 55 which you quoted. That was the total path length of all the QSOs that you surveyed, is that right?

Wolfe: Yes.

Felten: And how many QSOs?

Wolfe: We observed a total of 68 QSOs, so Δz is roughly 0.8 per object.

Felten: And where does that range, from what to what, typically?

Wolfe: I'd say $z = 1.8$ on the average to $z = 2.6$.

Felten: So that all lies at high z basically. Well, now that means that any calculation of the density and cross-section depends strongly on cosmological model. Is that right?

Wolfe: I agree. Jim, I just want to say, we don't know what's going on at those redshifts. The only comparison I made was with the least controversial set of assumptions that I could come up with. That doesn't mean that they're correct. But at least it's something. The model is conservative. When I did that calculation, it was the luminosity function that you recommended in your paper. (Laughter) I showed the results in my talk. For q_0 equal 0 versus 0.5, they differ by a factor of two. For q_0 equal 0.05, which essentially gives the most disks one would expect, at most 3 disks are expected to occur in our sample. If you take $q_0 = 0.5$, you'd expect 1.8 disks, which makes the discrepancy even worse.

Felten: And your conclusion from this is that the size of these dinner plates has to be roughly 3 Holmberg radii?

Wolfe: Yes, if the co-moving density of these objects equals that of spiral galaxies.

Felten: Looking at present-day galaxies, to what Holmberg radius would you have to go in order to find a column density of $2(10^{20})$ cm^{-2} in neutral hydrogen?

Wolfe: Less than 1.5 Holmberg radii.

Felten: Less than 1.5 Holmberg radii? Not even smaller than that?

Wolfe: No, $2(10^{20})$ corresponds to 1.5 Holmberg radii. That's what Bosma's result is, and that's what I used.

Felten: Okay, so that's just a discrepancy factor of 4 or 5.

Wolfe: I think it's less than that. Frank, in our Arecibo survey of the outskirts of spirals, what did you think the radius corresponding to $2(10^{20})$ would be?

Briggs: 1.2 to 1.3 Holmberg radii.

Felten: I thought you quoted Sancisi earlier as saying that the radius was smaller.

Wolfe: No.

Fall: I just want to ask how much you could change the size effect by assuming that the disk may have had a much higher gas to star ratio (or whatever these objects are). So, assume you're looking down to fixed column density which is how the objects are defined in your sample, but they're shrinking perhaps less than you think. Then, the radius to which you get fixed column density is changing more rapidly than the scale length of the disk.

Wolfe: I have to think about what you are saying.

Fall: Suppose you set out with these things being all gas, and they gradually turn themselves into stars. Then what we should be comparing with is not the column density of gas disks today, but with present-day galaxies where all the stars are in the form of gas.

Wolfe: I can say something about this. The column density that we see is typically 10 $M_\odot$ per pc^2. If that occurs at 3 Holmberg radii, what is that in scale lengths for an exponential disk? That's something like 13 or 14 scale lengths. That would mean that the star density at the center of the galaxy is 10^5 $M_\odot$ per pc^2 which is far too high.

Fall: But do we have to say that these objects are really 1.5 or 2.5 or 3 times bigger, or can we get away with some smaller factor if we take this other effect into account?

Wolfe: I'm not sure. I don't want to give a glib answer.

Fall: Because it seems to me your number is really an upper limit on the ratio of the size of the objects in the past to the size of these objects presently.

Wolfe: I thought I was being conservative based on the remarks I made in answer to Jim's question. If you assume the deceleration parameter is 0.5 (an $\Omega = 1$ universe), then the sizes have to be something like 6 or 7 Holmberg radii.

Sancisi: Two questions: One is, could you do as well with a larger number of smaller objects instead of large disks? Second is, suppose you have the dark matter that we observe now in spiral galaxies, which is about typically 3 or 4 times the mass of the luminous matter, in a flat distribution instead of a halo. Would it help in some way, or how would it affect your dynamic interpretations? Would it be conceivable, for instance, that your objects are what we now see in galaxies, including everything dark and luminous? I'm thinking of a flat system which is dark and luminous as might be seen today.

Wolfe: What was the first question? (Laughter — question is re-stated) Okay, all the evidence is consistent with a large number of small objects, except for two observations. One is the stuff that Frank discussed. The VLBI really rules out, at least in the case of 0458-020, a dwarf galaxy. I also think that if Tony Tyson is really looking at galaxies that are at a redshift from 1.5 to 2, his things really have to be intrinsically big. So, I think the answer to your question relies on that. Maybe Allen Schiano would like to comment on the answer to the second question.

Schiano: If the dark matter were planar rather than spherical, it would not affect things by a large factor. You can put a flattening parameter in the calculation, and it only makes a small effect.

Wolfe: But, for a given amount of mass, the gravitational potential is stronger, right?

Schiano: Yes.

Sancisi: But the other part of the question was really: Is it consistent with your results that all the mass that you see in these objects at high redshift is all the mass that we see now in the galaxies? Is it possible that your HI has turned into the dark matter?

Wolfe: The Bahcall missing matter, is that what you mean?

Sancisi: No. Evidence for dark matter from the rotation curves.

Wolfe: No. There is not enough for that.

Norman: You mean't to say Oort missing matter.

Wolfe: I'm sorry. Yes.

H. E. Smith: Let me just follow up on your plug for our poster paper and add one other constraint here, and that is that these things have formed stars in the past, because we see heavy elements, but they are not strong Ly-α emitters. From the Ly-α emission limits we set star formation rates on the order of a solar mass per year per object. You can destroy Ly-α or what have you, but they are not very actively forming stars of the type that some models would require at these epochs.

Wolfe: That's a good point.

Savage: You very nicely demonstrated the difficulty in getting the abundances out of metallic lines, but there is a lot of kinematical information that is relevant. Could you make some comments on that?

Wolfe: Okay, what I would say is that there seem to be two types of objects in our sample. In one type (there are 3, 4, or maybe more objects), the velocity dispersion we see is less than 10 km s^{-1}. I can't believe it's any more than that. This evidence is from saturated FeII, SiII and CII lines, and these are effective b-values, so the velocity spread is buried in there. I think whatever we're looking at, in those cases, they're very quiescent systems. However, there are systems that do look like the normal Ly-α metal systems that have b-values of 40 km s^{-1}, something like that. I guess the interpretation would be that there is an extra, more turbulent, component in those cases. Maybe that's the halo. That stuff looks like what you're seeing along the line-of-sight to the Magellic Cloud. So, I think there are both types of systems. I don't think we have enough information yet to say how things are distributed. There are only about 15 objects.

Tyson: Were you able to reconcile this infall and collapse and flattening with such a low velocity dispersion?

Wolfe: Yes.

Tyson: I want to get back to this bit with 10^{11} $M_\odot$ coming down to a disk after you already had a disk. I'm still uncomfortable about that. If that occurs after the formation of the disk, from mass loss and red giants and what have you, why wouldn't you expect every elliptical galaxy to have a disk?

Wolfe: Well, Jerry, it's your model. How much mass came out of the halo?

Ostriker: The principle idea is that it's environmental. Given that time scale, that's the same time scale in which groups and clusters form, and it just gets swept out as it gets brought down. That's why you find ellipticals in groups and clusters and not in the field.

Wolfe: I guess that's the answer. (Laughter)

Ostriker: It has to be environmental or otherwise you wouldn't find the ratio of ellipticals to spirals to vary between clusters and the field.

Antonucci: I wanted to ask if you could constrain either the number of Bothun type galaxies or the number of damped Ly-α systems at low redshift from the 21 cm studies.

Wolfe: I'm not sure how to answer that. As Bob Brown said, 1 % of the objects in his survey with Mort Roberts exhibited 21 cm absorption. Their survey was sensitive to $N(HI) > 10^{21}$ cm^{-2}. The Bothun column density is also something like 10^{21} cm^{-2}. That's one argument why I don't think the Bothun galaxies are that common right now, unless they're mainly stellar, because I think they would have shown up in Bob's survey.

Antonucci: And similarly, what about the damped Ly-α systems?

Wolfe: There has not been an adequate survey for the damped Ly-α systems at low redshift.

Antonucci: I mean, would they have shown up in the 21 cm survey?

Wolfe: Oh, definitely, that's what I'm saying. The strongest systems with $N(HI) > 10^{21}$ cm^{-2} definitely would have shown up. I don't think they're that common at low redshift.

Antonucci: So you don't really have to wait for Space Telescope to look for damped Ly-α?

Wolfe: Well, the problem with the 21 cm line is that there are all kinds of selection effects, and if the spin temperatures are high or if the radio sources are too large, we'd miss them. Moreover, the weaker systems with $N(HI) \simeq 10^{20}$ cm^{-2} are difficult to get in the 21 cm line. I trust the Ly-α more than the 21 cm line to get the statistics. The 21 cm line is beautiful for getting the details.

Jauncey: Artie, I think there is one thing that worries me. You see three 21 cm systems out of four damped Ly-α systems down to about the 1 % level. I mean, there is a real dichotomy with Brown's results. Is that cosmological, or is that purely circumstantial?

Wolfe: It could be cosmological, if you take those numbers literally. But remember that the Greenbank survey is sensitive to 21 cm optical depths of about 0.1–0.2. The problem with that is the line we found for 1157+014 had an optical depth of 2 %. That's a problem. To really make sense of this you have to do a survey down to a 4σ limit of 1 %, and I think that's impossible from Greenbank.

Jauncey: It would also make a lot of sense to continue your survey out in redshift. In other words, there are getting to be enough regular QSOs that you can get the 21 cm as well, to take it from $z = 2.8$ to $z = 3.8$.

Wolfe: Haven't we looked at most of them, Frank?

Briggs: Well, going out in redshift, you're making the assumption, when you think about using radio techniques, that you'll find something was in front of the extended radio lobe that wasn't in front of the optical QSO.

Jauncey: All redshift greater than 2.8 radio QSOs are compact, none of them are extended.

Briggs: Then there is no sense to looking at them in the radio, because you have a better measure of whether there's something there that has a lot of neutral hydrogen in it if you look at Ly-α.

Wolfe: I think we looked at most of the bright ones anyway, but the trouble is, the redshift path from Arecibo is so small. Again, I'd like to advertise the Arecibo upgrade, because if it's done, we would get continuous redshift coverage all the way out to high redshifts in every case. I think if the upgrade ever gets done, we should do that type of survey. 3C196 is a beautiful example of an object we would have missed optically. I think 3C286 would have also been missed optically because the Mg II equivalent widths were incredibly weak. Spinrad had to spend something like 20 nights, I'm not sure, at Lick to get those equivalent widths.

Kronberg: One interesting diagnostic, I think, of what the evolution of size and distribution of sizes is as a function redshift is the fraction of non-intersection by whatever means you investigate as you go to successively larger redshift intervals. Gary Welter and Judith Perry and I tried to model this and got some interesting results in the sense that up to redshift of 3, you do expect over quite a range of cosmological models and interveners, some detectable fraction of non-interceptions. You can actually, then, look for that. So it might be that the fraction of non-interceptions, by whatever criteria you're searching, is very interesting. That's one of the things we hope to do with the next set of rotation measures at high redshift. I'll try to get the noise down on this and see how this goes with redshift. We've actually run some models, and so just to throw some numbers in, a 50 kpc global size and keeping that constant, is roughly consistent with a q_o close to 0.1 and in our case a Faraday strength evolution of $(1 + z)^1$, which is mixing the issue now. However, you can also do this for optical absorption line systems, too.

Wolfe: I don't think we have enough data yet, but eventually we will.

Disney: You made an interesting remark about the amount of continuous absorption. Presumably if it was the amount of absorption you'd get from a normal dust-to-gas ratio, the quasar itself would be highly dimmed and reddened. Would it not just drop out of your sample, so in other words, the only guys you're going to see are the ones that aren't dimmed?

Wolfe: That's right.

Ostriker: I will be addressing exactly the question you asked in my talk this afternoon.

Stocke: Relevant to your point about QSOs turning-on inside such a gas cloud, Paul Hintzen and I found an object with a redshift of about 0.6 that appears to be in a 100 kpc long cloud that primarily has OII emission. It has a solid body rotation out 50 kpc on either side of the QSO. This, in principle, could be such an object.

Wolfe: Is there any evidence of dust in that system?

Stocke: We have not looked specifically.

Wolfe: I think dust is crucial when you consider the dynamics.

Stocke: The difficulty for this object is, of course, the lower redshift.

Wolfe: Yes.

Tyson: While the morphological class of dwarf spirals exists, the fact is that the range in size of spiral galaxies is very small compared to other morphological types of galaxies. I'm curious, if the rate of infall from the halo is some determinate as to what the final resultant Holmberg size of the galaxy will be, then certainly we should find a greater dynamic range of Holmberg radii for spiral galaxies, unless we believe that all halos have the same amount of mass available to descend.

Wolfe: I don't think that's a problem, but I have to be honest with you, I haven't really constructed a model. It's sensitive to the density distribution and where that cuts off. We know nothing about that. So I can't answer your question. If we really knew how the dark matter was distributed, then I could answer your question. But we really don't know that.

Tyson: How big are these things Bothun is finding? Are they bigger than an average spiral?

Disney: They about 10 times larger.

Barthel: From our combined VLA and Palomar studies, we know of quite a high number of small distorted radio QSOs between redshifts of 2 and 3, and they have very strong associated absorption systems. These would be very good candidates to look for HI. Both the radio and the optical suggest that the clouds are from 0 to 10 kpc.

Wolfe: Well, one problem is that in those objects the spin temperature might go through the roof.

Ostriker: Artie, you say your objects have to be so and so big. How big?

Wolfe: Well, I said 50 kpc, but that's one object. I have to be careful.

Ostriker: Could that object be bigger?

Wolfe: Yes. We have no upper bounds.

Felten: In the lecture you gave, you had to deal rather quickly with the question of whether Tony Tyson's observations put some constraints on the contribution that dwarfs can make to this problem. You spoke to me privately about this, and I think you mentioned that this constraint applies to $z = 0.1$. Do you have more to say about this?

Wolfe: Well, Tony should really defend himself. Tony has an argument from the two-color plots that he is not looking at dwarfs at a redshift of 0.2. He's also gotten redshifts recently, of a lot of his objects. They're all brighter members of his population, but they seem to be objects that have redshifts of 1.0. That's his other argument against them being dwarfs.

Felten: So, he could put some constraints on the number of dwarfs at a redshift of 0.1 or 0.2, or perhaps at 1.0 or 2.0.

Wolfe: No, not at redshift of 1.0 or 2.0, but I think he would say that his observations ruled out a very significant population of dwarfs at redshifts of 0.1 and 0.2. However, I don't want to quote him without getting a little bit more in touch with him.

Tyson: I spoke with Tony Tyson in Pasadena about the color distribution of these objects he's finding in his deep fields. That color distribution is inconsistent with the color distribution of the DDO dwarfs already cataloged by color distribution that we have. So if there's a selection effect that permits us to see those and not others, it cannot be used as a steadfast reason against there being a population of dwarfs with another color distribution that is a product of them being in a lower state of star formation.

Wolfe: That's right. That's always possible. You can do anything you want at high redshift.

Felten: Do any of those constraints really apply at a redshift of 1 or 2? That's what we're interested in.

Wolfe: No, I don't think so. There could be a high population of dwarfs at a redshift of 1 or 2. I agree.

Felten: Or other things as well.

Wolfe: Let me say, if Tony's right that he's looking at objects that are at a redshift of 1.0, they can't be dwarfs, because they're extended.

Felten: No, but he probably wouldn't see the dwarfs at a redshift of 1.

Wolfe: No, but I'm saying that Tony's seeing objects which are not dwarfs; that's a better way to put it. The objects that he's seeing that fill 20 % of the sky are not dwarfs.

QSO ABSORPTION LINES AND THE UNIVERSE AT REDSHIFTS BETWEEN 1 AND 4

J. P. Ostriker
Princeton University Observatory

1. INTRODUCTION

Present day cosmological studies bear a certain resemblance to the debates of the pre-Socratic Greek philosophers. A great deal of intelligence and some guesses, which in retrospect will seem to be fairly shrewd, combine with wild speculation to produce a heady brew. The problem, obviously, is that our theories have been under-constrained by the discipline afforded by observation. Galaxies, although numerous, are very difficult to observe at redshifts greater than unity, with only a handful of presumably unrepresentative examples known (cf Djorgovski, Spinrad and Dickenson 1987) having measured sizes, luminosities and redshifts. Evidence, such as we have, shows mild evolution in the counts and a significant evolution of certain special types (e.g. Dressler, Thompson and Shectman 1985), but the power of our instrumentation does not really reach to the epochs where evolutionary effects are expected to be dramatic. QSOs on the other hand have been seen to redshifts $z > 4$, but the number of objects seen at high redshift is so small and the prospect of seeing a much larger number so poor that, although giving us uniquely important information, they do not provide a powerful tool for studying, say, the growth of perturbations.

QSO absorption lines have some of the virtues of both of the above. They can be seen to the same large redshifts as the QSOs, but they are far more numerous with roughly 1 metal line system per unit redshift and 100 Ly-α line systems detectable per unit redshift in each QSO spectrum, using present techniques.

Observation of absorption lines gives us information concerning chemical composition, the ionizing radiation field and the fluctuation spectrum of density inhomogeneities. At the present time, both the available data and the analytical tools developed allow only a rudimentary modelling to be attempted, but we can begin to address some of the obvious questions. In this presentation I will not, except tangentially, discuss chemical composition, considering primarily the Ly-α clouds which are, in the majority, so metal poor as to permit the assumption of a primeval composition to be satisfactory. But I will address the questions of temperature, size, spatial distribution, source of ionization and some of the implications in terms of the origin, physical state of the clouds, and how they inform us about the conditions attendant upon galaxy formation. I will also comment briefly on how the metal line systems may be used to tell us something about the possibility of cosmological dust obscuration.

2. THE Ly-α CLOUDS

First observed by Lynds (1971) and first analyzed in depth by Sargent et al. (SYBT, 1980), the 'forest' of lines seen in high dispersion spectra shortwards of the QSO Ly-α emission line are customarily analyzed in terms of a broadly distributed intergalactic population of metal poor (H, He) 'clouds.' For the moment it is not necessary to distinguish between the possibilities that these are truly quasi-spherical 'clouds' or better approximated as quasi-cylindrical 'filaments' or quasi-planar 'sheets' of matter. High resolution spectral observations show directly that the lines are distinct, that the filling factor of clouds (C) along any given line-of-sight is quite small ($f_C \lesssim 10^{-2}$), and that the intercloud medium (ICM) must have a very low density of neutral hydrogen. A recent study by Steidel and Sargent (1987b) gave for the intercloud Gunn-Peterson test a limit on the density of neutral hydrogen n_{HI} (ICM) $\lesssim 10^{-13}(1+z)$ cm^{-3}. The simple fact that, at a redshift of $z \approx 2.5$, the intercloud gas produces $\lesssim 1/10$ of the Ly-α background, as compared to the smoothed out clouds, indicates that the neutral hydrogen density of the intercloud medium must be less than that in the clouds by a factor:

$$\frac{\bar{n}_{HI}(ICM)}{\bar{n}_{HI}(C)} = \frac{d\bar{\tau}_{Ly-\alpha}(ICM)}{d\bar{\tau}_{Ly-\alpha}(C)} f_C \lesssim 10^{-3} \tag{1}$$

where $\bar{\tau}_{Ly-\alpha}$ is the mean Ly-α optical depth. This requirement of a high contrast between the clouds and the ICM lead SYBT to postulate confinement of the clouds by a higher temperature lower density medium. Analogous but somewhat more detailed arguments by Ostriker and Ikeuchi (OI, 1983) and Ikeuchi and Ostriker (IO, 1987) further developed this picture. It is important to note that the term 'confined' should not be taken to imply pressure equilibrium between the clouds and the ICM. 'Confined by' should probably be replaced by 'surrounded by' or 'immersed in,' because the clouds may have a significantly larger pressure than their surroundings and spend a large part of their lifetimes in essentially free expansion.

The observations also indicate fairly directly that the typical detected cloud has an HI column density of $N_{col}(HI) = N_{14} = 10^{14}$ atoms cm^{-2} and a Doppler parameter of $b \approx 25-40$ km s^{-1}. This is slightly in excess of the values expected for photoionization ($T \approx 10^{4-5}$ K) and indicates mildly supersonic motions. If the observations of Foltz et al. (1984) of Ly-α lines along the line-of-sight to the lensed QSO 2345+007AB are interpreted in terms of a cloud size, then a radius of $R_C \approx 10h$ kpc is indicated, where h is the Hubble constant in units of 100 km s^{-1} Mpc^{-1}. Adopting this, and a fiducial value for the ionizing flux of $J_{21} = 1$, where

$$J = J_{21} \times 10^{-21} \text{ ionizing photons cm}^{-2} \text{ steradian}^{-1} \text{ s}^{-1} \tag{2}$$

at the Lyman limit, then the density in the clouds is approximately 10^{-4} atoms cm^{-3} and the density of neutral atoms smaller by a factor 3×10^4 so that $n_{HI}(C) \sim 10^{-8.5}$ cm^{-3}. It may be of interest that, if the Universe had a mean density equal to the internal cloud density, then at the reference epoch of $z = 2.5$ this corresponds to $\Omega_{b(C)} \approx 0.06\ h^{-3}$, somewhat in excess of the mean baryon density as determined by light element nucleo-synthesis (Yang et al. 1984). This relative overdensity is still larger when one allows for the baryons in galaxies and in the intercloud medium. Thus the clouds must have been produced by a process which compressed, by a significant amount, the preexisting gas. Quantitatively, we have

$$\Omega_{b(C)} = \frac{2.0}{h^{5/2}(1+z)^{2.75}} \left(\frac{N_{14} J_{21}}{R_C/(10kpc)}\right)^{1/2}, \tag{3}$$

with a filling factor

$$f_C = 4/3(R_C/\lambda) = 6 \times 10^{-3} h(R_C/10 kpc), \qquad (4)$$

where we have taken the cloud density, λ, to be 60 per unit redshift at $z = 2.5$ and $N_{14} = 1$.

The small cloud size also has important implications for the question of containment by gravity. The ratio of the gravitational energy to the thermal energy of a spherical cloud is

$$\left|\frac{W}{U}\right| = \frac{GM_C m_H}{5kTR_C} = 9 \times 10^{-4}(1+z)^{1/4}(N_{14}J_{21})^{1/2}(R_C/10 kpc)^{3/2}, \qquad (5)$$

where m_H is the mass of the hydrogen atom. Therefore, for the conditions we are considering, $|W/U| \approx 10^{-3}$, self-gravity is unimportant. Correspondingly, if the clouds are confined by dark matter halos, then, for the adopted parameters, the dark matter halos, interior to the observed cloud, would be required to be 10^3 times more dense than the baryon component which is consistent with the known value of Ω_b.

Since all of these arguments depend critically upon the cloud size as estimated from one observation, it is extremely important to determine that parameter more securely. The estimated typical cloud size is however consistent with other less quantitative methods based on the ability to cover a QSO's broad emission line region, ($R_C \gtrsim 10$ pc), the Doppler width ($R_C < 100 h^{-1}$ kpc) and evaporation limits (see SYBT and IO).

With this background we can ask how the clouds may be maintained. Black (1981) proposed self-gravitation. If the size estimates given above are roughly correct, gravity will be unimportant in general, and furthermore, a self-gravitating isothermal object is unstable to collapse or expansion. Rees (1986) and Ikeuchi (1986) considered the possibility that the clouds could be contained by the potential of dark matter halos. This is quantitatively more plausible than self-confinement, but appears inadequate if the clouds are as small as 10 kpc. There is also the issue of whether the observed trend of decreasing cloud numbers with decreasing redshift could be produced in this model, a question we will address more quantitatively in the next section. In any case, this model is quite testable since detectable correlations will be built up gravitationally between adjacent clouds. Correlations have now been detected (Carswell and Webb 1987), or new limits set, which are lower than those obtained by SYBT, allowing for confirmation or rejection of this model.

One can ask the question of whether any confinement is even necessary. Could clouds have freely expanded from an earlier epoch? For adiabatic expansion, the final expansion velocity is $\sqrt{3}$ times the initial isothermal sound speed or about 35 km s^{-1} which is within the range observed. But energy is continually pumped into the expanding approximately isothermal cloud by the background radiation with the consequence that a cloud formed at radius $R_{C,0}$ and epoch z_0 will have a velocity $v_C^2 = 6c_s^2 \ln(R_C/R_{C,0}) \simeq 9c_s^2 \ln[(1+z_0)/(1+z)] \approx 50$ km s^{-1} where c_s is the sound speed, which is significantly larger than permitted by the observed Doppler parameters. For clouds confined by an adiabatically expanding intercloud medium for which pressure $P_{ICM} = P_{ICM,0}[(1+z)/(1+z_0)]^5$, the internal density will scale as $(1+z)^5$ and consequently the velocity is always fixed in terms of local Hubble velocity. Since $R_C \propto (1+z)^{5/3}$, therefore

$$\frac{\dot{R}_C}{R_C} = \frac{5}{3}\frac{\dot{z}}{(1+z)} \quad \text{or} \quad v_C(t) = \frac{5}{3}H(t)R_C(t). \qquad (6)$$

For an estimated radius of 10 kpc, this gives $v_C(t) \approx \frac{5}{3}h(1+z)^{1+q_0}$ km s^{-1} or $v_C \approx 6$ km s^{-1} at $z = 2.5$, which is quite consistent with observations given the added thermal broadening. Direct, if simplified, numerical integrations show (IO) that clouds which are initially pressure confined will ultimately expand freely as the external pressure drops to the point where the velocity calculated above exceeds the sound speed.

What is the origin of the Ly-α clouds? This question cannot be addressed without determination of the previous issue of confinement of course. The obvious possibilities are:
1. through a thermal instability,
2. via gravitational condensation; or
3. in shocks caused by either gravitational or hydrodynamic processes.

Since the shocks must be thermally unstable to result in $\sim 10^{4.5}$ K clouds, the third possibility may be considered a variant of the first. A pure thermal instability, as treated by Field (1965) and Kondo (1970), is unpromising, simply because the postulated initially uniform but unstable state is intrinsically implausible. Two variants of a gravitational origin have been proposed, as noted before; gas which collects in the potential wells of dark matter perturbations (Rees 1986; Ikeuchi 1986) and initially collapsing lumps of dark matter plus baryons within which the baryon motion is reversed to expansion after ionization and reheating (Bond, Szalay and Silk 1987). In both of these pictures, cloud-cloud correlations should be detected and in the latter a correlation between expansion velocity, column density, and epoch is implicitly predicted. In the latter model the contrast between the cloud and intercloud density is relatively low, and, although the contrast in neutral fractions is amplified by consideration of ionization balance, it is not clear in this picture if it will be as large as is observed. Alternative (3), shock initiated thermal instabilities (OI; IO; Vishniac and Bust 1987; Chernomordik and Ozernoy 1983; Hogan 1987) has many attractive features. Since the cool gas is surrounded by much hotter gas, there is little difficulty in achieving either confinement or a sufficient contrast between cloud and intercloud neutral gas. The origin of the shocks could be either thermonuclear explosions associated with star/galaxy formation or merely the largely ignored hydrodynamical consequences of gravitational fluctuations which will occur on small scales in the cold dark matter picture, just as they occur on large scales in the Zeldovich 'pancake' picture.

Certain regularities have been discovered in the distributions of the column densities. Both the variation of the cloud number, N, with column density at fixed redshift and the variation of number with redshift at fixed column density can be fitted to power laws:

$$\left(\frac{\partial N}{\partial z}\right)_{N_{col}} \propto (1+z)^\gamma \quad \text{and} \quad \left(\frac{\partial N}{\partial N_{col}}\right)_z \propto N_{col}^{-\beta} \tag{7}$$

with $\gamma = +2.4 \pm 0.4$ (Bajtlik, Duncan and Ostriker 1987) and $\beta = 1.7 \pm 0.2$ (Carswell et al. 1984; Atwood, Baldwin and Carswell 1985). What has not, to date, been exploited is the fact that a necessary relation between β and γ exists for each theory proposing to explain the origin and maintenance of the clouds.

What is unknown, of course, and in principal very difficult to predict theoretically, is the intrinsic distribution of cloud sizes. One may suppose that the intrinsic distribution has the form $(dN/dM_c) \propto M_c^{-\delta}$. Then δ is determined by the distribution in observed column densities:

$$\delta = \frac{1}{3}\beta + \frac{4}{3}, \tag{8}$$

and, with δ fixed, the evolution to another epoch can be predicted on the basis of a specific theoretical model. That is, one can determine γ to be

$$\gamma = \left(\frac{5}{3}\sigma - \alpha\right)\beta + \left(1 - \frac{1}{2}\Omega_{b(C)} + \alpha - \frac{7}{3}\sigma\right) \quad (9)$$

where the ambient pressure is assumed to vary as $P_{ICM} \propto (1+z)^\sigma$, and the ambient ionizing flux as $J \propto (1+z)^\alpha$. Other observations, which we will shortly describe, show a relatively constant ionizing flux: $\alpha \approx 0$. Then one can characterize models in terms of the variation of the pressure with (for $\Omega = 1$): $\sigma = 0$ for gravitationally confined clouds; $\sigma = 3.6$ for ram pressure confined; $\sigma = 4.5$ for freely expanding models; and $\sigma = 5$ for confinement by an adiabatically expanding medium. Detailed analysis, to be presented elsewhere, tentatively provide evidence for $\alpha = 0$ and $\sigma = 4 - 5$.

Accepting this, one arrives at a picture in which the background intercloud gas has been shock heated to a pressure $P_{ICM} \propto (1+z)^5$ and has a density corresponding to $\Omega_{b(ICM)} \approx 3 \times 10^{-2}$ with the clouds having a density corresponding to $\Omega_{b(C)} \approx 2 \times 10^{-2}$ and a filling factor (at $z = 2$) of $f_C \approx 1 \times 10^{-2}$. Even within the context of the adopted model, it must be realized that these estimates are uncertain by approximately $10^{\pm 0.5}$.

Clearly the intercloud medium cannot be homogeneous, since we know from observation of X-ray emitting gas that the ambient pressure in the intergalactic medium varies by several orders of magnitude from place to place. A model within which there is a relation between filling factor, f, and pressure, P, of the form $P \propto f^{-k}(k \sim 2)$, (analogous to that found by McKee and Ostriker (1977) for the shock-heated interstellar medium) seems not implausible. Adoption of such a picture changes the conclusions concerning confinement surprisingly little but does indicate that the higher column density clouds may preferentially arise from regions having a higher pressure.

Finally it is worth noting, as has been done by several previous authors, that only one *expects* to see Ly-α clouds having a limited range of masses. For the adopted conditions, those with masses $\gtrsim 10^9$ $M_\odot$ would be expected to be gravitationally unstable to collapse, making small galaxies, and those with masses $\lesssim 10^6$ $M_\odot$ would, at earlier epochs, be subject to significant evaporative losses. Thus it is interesting that the galaxy mass function seems to reach a maximum at $\sim 10^9$ $M_\odot$ with dwarf, gas-rich, irregular galaxies and the intergalactic HII regions (Sargent and Searle 1970), at the lower end of the galactic mass function seen in the mass range $10^8 - 10^9$ $M_\odot$.

3. THE PROXIMITY EFFECT

Whatever the origin, confinement, and geometry of the Ly-α clouds may be, those finding themselves in regions of higher ionizing flux will show a lower column density of neutral hydrogen. This combined with the intrinsic distribution observed in column density indicates that, at a given column density, fewer clouds should be detected in regions of higher ionizing flux. Thus, near QSOs, especially, bright ones with strong UV 'bumps,' a decline in the number of Ly-α lines should be seen. An analogous effect has been detected by several observers and termed by Murdoch et al. (1986) the 'inverse effect' (see also Tytler 1987), since it causes a *decrease* in the number of Ly-α lines with increasing redshift as one approaches a QSO. In a recently completed study, Bajtlik, Duncan and Ostriker (1987) found that there was a statistically significant decrease which was correlated with QSO UV luminosity in just the fashion expected for the proximity effect. Furthermore, observation of the decline allows one to measure the UV background flux, since the relative decline depends on the ratio of the QSO light to the background UV, while the light from the specific QSO is directly observed.

The results are perhaps slightly surprising. Of the three models for the UV background presented by Bechtold et al. (1987), model I, with a relatively high and constant value of $J_{21} = 1$ seems best. Their model III, which at $z > 3$ is a factor of 10 lower seems definitely to be ruled out. This is interesting because it is model III which, according to their calculations, provides a best fit to the available data on high redshift QSOs. In support of this, Shapiro and Giroux (1987) have found that a relatively high ionizing flux is required to understand the Gunn-Peterson effect for a photoionized medium. There are two possible explanations for this contradiction. Either some other source of UV dominates over the QSOs at $z > 2.5$ or we are undercounting the numbers of high redshift QSOs. The former might be difficult to arrange since, if it were so, the postulated sources would be brighter and/or more common than QSOs in the $z = 0$ visible band, and it is surprising that they have gone undetected. Alternatively, obscuration due to the passage of light through intervening galaxies as proposed by Ostriker and Heisler (1984) and Heisler and Ostriker (1988) may have removed from the observed sample those QSOs responsible for most of the UV background flux at $3 < z < 4$. Even allowing for the dust absorption of UV, Heisler and Ostriker's models *with* obscuration are close to the model I of Bechtold et al. which matches the proximity effect.

4. SHELLS, VOIDS AND CORRELATIONS IN THE Ly-α CLOUD DISTRIBUTION

Do large voids exist in the Ly-α cloud distribution similar to those found in the nearby distribution of galaxies? Preliminary studies of one well-studied QSO, Q0420–388 by Carswell and Rees (1987) and Crotts (1987), are contradictory, the former finding no evidence for voids and the latter a $50h^{-1}$ Mpc void at the 99% confidence level. We (Ostriker, Bajtlik and Duncan 1988) confirm the results of the first paper and in a more detailed study set rather low limits on the fraction of space filled with voids as large as $50h^{-1}$ Mpc. But we do find a significant excess of line splittings in the 200–500 km s^{-1} range. This can possibly be attributed to two distinct causes: gravitationally induced correlations as would be expected in the models of Rees (1986), Ikeuchi (1986), and Bond, Szalay and Silk (1987), or as evidence for an origin in fragmenting shells as proposed by Chernomordik and Ozernoy (1983) or OI (1983). The former models predict that the feature should decrease with increasing redshift and decreasing column density while the latter (shock models) predict the opposite. Analysis is underway at present to determine which is more nearly correct.

5. METAL LINE SYSTEMS AND DUST OBSCURATION

If a line-of-sight to a distant QSO passes through the disc of a typical spiral galaxy, one would expect a significant amount of obscuration to occur. Taking our galaxy as an example with $\Delta M_V = 0.2$ appropriate to the cosecant obscuration law, then, $\Delta M_V = 0.4$ locally for a line-of-sight through and perpendicular to the local disc and $\Delta M_V \approx 1$ seen looking through the local disc at a random angle. At a fixed observed wavelength the obscuration would increase as $(1 + z)$ as we imagine looking through our own galaxy at larger redshifts simply due to the linear frequency increase in dust opacity. If our galaxy had a lower dust-to-gas ratio at an earlier epoch, one would of course scale the above noted dependence by $(Z/Z_\odot)$. A very crude treatment indicates that the typical obscuration $\bar{\tau} = \text{const}[(1+z)^{5/2} - 1]$ will increase with the number of QSOs observed proportional to $N_{true}\exp(-\bar{\tau})$, so obscuration has a sharp onset. It is also worth noting that, if the dust distribution is lumpy (as it would be if it were mainly in galaxies), it would be difficult to detect from observed objects. First, there would be little reddening of *observed* sources; obscured objects are just removed from the sample; Ostriker and Heisler (1984) find that the color of detected objects actually becomes bluer with increasing redshift. Second, any other measurements of obscuration, such as detected line ratios in observed samples, must be interpreted with caution, since those which are observed typically have been seen along favored lines-of-sight. In Heisler and Ostriker (1988) a ratio of $[\tau_{obs}(z)/\bar{\tau}(z)] \approx 1/2$ is computed.

Aside from such statistical approaches, is there a more direct test for dust obscuration? A signature for metals along the line-of-sight to QSOs is the detected metal line systems such as those due to CIV or MgII. In nearby systems (Galaxy, LMC, SMC) the ratio of dust obscuration, E(B-V), to the equivalent width in CIV absorption along various lines-of-sight is fairly constant, independent of metallicity. One could thus empirically determine a calibration for dust based on QSO metal line systems to estimate obscuration to individual QSOs. Fitting a synthetic spectrum to that observed by Steidel and Sargent (1987a) for Q0805+0441, and using that with a large data base of QSO absorption lines (Bergeron and Boisse 1984) gives $\bar{\tau}(z=2) \approx 0.25$, which would make the true $\bar{\tau}(z=2) \approx 0.4$ very close to the value taken by Heisler and Ostriker [$\bar{\tau}(z=2) = 0.5$] in their approach.

In summary, a number of different lines of evidence seem to indicate that dust associated with the metal line systems may obscure distant QSOs with $\bar{\tau}(z=2) \propto 1/2$ with a fairly rapid increase in $\bar{\tau}$ beyond that distance.

This work was supported by NASA Grant #NAGW-765.

REFERENCES

Atwood, B., Baldwin, J. A., and Carswell, R. F. 1985, *Ap. J.*, **292**, 58.
Bajtlik, S., Duncan, R., and Ostriker, J. P. 1987, *Ap. J.*, in press.
Bechtold, J., Weymann, R. J., Lin, Z., and Malkan, M. A. 1987, *Ap. J.*, **315**, 188.
Bergeron, J., and Boisse, P. 1984, *Astr. Ap.*, **133**, 132.
Black, J. H. 1981, *M.N.R.A.S.*, **197**, 553.
Bond, R., Szalay, A., and Silk, J. 1987, preprint.
Carswell, R. F., Morton, D. C., Smith, M. G., Stockton, A. N., Turnshek, D. A., and Weymann, R. J. 1984, *Ap. J.*, **278**, 486.
Carswell, R. F., and Rees, M. J. 1987, *M.N.R.A.S.*, **224**, 13p.
Carswell, R.F., and Webb, J. 1987, talk at NATO Conference, Cambridge University.
Chernomordik, V. V., and Ozernoy, L. M. 1983, *Nature*, **303**, 153.
Crotts, A. P. S. 1987, *M.N.R.A.S.*, **228**, 41p.
Djorgovski, S., Spinrad, H., and Dickenson, M. 1987, *Ap. J.*, in press..
Dressler, A., Thompson, I. B., and Schechtman, S. A. 1984, *Ap. J.*, **288**, 481.
Field, G. B. 1965, *Ap. J.*, **142**, 531.
Foltz, C. B., Weymann, R. J., Roser, H. J., and Chaffee, F. H. 1984, *Ap. J. (Letters)*, **281**, L1.
Heisler, J., and Ostriker, J. P., 1988, *Ap. J.*, in press.

Hogan, C. 1987, *Ap. J. (Letters)*, **316**, L59.
Ikeuchi, S. 1986, *Ap. Sp. Sci.*, **118**, 509.
Ikeuchi, S., and Ostriker, J. P. 1987, *Ap. J.*, **301**, 552.
Kondo, M. 1970, *Pub. Ast. Soc. Japan*, **22**, 13.
Lynds, R. 1971, *Ap. J. (Letters)*, **164**, L73.
McKee, C. F., and Ostriker, J. P. 1977, *Ap. J.*, **218**, 148.
Murdoch, H. S., Hunstead, R. W., Pettini, M., and Blades, J. C. 1986, *Ap. J.*, **309**, 19.
Ostriker, J. P., Bajtlik, S., and Duncan, R. 1988, *Ap. J. (Letters)*, in press.
Ostriker, J. P., and Heisler, J. 1984, *Ap. J.*, **278**, 1.
Ostriker, J. P., and Ikeuchi, S. 1983, *Ap. J. (Letters)*, **268**, L63.
Rees, M. J. 1986, *M.N.R.A.S.*, **218**, 25p.
Sargent, W. L. W., and Searle, L. 1970, *Ap. J. (Letters)*, **162**, L155.
Sargent, W. L. W., Young, P. J., Boksenberg, A., and Tytler, D. 1980, *Ap. J. Suppl.*, **42**, 41.
Shapiro, P. R., and Giroux, M. L. 1987, *Ap. J. (Letters)*, **321**, L107.
Steidel, C., and Sargent, W. L. W. 1987a, *Ap. J.*, **313**, 171.
—————. 1987b, *Ap. J. (Letters)*, **318**, L11.
Tytler, D. 1987, *Ap. J.*, **321**, 69.
Vishniac, E., and Bust, G. S. 1987, *Ap. J.*, **319**, 14.
Yang, J., Turner, M. S., Steigman, G., Schramm, D., and Olive, K. A. 1984, *Ap. J.*, **281**, 493.

DISCUSSION

Wampler: Jerry, when you're looking for the 2200 Å bump in redshift systems that are close to the redshift system of the emission line redshift of the QSO, I think it is perhaps worthwhile voicing a warning that the Fe lines have a minimum gap between the Fe bands at about 2200 Å. That is, UV multiplets 3 and 4 and so forth lie to the red of that. There seem to be possibly some FeIII bands that lie to the blue, but in any case the intrinsic spectrum of the quasar may have a dip there.

Ostriker: There were two absorption line systems, one near, which we therefore disregarded, and then one far away, and we used only it, for just the reason you mentioned.

Wampler: The gap in the rest frame wavelength is 300–400 Å, and so all I'm saying is just for people who do this to be careful.

Heckman: If there is a population of QSOs at high redshifts that were obscured by dust, why wouldn't there be a large population of flat-spectrum radio sources that were blank fields optically?

Ostriker: We addressed that subject in our paper. The easiest way to answer that is by way of metaphor. Should there be a large number of x-ray QSOs which are not seen optically by the same argument? No. Why not? Because the redshift of the observed quasars is too small for dust to be at all important at that redshift. It's just that the magnitude limit isn't faint enough. For the flat-spectrum radio sources what we did is this: we essentially took the observed sample and tried to imagine what would happen if you pushed it further away. Would the more distant QSOs have gotten into the sample and would they have been observed in radio? We found that the observed ones are too close to the radio limit typically, so that if you pushed them further away, so that they were far enough away so that if dust obscuration would be important in missing the optical, then you would also miss the radio.

Heckman: That's assuming something about how the luminosity evolution goes?

Ostriker: No. Whether or not they're there, I'm just saying you couldn't tell, because you can't see them far enough away.

Stocke: What about the x-ray background?

Ostriker: Julia, do you remember the latest number?

Heisler: You get 82% of the x-ray background from quasars with absolute magnitudes brighter than -21.5 for $H_o = 100$ km s^{-1} Mpc^{-1}. This is for our best model which has dust.

P. Shapiro: You must have a problem with the spectrum then?

G. Burbidge: Yes. What do you do about the shape of the background?

Heisler: Well, all we did was assume that the spectral index of the quasars in the x-ray regime was -0.5, which is the number that you see in the literature for models.

Ostriker: What wavelength was that calculated at?

Heisler: 10 keV.

Giacconi: I think that is perfectly alright if you assume that the spectrum of a quasar is evolving with redshift in the appropriate way to give you the right answer. That's alright, but there are implications.

Ostriker: Well, it's better than our first try when we had 300 % of the background. (Laughter) We're moving in the right direction.

Giacconi: But I think you must allow for the evolution of the spectrum. In fact, as deep as we see, the spectral index is more like 1 rather than 0.5. I also wanted to ask you a question.

Ostriker: So, Riccardo, you would agree that at 80 % of the background, it's not a big problem, one way or another?

Giacconi: No, it's becoming a big problem, a huge problem. The remainder is a huge problem.

Ostriker: No, but you don't think 80 % is too much?

Giacconi: No, if this is too much you must be changing the spectrum already. The other question had to do with dust. You know there were those calculations about trying to detect dust scattering due to x-rays from the Crab nebula and you could predict there was a halo due to scattering. In fact, there were searches for that. The searches were marginal because the optics were not very good at that time. So the question is, should we see scattering? Have you asked that question?

Ostriker: I just put the question you asked me to somebody during a thesis exam last week. (Laughter)

Giacconi: Oh. But shouldn't you see it, and hasn't it been looked for?

Ostriker: Yes you should see it. What Riccardo was saying—you've just made the same wonderful invention that I did—is that if there is dust obscuration due to an intervening galaxy, then when you look at that galaxy you'll see a 1 arc second halo in x-rays.

Giacconi: It is 1 arcsec?

Ostriker: Yes. Maybe its a few arc seconds; I don't remember the exact numbers.

Giacconi: If it is a few arc seconds, you could even go back now and look to see if its there.

Ostriker: We should look back at McCray's papers.

Shull: The papers were more directed toward the point sources.

Ostriker: No, no, but I'm saying that it is a point source. A quasar is essentially a point source.

Shull: Yes. And they've been looked at and it was detected.

Ostriker: Well, what's the angular size?

Shull: It depends on the energy and the grain size distribution.

Ostriker: We can assume it's the same as our galaxy. It's a few seconds and it should be seen. It would be an immediate test for dust.

Giacconi: The close-by QSOs and the far-away QSOs, you look at the distribution of counts, and you find if the close-by ones look wider.

Ostriker: Exactly.

Jauncey: Martin, you've been asked to ask a question. (Laughter)

Rees: I don't have a question. I just wanted to respond to your criticism of my model. You said first that it required fine tuning to get the potential wells right. To that I'd say that you have a hierarchy and for very small masses the potential wells are too shallow and for very large masses the potential wells are too deep. Therefore, there is some theorem which says it must be right for some intermediate range. (Laughter) The second point is the destruction between a redshift of 2 and the present. That depends on the extent to which they merge. It can't be done yet with the existing n-body computations, but there's no free parameter there and that factor is entirely predictable. The third point, about why the clustering should not be the same as for galaxies, has precisely the answer which you gave in your response to Don York when he asked why you didn't get the clustering in your shock model.

Ostriker: Martin, I didn't say it should be the same as for galaxies. I just say that it's predictable, what it should be, and it can be tested for, that's all.

Rees: And that is exactly what I would say in this model, too.

Jauncey: Can I put a suggestion to the audience. Those with serious questions, come out and discuss them with Jerry in the afternoon tea, and that leaves only those trivial questions, so if anybody has one they can ask it. (Laughter)

York: It's a matter of definition. From reading the cold dark matter paper and listening to Martin and listening to you, I conclude that people use the term "galaxy formation" in grossly different ways. To Martin it's one thing and to you its only those galaxies that we see today because you want your shocks and you want your stars and so forth, but somehow you say that Ly-α from them isn't associated with galaxies because there aren't big enough wells there. In the cold dark matter, paper galaxy formation is when you start the stuff down, but you don't worry about the stars. So, I just wondered what galaxy formation means; how do you define galaxy formation? (Laughter)

G. Burbidge: They don't know.

Ostriker: All you see is computer generated graphics. (Laughter)

Jauncey: Any more trivial questions? (Laughter)

FINAL PANEL DISCUSSION

Wolfe: I thought in this final panel that we should focus on the big questions, not the smaller issues. One of the points Dave Jauncey raised, which is clear from the talks of Bob Carswell and Richard Hunstead, is that it would seem that the only way a lot of these questions can be solved is by getting a large body of extremely high quality data, say at resolutions of ~ 0.1 Å. All the people who do this work know how difficult it is to get such data. The question is: how are we going to make progress in this area. I wonder if anybody has any suggestions along these lines. There's little doubt that ultimately many of the issues that we have discussed are only going to be decided with extremely high quality data which take a very long time to acquire.

E. M. Burbidge: One suggestion is to get groups together. They could put in joint applications for time on the best spectrographs and telescopes. You want wide coverage, obviously, and that means getting good coverage in the red as well as the blue.

Giacconi: I am concerned with this same type of problem, but I was introduced to the problem slightly differently, so forgive me for the approach I take. I saw a lot of data presented here that were taken under different conditions. It wasn't clear what the signal-to-noise ratios were, what the selection criteria were, and on and on and on. It seemed to me that it would be very difficult to put such disparate sets of data together. Each one seems to have done their own statistics. So that immediately made me leap to think about key programs on Space Telescope which had mixed reviews in the community and therefore I don't know how it would work. I think the approach would have to be to design some standards which people would entirely agree to and to develop some plans to join together disparate observations so that under certain condition you could use all of the data. This seems to me to be important if you want to get to the physics of these objects.

Wolfe: So you suggest, for example, that people publish the error arrays per pixel along with their data and so on?

Giacconi: Well, I think there should be better coordination so there won't be sub-standard data.

Shull: I'd like to comment on this problem from the point of view of an interstellar medium observer. We dealt with the problem of multiple components for years, and it really is important to have a high signal-to-noise ratio. If there's a trade-off, don't sacrifice signal-to-noise or coverage to get ~ 0.1 Å resolution. Signal-to-noise ratio should be 40 or 50, at least 20.

G. Burbidge: But you know basically what you're looking at. These people don't.

Savage: I'd also like to comment from the perspective of the interstellar medium observer. The way you see different fields develop is a standard to start out in ways of doing surveys, for hundreds of objects, for hundreds of systems, and so on. However, there should come a time when you decide to zero in on a few representative absorption systems and do a real job on them. I guess I've been a little disappointed in what I'm hearing here. I'm not seeing that happen, and I think it would be very worth while to try to pick out 3, 4 or 5 representative systems, maybe one object might contain these, and do like you say: a real job of physical analysis of what the spectra are telling you. However, you need high signal-to-noise ratios and relatively high resolution to do it.

Wolfe: Maybe Richard Green would like to comment. Its interesting that both Richard and I are involved in programs that obtained echelle data on the same object. We didn't know about each others programs.

Green: I wouldn't be as optimistic as Mike Shull. We can obtain a signal-to-noise ratio of about 20 in 3 nights duration, so 40 isn't a snap of the finger unfortunately. That's only if the object is as bright as 16th magnitude. I think the direction is just as has been wisely suggested. There's a consortium of York and Weymann and Foltz and I can go on, and we're all trying to pool efforts at some level to get the data. We may analyze it somewhat differently. However, we want to concentrate on a few objects and really study systems at the highest possible resolution to start answering these questions.

Sancisi: I would like to point out that we have been doing this in radio astronomy for a long time now. For instance at Westerbork we sometimes dedicate one month to the observation of one object if its worthwhile. It is possible to get time, if it's an interesting and important study, for instance for one week. Yesterday I saw observations that took one week to obtain. So why isn't this done with the optical people?

Wampler: To get this work done, it seems to me that three time scales are relevant. On a short time-scale there are at least two of these objects that people have worked on that are extremely variable. If one monitor, for instance, PKS 0215+015 so that we know when its bright, that particular one could be observed from both hemispheres, and there are four 4-m class telescopes in either hemisphere that could be used, and so that work could be done fairly quickly. On a somewhat longer timescale, we could look for more bright quasars; there must be still some around. On a longer time scale it's going to take the 10-m to 15-m class telescopes to do the job properly.

Wolfe: Cyril, do you have any comments about more bright quasars?

Hazard: Yes, I think Dave Jauncey had a poster paper on the importance of finding such quasars. It's pretty obvious that a lot of this work is being done inefficiently in the sense that people go to the catalogs and pull out what object they can find. Before the latest Hewitt and Burbidge catalog came out, I think there were only about 40 objects with redshifts of over 3. Now, these are not rare objects in the sky. I mean we find 10 to 15 objects with redshift over 3 per UK Schmidt plate. I think that it would be possible to do a detailed search and certainly come up, I think, with lots of objects over redshift 3. Something like 10 objects with redshifts greater than 3 between 16–17 magnitude wouldn't be unreasonable.

G. Burbidge: Cyril, how long has it taken you to get what you've gotten?

Hazard: Once you've got the plate, it would take about 2 hours to search.

G. Burbidge: No, I mean from the beginning. Is it 5 years or more?

Hazard: Well, it wouldn't take that long now. No one wanted to take the Schmidt plates we needed at first because everybody was convinced that very high redshift objects weren't there. But now that we've got them, there's no reason why we can't do it quickly.

Wolfe: I'd also like to make a point in this area. If we want to do abundance studies, which I think a lot of us are very interested in, I think the only objects that we should concentrate on are those with 21 cm absorption lines. This is because it's very difficult to get out abundances when you just look at the optical data on these systems in detail. There are lots of velocity components, and we don't know which one of those carries the bulk of the neutral hydrogen. I would like to ask the radio astronomers if there is any way of getting a few more of these very bright high redshift 1 Jy radio sources in the Arecibo declination range so we can use 21 cm information as well as optical data.

Jauncey: It's difficult Artie, depending on where you want to put your redshift cutoff. Observationally what we find is that at redshifts up to about 2.5 there are some steep spectrum quasars, the remainder are flat or mostly inverted. For example, Q0528 is an inverted spectrum. We might have a couple to go into your list. One in particular has a redshift of 3.6 and its between 1 and 1.25 Jy, but there are not many of them because they are flat spectrum sources. I think more to the point there are objects like 3C196 that Bob Brown had, where instead of them being a point source, you have extended sources. I think they're crucially important, because with instruments like Merlin, you can actually map the absorption. We can argue all we like about how big these clouds are, etc., etc. If you can map them with 0.1 to 0.01 arcsec resolution, then you can see exactly what they are. So I think there's one real thrust to finding the 21 cm absorbers, and I would say, by all means look for the extended ones as well as the compact ones. That should be a real push. You have to go in and make observations, certainly in that area, but finding them is hard.

Norman: This is a crucial point. It also comes back to this business of being somewhat collective in our approach. I think there is real need to (1) upgrade the Arecibo telescope to search for these things and (2) be able to map these things, not just with the Merlin, but with the VLA. So on a 5 - 10 year timescale we really do want to think about having tunable receivers that can make maps at high redshift. It's one of the few ways that you can actually do dynamics at a redshift of 3. If you have an extended radio source, you have a very high probability of seeing one of these Wolfe-type objects across an extended radio jet, and you will be able to get some form of a rotation curve. It's not a crazy idea. It will work.

Wolfe: I'd like to ask the big pundits in this room to make these attitudes known, because when we have suggested things like this in the past, we've gotten little response.

Jauncey: I think that one of the difficulties that we as observers on national facilities have is justifying something like 7 or 10 nights on a single object, or maybe two objects. Most time assignment committees look at those proposals, and they see in 6 nights, they've got three 2 night proposals. I mean, three 2 night proposals is three times as many papers. That basically is, I think, how a lot of the judgement is done on national facilities. One of the things I think that this whole community can voice is the value of agreeing on objects, agreeing on observing procedures and resolutions, but then make it very clear to the judgement committees who assign the time, that these really are crucial observations.

G. Burbidge: I think I have to point out the contradiction in what you say, because I have had experience with exactly what you're talking about. I've heard people say exactly what you say. The same people on the time assignment committee come in and take a very different view. You're talking about yourselves. (Laughter)

Ostriker: Geoff, I think the question is one of guidelines. In other words, if the time assignment committees were told to give one-third of their time to larger projects, one-third to medium projects and one-third to small projects, on some logrithmic basis, that might work.

G. Burbidge: That would help.

Ostriker: They would then do something different from what they now do.

G. Burbidge: I think you also have to be remember that in optical astronomy especially, maybe only, there's a large variety of important problems. So, even if you said you'd spend a large fraction of time on large projects, there would be a huge debate. I can remember a time when I was accused very strongly of giving no time to radio astronomers, and I had a time assignment committee which was very strongly pro-radio astronomy. In fact they were getting a very fair share, but they felt they had to have a great deal more time. The oversubscription is so high that you're bound to turn people down. That's the whole problem. It's very hard to give a single program a lot of time in the face of the combined opposition. That's the real problem.

Jauncey: The other point is you don't tell time assignment committees what to do.

G. Burbidge: Yes, but you also don't have to take their advice. (Laughter) In my period, they made recommendations and there was some discretion.

Wolfe: I think we should go on to some theoretical questions. Martin might want to say a few things.

Rees: I'd like to just make two remarks and show two viewgraphs. First, Don York said quite rightly that there has been lots of confusion about what was meant by galaxy formation. It has been discussed in the context of small galaxies or dwarfs and in the context of big galaxies that might be hundreds of kiloparsecs across. I thought I'd just show a viewgraph which summarizes galaxy formation in the cold dark matter picture. It emphasizes how in practice the luminosity function of galaxies depends on various steps where the physics is uncertain. In this picture, we expect gas to condense in dark matter potentials over the mass range of 10^8 (that's the lower limit which I discussed for the mini-halos) up to 10^{12}, above which the gas is so hot it can't cool. The luminosity function of the galaxies you end up with depends first of all of the mass distribution of these halos: how may small ones, how many big ones, and there are mergers going on all the time. However, even if you know that, you've got to know how efficiently gas falls into these potential wells which will depend on how deep they are and other parameters; and you've got to know the IMF of stars and everything else. So, in principle, there are these various steps of the argument, and it's the high masses which are relevant for the heavy element systems and the low masses which are relevant, of course, for the dwarfs and the halos.

That's a digression, but the one point which I want to bring up, which I think does relate to another of the issues, is the extent of the big (10^{12} $M_\odot$) galaxies, at a redshift of 2. We've heard that the cross-section may be something like 100 kpc squared and therefore this may be relevant to the big disks and the heavy element systems. It would be nice to know how we can perhaps get further evidence on whether there is indeed a lot of gas out 50–100 kpc in big galaxies at a redshift of 2. One way of attacking this question, which was already mentioned in discussion earlier, is to perhaps consider the special subset of these galaxies that actually have a quasar in them: the host galaxies of quasars at a redshift of 2 or 3. Because if a galaxy at a z of 3 does have a lot of gas extending out 50–100 kpc radius, then this gas may be in the clouds I discussed on Tuesday. It may be in the form of caustic sheets or surfaces, but it will extend out to large radii. If there's no central UV source or no large population of young stars, this gas will remain neutral; we will see it in absorption if there is a quasar behind it, but we can't see it in emission. However, if a quasar were to form in the middle of a galaxy answering that description, then one can easily calculate that the emission measure of all this gas, were it ionized, would be sufficient to radiate most of the quasar luminosity. In other words, if the covering factor for these clouds (or sheets, or disks, or whatever they are) of gas at $10^{4\circ}$K were sufficient, and if there were a quasar in the middle, then much of the quasar UV light would in fact be re-processed and radiated by these clouds. Therefore, if this were the case, then you would see extended emission line fuzz around the quasar. Now, there are of course limits on this emission line fuzz which I believe are not trivial which put constraints on this whole class of objects, depending on the filling factor, the covering factor, and all the rest. So, it is very important to set the best possible limits on fuzz; not just around low redshift quasars where you'd expect to see a normal galaxy, but also around high-redshift quasars where it is reasonable to believe that there is more gas out at larger radii. This is a way to probe the galaxies, and also relates, in a different way, to other problems we've discussed at this meeting. For instance, the sizes quoted for Ly-α clouds are of course constrained by the fact that if you get a deep line it must be large enough to cover the continuum source, and indeed perhaps the broad line emitting region. Were it the case that some of the light from a quasar were coming from a more extended region, then of course to get a deep line, you have to have a much larger cross-section, and much larger dimensions for the clouds. Therefore, in order to firm up the lower limits on cloud sizes, one needs to know the effective size from which the emission in a background quasar is coming.

I also want to mention another issue. George Miley told us the very interesting result that in radio sources at high redshifts, the optical axis is aligned closely with the radio axis. This is quite the opposite of the situation for radio sources at small redshifts. This suggests that the optical emission which one is seeing associated with these high-redshift radio sources is perhaps not star light, but is somehow related to the jet itself. It could be that there is some optical emission associated with gas going out along the jets; it could be that (as in the low redshift, low-power jets discussed by Whittle and collaborators) the gas pressure was raised by the jet, which enhances emission from gas; or it could be that there is gas in all directions, and there is optical beamed radiation which somehow excites, not all the gas, but just the gas in the direction of the jet. These things could explain the correlation. It's very important to understand the nature of the optical emission which give the preferred orientation, because this may be another handle on the distribution of gas at large distances from the typical proto-galaxy. As I said, if the galaxy is not active, then unless it has rapid star formation, you would just see these things in absorption. Where there happens to be an active nucleus in such a galaxy, it serves as a probe of the gas that is directly relevant to the absorption lines we've been discussing.

Boksenberg: What are the absorption line properties of this system you've described?

Rees: Well, again, this relates to what Don York was discussing. I think one question is whether the gas is in a lot of clouds or whether it's in sheets or caustic surfaces; this relates to the issue of whether or not Artie Wolfe's systems do have to be rotationally supported disks or just sort of sheets, possibly resulting from the general organized collapse. The relation of the Ly-α clouds to large galaxies, I think, is uncertain. You could say that at large radii there are clouds of low column density which don't have heavy elements, and these are Ly-α systems. Then it would be hard to avoid having the Ly-α clouds clustered like galaxies. Therefore, I would argue that big galaxies may be relevant to Artie Wolfe's systems and to the heavy element systems. The Ly-α clouds are more likely related to either dwarf galaxies or to the still smaller systems where the potential well is not deep enough to make gas gravitationally unstable. So I think large-scale protogalactic collapse is relevant to what Artie Wolfe was talking about, but of dubious relevance to the Ly-α systems.

Boksenberg: But in the quasar spectrum, would you expect to see a metal line system at the emission redshift of the quasar? You don't see this in general.

Wolfe: That's what I was going to ask. You do tend to see absorption there if the quasar is a radio emitter.

Rees: In most cases, of course, where there's not an active nucleus, you will only see this gas through its absorption effects on a background object.

Boksenberg: What I mean is that the quasar itself instrinsically, given all this gas around it, should have absorption. Don't call it a broad absorption line, but it should be narrow absorption.

Rees: Yes. Sure. It may have narrow absorption and fuzzy narrow-line emission and therefore I believe that the existing upper limits are perhaps of more relevance than is commonly realized in this subject. They are already, I suspect, constraining the density, covering factor and extent of the gas. Therefore, it's very important to look hard for fuzz around high redshift quasars.

Boksenberg: One more thing. If you don't worry about the fuzz, then you still would expect to see absorption which you don't usually see, so doesn't that constrain things?

G. Burbidge: Well, you do see it sometimes.

Rees: Well, that will constrain the covering factor.

Wolfe: You especially tend to see absorption in cases where the quasar is a radio emitter. That's interesting.

Boksenberg: Yes, but I'm just saying in general you don't see it.

Heckman: I think one could also make the point that in radio emitters there is an excess of absorption seen near the emission redshift, so therefore we better see fuzz around those objects since the covering factor is large.

Felten: Martin, if we do indeed succeed in finding one of these fuzzy objects, like your picture on the viewgraph, then well and good. However, if we don't find any, that might suggest that there are none of these protogalaxies. However, it might just be that the quasars don't form in the galaxies in the young protogalaxy state. They may only form in ones which are a little older and have collapsed to normal galaxy size. I would suggest that, therefore, this not be a crucial experiment. If it works that's fine, but if it doesn't work, then that's fine.

Rees: I agree. It would constrain only part of parameter space. No one really knows what sort of galaxies you need to have, if you need any at all, to get the quasars. Also, you don't know the covering factor, because if Artie Wolfe is right, and we are seeing the gas in a thin disk, then that, of course, doesn't soak up much of the UV radiation.

Felten: And there is so much parameter space.

Rees: Yes.

Wampler: I think there's another way of looking at this constraint and that is to look for a narrow component on Ly-α emission. Now, in the quasar Ly-α there's some evidence that Ly-α is being re-processed, because it appears to be about a factor of 10 less intense than you would have expected.

Rees: But that's close in.

Wampler: That's close in. That's broad Ly-α emission. In any case, the close in broad Ly-α emission appears not generally to come from an optically thick region, so it's not taking out a lot of the quasar UV continuum. Now, if you want to take a lot of the UV continuum out by these little clouds, you are going to produce a very strong narrow absorption line that sits on top of (narrow) Ly-α emission. You see those strong lines on the hydrogen series in Seyfert galaxies and things like that, but in quasars you don't generally see them, particularly in powerful quasars. There may be a few exceptions. I'm thinking of narrow emission from this halo which is reprocessed UV radiation below the Lyman limit. The way around seeing narrow absorption is just to make these clouds more optically thin.

Rees: But even then, you see a narrow emission line, and it would be very important to know if the narrow core is sort of cut in half.

Wampler: Yes, but the point is: I don't think you can take much out of the UV continuum, because if you did, you'd see a very strong spike in emission and on top of that absorption.

Hazard: I'd like to make a comment about that. We do see these very, very strong spikes on lots of quasars at the moment; they are very, very strong indeed. Equivalent widths are several hundred angstroms and we now have a sample of these objects. However, I just want to make a word of caution in all of this because, if you want to go out and look at a catalog of quasars and look at the fuzz around them, I'm afraid you won't find it. There are very good reasons for that. I think this type of problem you have to be terribly concerned about, because it may be relevant to the other types of problems we've been discussing here. I think you should realize how quasars are chosen. For example, let's say you want a big sample of optically selected quasars. Unlike a radio selected sample, the first thing that is done in a machine selected sample is to throw out all the objects with fuzz around them. (Laughter) You're not going to get a sample of objects with fuzz. Now the same type of thing tends to happen in objective prism surveys. When you see a fuzzy object you say its a galaxy. Now, I think I've talked to Martin about this before. There is a set of objects on the objective plates which have spectra which look like objects having redshifts over 2, but which are fuzzy. We have to put a sample of those aside to look at, and we haven't had time to do that yet. What I want to emphasize is that one shouldn't take any notice of the fact, necessarily, that there are differences between optical quasars and radio quasars at the moment or the fact that you can take a sample of objects, look at them, and see no fuzz around them. That's almost certainly built into the system at the moment. One has to be very careful. People that use samples of quasars should really talk to people who find them to really find out what the selection effects are. Let me just point out one thing, for instance. Say you have an absence of Ly-α absorption lines or the Ly-α emission line itself. Now I'm not saying this is a very big effect, but if you have an emission line which has got very strong absorption lines running over it, that emission line appears weaker than it would in another quasar that didn't have the absorption. Therefore, there's a bias in your sample immediately if your selecting on the basis of that strong emission line. All surveys, whether anyone thinks this or not, and most people don't realize this, that use broadband colors also select objects in many cases on the basis of strong emission lines. If its not based on strong emission lines, it's certainly based on the continuum colors. So all surveys are riddled with selection effects. These selection effects are not even well understood in most cases by the people who do the surveying. Certainly just going to a sample of quasars and saying this is an unbiased sample is almost certainly incorrect.

Giacconi: But you could take, for instance, all x-ray detected quasars and then go and study them all, right? The problem right now that I see is that the x-ray surveys are too deep. They're too small a field and too deep. For the all sky survey that ROSAT will carry out, there will be 50,000 objects spread all over the sky. They would tend to be nearer though. That would be a fantastic source of objects, even though its biased in a certain way.

G. Burbidge: You've still got 1000 or 1500 radio selected objects where bias doesn't occur. Here people look for fuzz, but only in a selective way, because they only tend to look for it largely at low redshifts, but you do have that sample, Cyril.

Hazard: I'm just saying you have to be careful.

Stocke: I'd like to re-iterate a comment that I made during Artie's talk. 3C275.1 is a radio selected object. It has extended 100 kpc fuzz primarily of narrow emission lines of oxygen. In fact, if you push it to redshift of 2, of course, given most cosmological models, it would be larger than the spectrograph slit and therefore excluded from the spectrum. So I think there is reason to believe that you wouldn't necessarily see narrow Ly-α emission in a normal slit spectroscopy survey of quasars. Also, you're spreading the object even more at redshift 2 than you are at a redshift of 0.6 where we found this one which is relevant if your fighting faint sky limits.

Wolfe: What about going to interference filters?

Rees: Could I just comment that the "cartoon" I drew applies to your object and to any intense cooling flow where there is filamentary structure out to large radius.

P. Shapiro: I'd like to object to the assumption that because Ly-α forest clouds appear to be less clustered than the C IV metal-line systems, that the Ly-α forest clouds are not likely to be associated with galaxies or at least protogalaxies. Why isn't it possible that the systems that are metal-line systems are associated with objects that simply have been around longer, and are therefore in regions that were clustering for a longer period of time, while the Ly-α forest clouds may be associated with protogalaxies that are younger and have had less time to cluster?

Boksenberg: You can say something else. Nothing need be clustered, particularly. If you have a core which is where the stars are with a relatively high velocity dispersion, and a halo where the hydrogen is with a velocity dispersion somewhat less, you get exactly what you see.

P. Shapiro: In that picture you put the Ly-α forest clouds out in the halo?

Boksenberg: It's a bigger cross-section so you see more of them.

P. Shapiro: So your agreeing with me essentially. We should be careful not to make the assumption that the Ly-α forest clouds cannot be associated with galaxies of hydrogen.

Boksenberg: I completely agree.

Wolfe: I think this brings up a topic that has been discussed a lot at this meeting and that is David Tytler's suggestion of a single population giving rise to the column density distribution. What do people think about this?

Turnshek: Let me show you what people think about this by showing the results of the informal survey we took a few hours ago. There were only 27 responses, so many attending chose not to respond. There are some people of those responding who would not answer especially the last 2 questions. The percentages are based on 27 responses.

From 27 Responses	Ejected	ORIGIN Origin not related to ejection, but material is intervening and resides in the cluster containing the QSO	Cosmologically Intervening	"It's all wrong"
"TYPE" OF SYSTEM		PERCENT*		
Broad Absorption Line	76	15	< 1	8
Narrow z_{abs} Near z_{em}	24	52	14	5
Weak Ly-α Forest	3	4	78	4
Narrow, Highly Displaced Showing Metal Lines	9	5	74	5
Damped Ly-α	0	8	85	0

*The total is less than 100% because all questions were not answered.

<u>Ostriker</u>: Does ejected mean ejected from our galaxy? (Laughter)

<u>G. Burbidge</u>: No. Ejected from the quasars. Let me try to make this clear. It starts with the hypothesis that all of the absorptions are either intrinsic, in some sense coming from the quasar, or intervening. That's where we all started from. We have chosen to discuss through the majority of this conference, the intervening hypotheses. It is entirely possible that one can start discussing how many of all the systems that one sees, not only the broad ones but the narrow ones as well, can be associated with ejection, independent of the question of this distance to the quasar which we could raise if we wanted to. But this is where one should start the discussion, and not where it started at the beginning of this meeting. Of course, the idea here was that people wanted to explore the universe, and you can't do that unless you choose the right hypothesis to start with, independent of the data of course. (Laughter) Sorry.

<u>Felten</u>: It's a bandwagon!

<u>Burbidge</u>: Well, I didn't say that. (Laughter) It's fairly obvious though.

<u>Wolfe</u>: With that objection noted, Nolan Walborn would like to say a few things. He wanted to say something about the relevance of some local phenomenon and about BAL quasars.

<u>Walborn</u>: I wanted to show you a couple of local phenomena that I've been impressed by. I've seen some fascinating spectra in the last couple of days, and these local phenomena may be relevant. The first thing I'd like to show you is a local narrow absorption line system at a distance of about 2.8 kpc from the sun. These are interstellar line profiles from the Carina nebula, a region of very massive star formation. What you see is a very large velocity range in strong interstellar line components. This object has absorption spanning a wide velocity range from −220 to

+120 km s^{-1}, or 340 km s^{-1}. That was a MgII profile I just showed. Here is a CII profile in another object. The interpretation has a feature at -350 km s^{-1} and that's confirmed in MgII. The total velocity range is 430 km s^{-1} in that object. And as you saw, those features are strongest in the low ionization lines, but you can recognize some structure in some of the high ionization lines. This is CIV in that same star with possible features at -350 and 70 km s^{-1}. Finally, there's a summary of all of the velocity ranges and all of the high velocity components that there are in that object. The extremes are -350 km s^{-1}, quite a strong feature I just showed you, and there's another strong one at $+200$ km s^{-1}. So there's a velocity range of 550 km s^{-1} in sharp, strong absorption features in a local giant HII region with massive stars in it.

The other thing I wanted to show you was some local BAL objects. These are a set of wind profiles in O stars, a luminosity sequence. You see very strong CIV P Cygni profiles with terminal velocities the order of a few thousand km s^{-1}. Here's that same sequence in SiIV, showing how the SiIV is a function of luminosity or density in the wind. Just to make a point that there are a lot of non-classical P Cygni-type profiles seen in O and B spectra, and quite a bit of progress is being made right now in interpreting those in terms of spherically expanding envelopes. There's just an overview there (shows diagram). You can't see the details. I remember, when I first showed this display, Pat Osmer said "Gee they look like quasar spectra" and my reaction to some of the quasar spectra is "Gee they look like O star spectra." Here's N V and Ly-α in a case where you have a fairly light reddening so that the Ly-α is able to recover. You see SiIV and CIV. You can see the SiIV weakening as you go to lower luminosity stars.

Finally, I'll show 3 more viewgraphs of such objects. These are some Wolf-Rayet stars, narrow-line WN stars. Here you see what happens when you have strong interstellar extinction. The damped Ly-α just completely obliterates the solar region of the spectrum and emerges with the blueshifted absorption from the NV. Here you see some very peculiar SiIV profiles somewhat similar to the ones I saw in quasars with relatively weak emission and somewhat detached very strong absorption troughs. Also I just wanted to comment on Wolf-Rayet stars and the abundance question. Most of us believe that these are objects showing the results of CNO cycling, enhanced nitrogen and helium and decreased hyrdrogen and carbon. But the abundance determinations are an extremely difficult thing to do, and what most people in this field do at the present time is only make a comparison of ratios of ionization states of, say, nitrogen and carbon, similar ionization state. The details of absolute abundances are very sensitive to the modeling, and very few people have the courage to do that even in Wolf-Rayet stars. I doubt whether quasars are easier. Here is the further region in the same star showing a strong CIV even in these WN stars. Here are some other peculiar profiles. These are not resonance lines, but excited state transitions. Well, the point is, considering the profiles you have seen, if you scale the velocities by a factor of ten, they're similar to quasar BAL profiles. But the last one here shows you one in which you don't even have to do that. This is SN1987A from South African observations and you see on the first day they even misinterpreted what they were seeing. This absorption trough is at about 6100 Å. They thought it was a SiII feature in a Type 1 supernova, but in fact its H-α shifted to 20,000 km s^{-1}. So, just in conclusion, the main point I want to make is that there is an enormous amount of progress in the literature that has been produced on understanding these kinds of profiles in terms of expanding envelopes in which the absorption arises along the line-of-sight to the continuum source of emission and the rest of the envelope is seen against the sky. This modeling started out 15 or 20 years ago with very simple approximations and has now become quite refined, and right now people are utilizing effects like shocks and x-rays in winds, the difference of assuming a finite disk instead of a point source for the continuum source and overlapping profiles. I think there are tremendous tools in the literature. There is a whole parallel literature on wind profiles from cataclysmic variable disks. I think there is a tremendous formalism and body of tools in literature which could be applied to interpreting some of these profiles which you're now finding in these BAL quasars. I'd like to suggest that a very fruitful cross-pollination might take place if the next quasar workshop brought together people here who are experts in those spectra, and people like Abbott and Hummer who are currently carrying forward the state of art modeling of these profiles in O and B stars, and other people I don't know as well in the cataclysmic variable field.

G. Burbidge: Well there have been attempts like this involving radiation driving for the quasars. For example, work done by Lucy.

Walborn: I have a feeling that was quite a few years ago, and quite a bit more has been done both observationally and theoretically.

G. Burbidge: There is some evidence for line-locking in these systems which is another subject largely ignored somehow this week which would suggest the intrinsic hypothesis, even for systems with quite large velocity and redshift differences.

Wolfe: I want to get back to David Tytler's suggestion of a single population. This subject came up constantly. Do people think this really is one single population, or is something else going on?

Rees: Could I raise the question: do we know anything about the heavy element abundances in the low column density systems. Suppose we wanted to have high heavy element abundances in systems with HI column density of 10^{14} cm^{-2} but a low ionization level, is there anything which would argue against that?

Bergeron: One of the programs I have right now is to search for the heavy elements in the systems with HI column density around $10^{16}-10^{17}$ cm^{-2}. I think if for different ions one is able to reach equivalent widths which correspond to a depression of 10% to 3% around CIV, then we may be able to answer your question. But again, that's back to the same point as before. We need very good signal-to-noise ratio, high resolution spectra.

Giacconi: I don't understand the question though, because if you admit that there is a gradient of metallicity as you go in, then the cross-section from the stuff which is metal poor is going to be enormous as compared to that which is very metal rich. So you'll always find metal-poor systems.

Rees: But that's a model dependent argument.

Giacconi: Right, I understand that. What I'm saying is that, if you believe in what you're saying, then you can't press that. (Laughter)

Boksenberg: There are objects with 10^{14} cm^{-2} column density where you see metals very clearly, of course. Those are the same column densities where you normally don't see metals. The point is that there are two populations, not one population actually, but two. The question is whether they merge together and what David Tytler has said is that in the ones where we don't see metals the gas may be neutral, but I don't really clearly see why that should be.

Wolfe: Is there any other way to get a handle on the ionization state?

Boksenberg: Remember, David clearly doesn't say that he sees a single population. That's only his proposition. However, if the gas is not neutral, then it is under abundant.

Tytler: The exciting point is: are we seeing any primordial material? That's what would interest me. What we want is the actual distribution of abundances and we're an awful long way away from getting that. For HI column densities of 10^{14} cm^{-2} I don't really see how we're going to get the abundances down there very easily. We might just scrape the top of it and find that one or two of them do have metals, and you'd immediately know they have high ionization, but that's not going to help you in any great detail.

Disney: I have just a kind of flippant comment. It's amazing how many things if you plot them in a power-law spectrum and you take account of the fact that they are divergent at the ends, you have a slope of like 1.5. For example, radio luminosity functions for large numbers of different types of radio sources are like that. I don't think that you could ever say that that tells you that there's only one population.

Wolfe: There are a lot of people here working on luminosity functions, do they have any comments about that? I remember in the Virgo cluster the different types of galaxies have different shapes of luminosity functions. Isn't that correct?

Sandage: That's right. You can put a fixed luminosity function through all the Hubble morphological types, but that's nonsense, because each type has its own luminosity function. Then you have to add them up with different percentages, and you add them with different percentages in different environments. So, to say there's a universal Schechter function I think is wrong.

Wolfe: So you need a crucial parameter like morphology to distinguish between them.

G. Burbidge: You need something more than you've got.

Shull: This is a point on abundances. From the damped absorption systems, you can use your upper limits to solve this curve of growth effect, because you can not slide your curve an unlimited amount to the right. I think the observers ought to do very good jobs on these weaker lines, for example of FeII.

Wolfe: The FeII is crucial.

Shull: That's crucial. There are very good lines there.

P. Shapiro: Was the comment about the gas being neutral to the effect that the column density would simply be too small in metals.

Boksenberg: Exactly.

Wolfe: Any other comments about this business of two populations, or about populations in general? Another issue is magnetic fields. I believe a lot of people were very excited about what Phil Kronberg had to say. Does anyone have comments.

Rees: Well, I'd just like to amplify the point you made in your talk this morning, Art. If it is true that the magnetic field in some of these systems as inferred from Faraday rotation is of the order of a micro Gauss as in a galaxy now, this is very important indeed for the origin of magnetic fields. This is because the popular view, as espoused at enormous length in Parker's book for instance, is that you start off with a galaxy having a seed field that may be 10^{-20} Gauss or thereabouts, and it's amplified by a dynamo mechanism whose growth time is of order the rotation period. Then, if we are seeing, as you imply, the outlying clouds of a young disk, there's certainly been not enough time for enough e-foldings to get from 10^{-20} Gauss up to 10^{-6} Gauss. Therefore, this has very important ramifications for popular ideas on the origin of galactic magnetic fields.

Wolfe: Do you think that this means that they're primordial then or something else?

Norman: You'd need something more like 10^{-9} Gauss for the seed field. Jerry would like that, but he's gone.

Giacconi: The other point is that they become extremely important when you start talking about time for dissipating the clouds, conduction of the heat, and all that. I would be very surprised if no one was worrying about it.

Wolfe: You were surprised that they were so big?

Giacconi: No, I was surprised that nobody was worrying about it, because concerning the conditions for survival of the clouds, obviously if you don't take into account magnetic fields, you don't know what the conductivity is.

Norman: I'd like to make a few points. One is that recent work by Field, Ikeuchi and myself on the dynamo problem indicates that Parker's large number of papers on the subject probably contain some erroneous statements, and that the dynamo action as he proposed it probably does not work. The second point is the pressures that you infer from these magnetic fields of 10 micro Gauss are truly enormous. They are of order 10 to 100 times what we see now in the interstellar medium. So you're talking about further out in radius at an earlier epoch where its much more fragile and you have a significant problem.

G. Burbidge: But you have that problem anyway because in the radio sources where you've got extended lobes, there is very good evidence, based on the arguments that go way back, that the fields are strong, like tens of micro Gauss in a region where you could certainly argue density's very low. It seems to me we've had the problem around for a long time.

Norman: That's right. Those things are probably dynamically expanding. These do not have characteristic expansion velocities of the order of 10 km s^{-1}.

Wolfe: One thing that should be mentioned, though, is that the kinetic energy density may be higher in these systems than it is locally because there is more gas per unit volume back then.

Norman: Right, but you still have problems.

G. Burbidge: Well, from time to time it's argued that the dynamo mechanism won't work. I gather that's what you're now saying. You're beginning to sound like Alfven to me. He may be right about some of these things.

Norman: No. Without going into details, let me just say that there are fascinating things to be thought about concerning the dynamo mechanism.

G. Burbidge: But you still need a seed field, right?

Norman: What Martin says is absolutely correct, of course, right. You need to get around at least once. Therefore, you need a seed field of the order of 10^{-9} Gauss, unless you can think of another way to do the dynamo.

Rees: Parker, Zeldovich et al. and the rest would conjecture that the seed field you need is 10^{-20} Gauss, and there are ways of getting that by either Compton drag as the galaxy forms on the microwave background or possibly from the first few stars that can make an internal field and then spew it out again. So getting a 10^{-20} Gauss seed field is not a problem whereas if it has to be 10^{-9}, then I suppose this would have to be some sort of cosmological primordial field.

Norman: Well, I can report that Mike Turner has tried very hard to do this from symmetry breaking effects, etc. and he can only get 10^{-20} at most. There is no obvious way to do this from the phase transition.

G. Burbidge: Well, that's the end of that. (Laughter)

P. Shapiro: If you really believe that these systems are large disks, then there must be some orientation effect also in the Faraday rotation measurement, where since it's the component parallel to the line of sight it contributes in the Faraday rotation, and the low column density systems are the ones that are seen face-on. If the magnetic field is anything like it is in our galaxy, that will be the case where its minimum.

Wolfe: The evidence is consistent with that.

P. Shapiro: And it will be a stronger effect than just the column density. The object with the lowest column density will have a proportionally lower effect on Faraday rotation.

Wolfe: The next topic I want to ask people about is galactic halos. Halos is a topic thats been around for a long time and I wondered what the people now thought about the evidence that the metal-line systems really are produced by halos.

Norman: I'd like to show a diagram (see page 90 in the poster publication). This diagram has been shown in bits and pieces throughout the meeting. This is work done by Laura Danly, Chris Blades and myself, following work by Art Wolfe. The black points are from all the quasars observed by Young, Sargent and Boksenberg and by Foltz et al. The open circles are due to the galactic halo studies by Danly, Savage and others. These new green points, which Danly added today, are from the new Mg II survey work of Lanzetta, Turnshek and Wolfe. These green points circled by red are damped Ly-α systems. It seems to me, generally speaking, if we really want to understand the nature of the quasar absorption lines, we need a framework in which to set things. We need to understand what the halos of nearby galaxies, including our own galaxy, would actually produce; and then we need more detailed physics, and perhaps we can understand. Here we try to do this for our own galaxy, and this is the result. We also hoped, which was about a year ago, we could get lots of IUE data and do this for nearby starburst systems, etc., etc., but, in fact, that's extremely difficult. We have to wait for, and I 'm quite serious about this, we have to wait for Space Telescope which will create a revolution in this type of work. And so, here we have the data on the systems. The green systems, as was mentioned yesterday, may be the ones that are most like our galactic halo. Roughly a third of these MgII selected systems may be somewhat similar. These black ones are CIV selected. Maybe when we think we understand the physics, just something simple, like the ionization structure of these things, we may be able to establish some sort of correlation. But even this question will require a lot more data.

Boksenberg: Can I say something on this? I think you have to be very careful with these things. I don't exactly know how you did this because it depends on what form the data have been published, but if you look at the profiles of these objects, they're complex in CIV usually and not so in SiII. In other words, there are components which may be one-to-one in ratio in SiII or CIV, and others which may only have CIV and no SiII. So there's a mixture of things here unless you've done that separation yourself. I don't know if you have.

Danly: I think that's the point, actually. There are clouds, or whatever, that are causing the absorption to be similar, that have both the SiII and the CIV or the CII and the CIV, and then there are those which are purely highly ionized and you don't see both.

Boksenberg: I'm sorry to butt in, but I meant in a given case you have a complex profile. It's not that case compared with another case. There may be a complex CIV and related to that is a less complex SiII, so only some of the SiII components line up with particular CIV components while others don't. So what kind of diagram you get just depends on whether you've done that deconvolution.

Norman: Well, this was done by your collaborator. The high-resolution CIV study is the best we have. We only took the really high resolution data.

Boksenberg: Well, the diagram says equivalent width, but I was wondering if that's the total equivalent width.

Blades: It's the total CIV equivalent width.

Boksenberg: If it is then there is some problem. It doesn't mean that you'll get the same correspondence to the galactic gas. On the other hand, I don't think this is the halo that one is looking at here which is just a few kiloparsecs up. So what you're comparing, I don't know.

Norman: Obviously one would want to have a lot more sight-lines through our halo and sight-lines through nearby starburst systems, etc., but this is the best we can do. This is really the only good data on sight-lines through our halo that we have and that's (the diagram) the answer. There's a lot more to do.

Boksenberg: That doesn't give you the cross-sections you're looking for.

G. Burbidge: No. You still have the cross-section problem. You've always got to be looking for an out if you believe this scenario.

Danly: A relevant point here concerns the LMC sight-lines. Some people have discussed this as being an anomolous line-of-sight. We've heard about that today and followed it in the literature, but a point that Blair made in his talk is that it's very relevant, I would say, because not only is it the full extent of our galaxy, but it also contains some sort of material that may be associated with the LMC—Milky Way pair, which at large redshifts is probably a better representation of what the field of galaxies and satellites and forming objects is like anyway. The important point here is that the material is predominately neutral. For example, you look at the supernova data and you see components all the way down, MgI and MgII, and all the way across in velocity space between the Milky Way and the LMC. Whereas the high ions, the CIV and SiIV, they're limited to the nearby galactic space, an LMC component and a Milky Way component. So I think this is representative of large distances, at least out to 55 kpc, and the large galactic environment.

Shull: This is very persuasive as is Don York's argument, but there are lethal things among them. One is, you look at η Carina, or the LMC, and they are full of molecular clouds and dust, none of which you see in quasar absorption line systems.

Norman: That's a theorist's problem.

Danly: In response to that, I don't know why that's lethal. I think we're in agreement. You have separate populations. The highly ionized species correspond to the majority of the quasar absorption lines.

Boksenberg: That's right! And that's telling you something. They don't have to be the same. In the halo there's a higher exposure to ultraviolet flux and that sort of thing. So, I wouldn't expect a direct correspondence. Now, it may be that the populations actually differ anyway, but even then the environments differ. So the thing is to see what it tells us rather than ruling out a proposition. I know you're not doing that, but I think one should look at it and see what it's actually telling us.

Danly: That's a point we made in our poster. You can extrapolate. If you know what the low density environment is like at 5-10 kiloparsecs above the plane, then you can also infer some things about large galactic radii where the gas densities also drop off to low densities and it's also exposed to the same extragalactic radiation field. If anything, the radiation field is softer than that near the inner galaxy because it is not subjected to the O and B hot star radiation field.

Boksenberg: At earlier epochs the quasar radiation field is present.

Danly: Yes. That's right. We're saying that's separate from all these things.

Boksenberg: Yes.

Wolfe: Well, we only have a few more minutes, so if anybody would like to say something that is very pressing they can.

Felten: I would like to ask if the moderator would like to add anything more regarding the question Renzo Sancisi asked this morning concerning whether there's a shortage of gas or mass available in galaxies today to account for the large extent and the large column depth of the disks that you need at large redshift to account for the damped Ly-α systems. We've had some private discussions on this and it seems as if all you want is mass, that there's no great shortage of mass.

Wolfe: The problem is how to go from something which may be big to something which is apparently small. There's not a problem with mass. That's all I wanted to add and that's it!